PERIODIC CHART OF THE ELEMENTS

Key: Atomic Number → 11; Name → Sodium; Symbol → Na; Atomic Weight → 23.0

Period	Group IA	IIA	IIIB	IVB	VB	VIB	VIIB	VIII	VIII	VIII	IB	IIB	IIIA	IVA	VA	VIA	VIIA	Noble Gases
1	1 Hydrogen H 1.0																	2 Helium He 4.0
2	3 Lithium Li 6.9	4 Beryllium Be 9.0											5 Boron B 10.8	6 Carbon C 12.0	7 Nitrogen N 14.0	8 Oxygen O 16.0	9 Fluorine F 19.0	10 Neon Ne 20.2
3	11 Sodium Na 23.0	12 Magnesium Mg 24.3											13 Aluminum Al 27.0	14 Silicon Si 28.1	15 Phosphorus P 31.0	16 Sulfur S 32.1	17 Chlorine Cl 35.5	18 Argon Ar 39.9
4	19 Potassium K 39.1	20 Calcium Ca 40.1	21 Scandium Sc 45.0	22 Titanium Ti 47.9	23 Vanadium V 50.9	24 Chromium Cr 52.0	25 Manganese Mn 54.9	26 Iron Fe 55.8	27 Cobalt Co 58.9	28 Nickel Ni 58.7	29 Copper Cu 63.5	30 Zinc Zn 65.4	31 Gallium Ga 69.7	32 Germanium Ge 72.6	33 Arsenic As 74.9	34 Selenium Se 79.0	35 Bromine Br 79.9	36 Krypton Kr 83.8
5	37 Rubidium Rb 85.5	38 Strontium Sr 87.6	39 Yttrium Y 88.9	40 Zirconium Zr 91.2	41 Niobium Nb 92.9	42 Molybdenum Mo 95.9	43 Technetium Tc (99)	44 Ruthenium Ru 101.1	45 Rhodium Rh 102.9	46 Palladium Pd 106.4	47 Silver Ag 107.9	48 Cadmium Cd 112.4	49 Indium In 114.8	50 Tin Sn 118.7	51 Antimony Sb 121.8	52 Tellurium Te 127.6	53 Iodine I 126.9	54 Xenon Xe 131.3
6	55 Cesium Cs 132.9	56 Barium Ba 137.3	* 57 Lanthanum La 138.9	72 Hafnium Hf 178.5	73 Tantalum Ta 180.9	74 Tungsten W 183.9	75 Rhenium Re 186.2	76 Osmium Os 190.2	77 Iridium Ir 192.2	78 Platinum Pt 195.1	79 Gold Au 197.0	80 Mercury Hg 200.6	81 Thallium Tl 204.4	82 Lead Pb 207.2	83 Bismuth Bi 209.0	84 Polonium Po (210)	85 Astatine At (210)	86 Radon Rn (222)
7	87 Francium Fr (223)	88 Radium Ra (226)	** 89 Actinium Ac (227)	104 Kurcha- tovium Ku	105 Hahnium Ha													

Series														
* Lanthanide Series 6	58 Cerium Ce 140.1	59 Prase- odymium Pr 140.9	60 Neodymium Nd 144.2	61 Promethium Pm (147)	62 Samarium Sm 150.4	63 Europium Eu 152.0	64 Gadolinium Gd 157.2	65 Terbium Tb 158.9	66 Dysprosium Dy 162.5	67 Holmium Ho 164.9	68 Erbium Er 167.3	69 Thulium Tm 168.9	70 Ytterbium Yb 173.0	71 Lutetium Lu 175.0
** Actinide Series 7	90 Thorium Th 232.0	91 Protactinium Pa (231)	92 Uranium U 238.0	93 Neptunium Np (237)	94 Plutonium Pu (242)	95 Americium Am (243)	96 Curium Cm (247)	97 Berkelium Bk (247)	98 Californium Cf (251)	99 Einsteinium Es (254)	100 Fermium Fm (253)	101 Mendelevium Md (256)	102 Nobelium No (254)	103 Lawrencium Lr (257)

CHEMICAL PRINCIPLES AND THEIR BIOLOGICAL IMPLICATIONS

CHEMICAL PRINCIPLES AND THEIR BIOLOGICAL IMPLICATIONS

RAYMOND F. O'CONNOR
Santa Barbara City College

Hamilton Publishing Company
Santa Barbara, California

Published by **Hamilton Publishing Company,**
a Division of John Wiley & Sons, Inc.

Library of Congress Cataloging in Publication Data:

O'Connor, Raymond F.
Chemical principles and their
biological implications.
1. Chemistry. 2. Biological chemistry.
I. Title.
QD31.2.023 540 74-3367
ISBN 0-471-65246-6

Printed in the United States of America

10 9 8 7 6 5 4 3 2 1

PREFACE

One of the most profound commentaries on the twentieth century might very well be this phrase from a song by Bob Dylan—". . . and the times they are a changin'." Nowhere is change more apparent than in science and education. Traditional boundaries between branches of science have all but disappeared. A scientist may be a biomedical engineer, a mathematical biologist, or an astrochemist, to name a few.

The philosophy and content of lower-division science courses is also changing. For example, one currently fashionable approach to the teaching of biology is to examine the way in which pollutants in soil, air, and water interact with living organisms, and to show how ecological relationships are upset by mismanagement of natural resources. An argument in favor of such an approach is that students tend to develop a positive attitude toward a subject when it is relevant to their own lives, and this attitude enhances their learning.

If this approach can be justified for one course, why not for others? Why not design a chemistry course for those who are not science majors but who need or want to know something of the role of chemistry in life processes? Why not, for example, design a chemistry course that nurses, physical education majors, and biologically oriented students would find relevant? These students are usually required to take chemistry, yet all too often they derive little profit, and less pleasure, from the experience. They are alienated by the emphasis on measurement and stoichiometry, or they become frustrated by an attempt to compress general, organic, and biological chemistry into one short course. There is a real need for a new approach—one that links basic chemical concepts with biological phenomena. This book is an attempt to meet that need.

Chemical Principles and Their Biological Implications was designed to be used in an introductory chemistry course. It assumes no previous knowledge of chemistry, and requires only the most basic arithmetic skills. Although the emphasis is on chemical theory and its applications, the quantitative aspects of chemistry are not ignored. Measurements, gas law calculations, and stoichiometry are included, but they are not unduly emphasized. The chief distinction of this book lies in its approach and in the examples that have been chosen to illustrate the relationships between chemistry and biology. Material from organic and biological chemistry is included wherever it is relevant, and not treated as a separate topic. This integration is unique in a text for beginning students.

Each chapter is preceded by a set of learning objectives designed to point out concepts and ideas that are particularly important so that the student will be especially attentive to them as he reads. The questions that follow each chapter can help a student to test his comprehension of these ideas. Answers to approximately one-third of these questions are found in the back of the textbook. A complete set of answers is contained in the teacher's manual.

A manual of relevant laboratory exercises is also available. Each exercise is designed to complement a particular segment of the text by requiring the student to relate his observations to chemical theory. Materials from plant and animal sources, including living cells and tissues, are used in the majority of exercises. Convenient report sheets simplify the instructor's evaluation of student performance.

Chemical Principles reflects the conviction that a textbook should be more than just an outline that must be filled in and interpreted by the instructor. It should contain a clear and logical sequence of ideas, and should be written in a style that is pleasing and easy to read. Hopefully, this textbook fulfills these requirements.

It is with pleasure that I acknowledge the editorial contributions of Nancy Marcus; the critical reviews of Professor Leonard Druding, Rutgers University, and H. L. Retcofsky, Shadyside Hospital School of Nursing, Pittsburgh, Pennsylvania; and the fine artwork of Dale Johnson. I would also like to express my appreciation to George Thomsen and Ron Lewton of Hamilton Publishing Company for their untiring efforts, and to Paul E. Harris, Jr., for his faith in this project.

Santa Barbara, California *Raymond F. O'Connor*

ABOUT THE AUTHOR

Raymond F. O'Connor is associate professor of chemistry and life science at Santa Barbara City College, Santa Barbara, California, where he has taught since 1965. He received his B.A. degree from Miami University at Oxford, Ohio, in 1951, and his M.S. degree from the University of Michigan at Ann Arbor, in 1954. Prior to his academic career, Professor O'Connor acquired a background of research and industrial experience through associations with Massachusetts Institute of Technology; Merck, Sharp & Dohme; and the University of California, Berkeley. He is the author of laboratory manuals in biology and in chemistry, and has contributed several articles to professional journals. His professional affiliations, past and present, include the Society of American Bacteriologists, A.A.A.S., Sigma Xi, the American Rocket Society, and the American Chemical Society.

CONTENTS

CHEMICAL PRINCIPLES AND THEIR BIOLOGICAL IMPLICATIONS

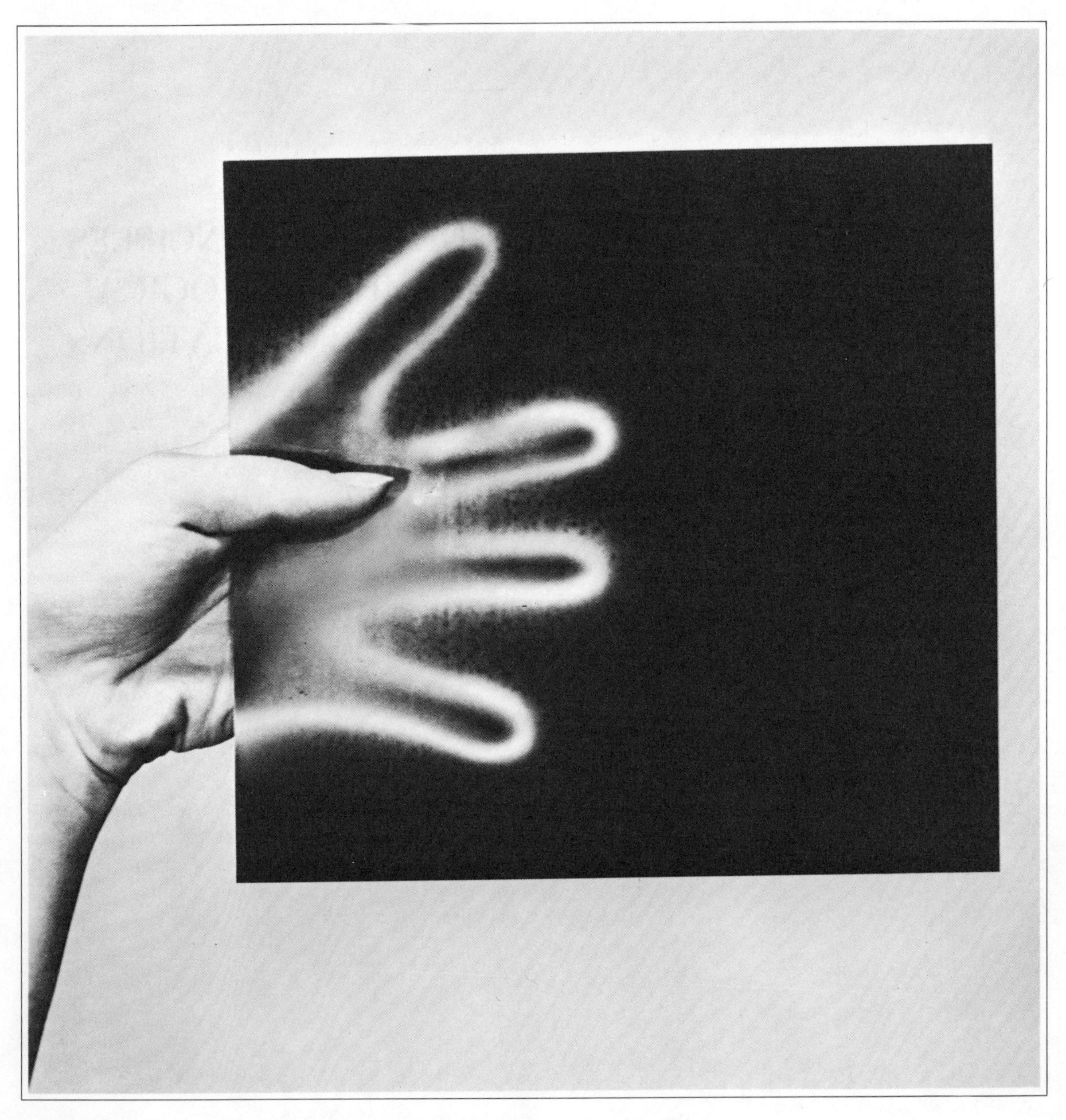

Liquid crystals on a sheet of plastic. The color of the crystals is related to their temperature. (Courtesy LiquiCrystal, Inc., York, Pennsylvania)

1

CHEMISTRY: THE THEME AND THE TECHNIQUES

LEARNING OBJECTIVES

1. Explain the observed differences between solids, liquids, and gases, using a scientific model.
2. Distinguish between a measurement and a measurement system.
3. Describe the metric system of measurements.
4. Distinguish between mass and weight.
5. Describe the relationship that exists between the Fahrenheit, Celsius, and Kelvin temperature scales.
6. Explain the principle behind the operation of a barometer.
7. Convert a measurement expressed in either a British or metric unit to its equivalent in any other related unit.
8. Distinguish between density and specific gravity.
9. Starting with a solid substance at its freezing point, describe its pattern of heat absorption until it becomes a vapor.

KEY TERMS AND CONCEPTS

calorie	matter
chemical property	meter
cryogenics	physical property
density	pressure
freezing point	scientific model
gram	specific gravity
heat of fusion	specific heat
heat of vaporization	temperature
liter	torr

Impressions of chemistry are almost as varied as definitions of love. Chemistry is regarded by some as a branch of technology which produces such agreeable and useful items as mouthwash and color television sets; by others as a monastic vocation practiced by bearded and bespectacled bumblers who are almost out of touch with reality; and by still others as an enterprise only slightly less noble than the search for the Holy Grail. To many students in introductory courses, it seems an incomprehensible jumble of numbers and symbols. In reality it is a body of knowledge that has taken form gradually over a period of about 25 centuries, growing out of man's search for an explanation of the nature and meaning of his universe and of himself.

1.1 THE NATURE OF CHEMISTRY

Chemistry is a collection of observations and theories concerning the composition of material things and the changes that they undergo. It is a vigorous and rapidly changing discipline with indefinite boundaries, encompassing such diverse goals as finding a cure for cancer and unraveling the composition of moon rock.

Chemists often refer to themselves as "scientists" and to their procedures as "the scientific method." To the layman these words carry the implication that scientific investigations are carried out on a different level from that of ordinary thought processes. The thought processes of a chemist, however, are not significantly different from those of a gambler in deciding which horse to bet on: luck, intuition, and a knowledge of past events figure strongly in both. Louis Pasteur underscored the relationship between luck, knowledge, and success in scientific discovery when he said, "Chance favors the prepared mind." A willingness to look for new meaning in familiar objects and events has played an important role in such scientific advances as Fleming's discovery of penicillin, Watson and Crick's model of the DNA molecule, and Newton's elucidation of the laws of gravity.

Penicillin was discovered by Alexander Fleming, a British microbiologist. A culture of *Staphylococcus* had been contaminated by a blue-green mold called *Penicillium notatum*. Fleming observed that the bacteria did not grow in the vicinity of the *Penicillium* and correctly surmised that the inhibition was the result of a chemical produced by the mold. Although this inhibition had been noted by others before him, Fleming was the first to recognize its importance.

1.2 MATTER

The material objects studied by chemists may be collectively referred to as *matter*. Matter may be defined as anything which has mass and occupies space; this means that matter is both tangible and real. Matter (which has mass) has two behavioral features which can be measured: (1) matter is attracted toward other matter, and (2) matter once set in motion tends to continue in motion in the same direction unless acted upon by an outside force. Although both of these features of matter will be discussed later in detail, it is worth noting that the attraction between two bodies increases as the mass of either body increases. This attraction becomes readily apparent only when the mass of one (or both) of the bodies is relatively large, as for example, the attraction between a raindrop and the earth. The attraction between two raindrops is not so obvious because their masses are so small.

The technique called weighing is, in effect, measuring the force of attraction which the earth exerts on the mass of the object being weighed. The same object would weigh less on the surface of the moon because the moon's mass is smaller, and its attraction lower, than that of the earth (Figure 1.1). However, the mass of an object is always the same, regardless of its weight.

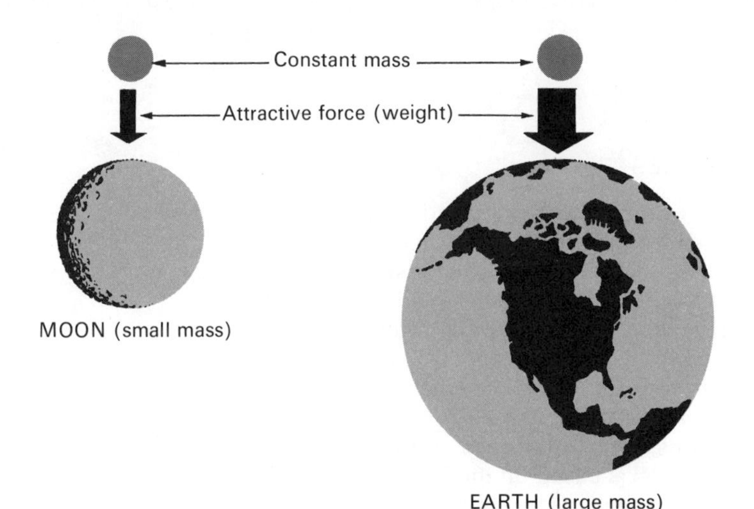

Figure 1.1
The effect of location on the mass and weight of an object. The mass remains the same regardless of its location. The weight varies in relation to other masses.

1.3 PHYSICAL STATES OF MATTER

Although the reality of earth rock in the liquid form becomes obvious enough during volcanic eruptions, earth rock is usually regarded as a rather substantial solid. However, the water which makes up a raindrop exists under ordinary conditions in three different forms or *physical states:* solid, liquid, and gas (vapor). An open bottle of carbonated beverage illustrates all three states of matter: the container is a solid, and the contents consist of a colorless gas (carbon dioxide) dissolved in a liquid (Figure 1.2). The differences among the three physical states may be determined by performing a few simple manipulations with the bottle and its contents.

A deflated plastic bag tied over the neck of the bottle will eventually be filled with a colorless gas which can be shown to be carbon dioxide. If the bag is detached and then held for a short time with its mouth against that of a similar bag containing air, appropriate tests will reveal that the contents of the two bags have mixed and are now identical (Figure 1.3).

If the liquid is poured into other containers having different shapes but with the same capacity as the bottle, the liquid adjusts to the shape of the container but the quantity of liquid does not change (Figure 1.4).

Repeated observations on the solid container show that, under the conditions of this experiment, it changes neither its size nor its shape. Table 1.1 summarizes the findings of this experiment.

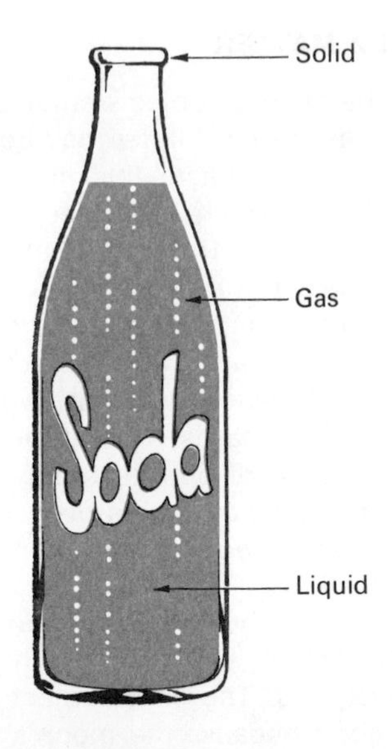

Figure 1.2
A carbonated beverage and its container. Together they represent the three physical states of matter.

TABLE 1.1 Observations Concerning the Physical States of Matter

	Solid	Liquid	Gas
Shape	Definite	Assumes shape of container	Assumes shape of container
Volume	Definite	Definite	Fills container

1.4 A MODEL OF THE PHYSICAL STATES

Because these findings are statements whose truth may be verified by repeated observations, they are referred to as facts or data. The facts do not attempt to answer such questions as "Why do solids weigh more than the same volumes of gases?" and "Why do gases distribute themselves uniformly throughout a container?" The answers to these questions are to be found in an explanation called a *model.* A scientific model is not a copy or a miniature representation of some material object, such as a model of a boat or a flower. Rather, it is a set of statements (and sometimes assumptions) whose purpose is to account for some phenomenon. Since a model is based on the facts that are available, it is subject to modification as new facts are brought to light.

If we review the observations made during the experiment, it may be possible to construct a model which will give us some insight into the physical states. The filling of a bag with carbon dioxide, and the fact that carbon dioxide and air were found in both bags, suggests that movement is a property of gases. The transparent nature of the bottle, liquid, and gas suggests that they may all be composed of particles invisible to the unaided eye. The failure of the solid or liquid to increase in volume could be the result of forces holding them together. Since the liquid underwent rather drastic changes in shape,

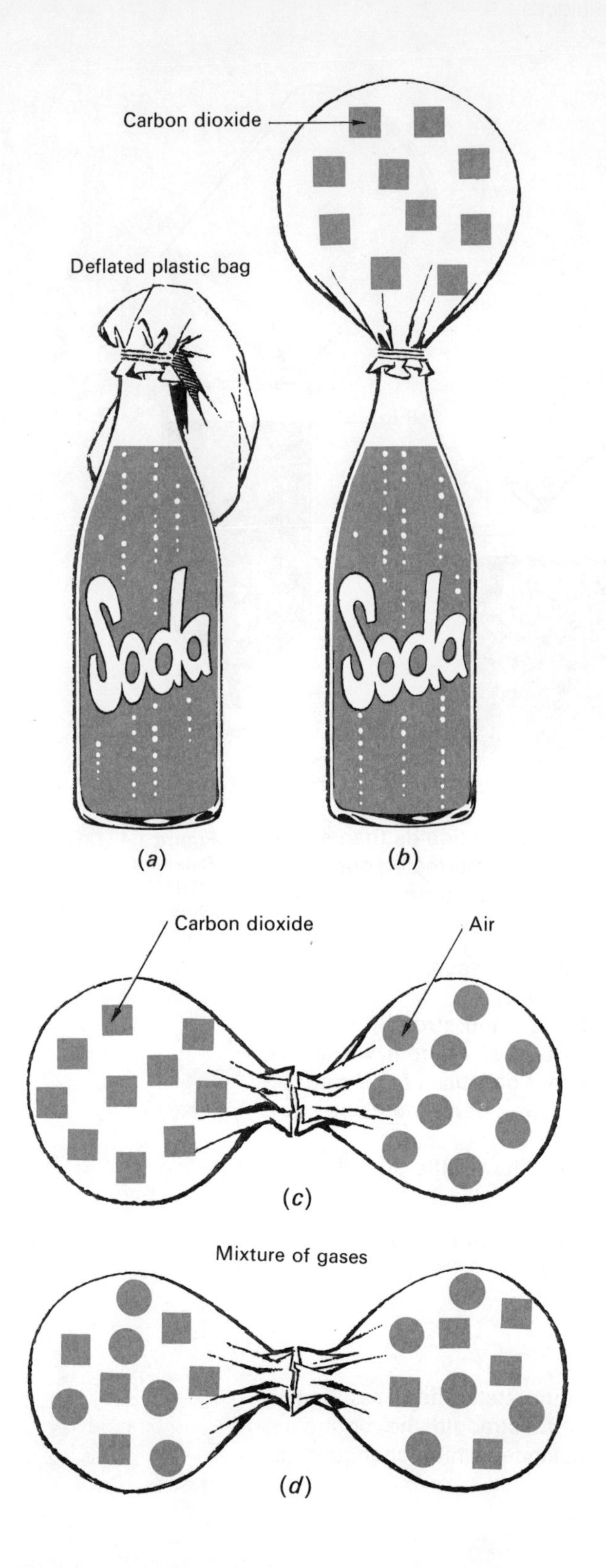

Figure 1.3
A demonstration that gases assume the shape and volume of the container. (a) Plastic bag tied over neck of open bottle. (b) Bag fills with escaping gas. (c) Mouths of bags are held together. (d) Appropriate test shows contents of bags are identical.

Figure 1.4
A demonstration that liquids adjust to the shape of the container but do not undergo a change in volume.

the attracting forces would appear to be weaker in liquids than solids. Although we would be rash to insist that we have perfected our model after only one experiment, the following model does provide a satisfactory explanation of the data in Table 1.1.

1. All matter consists of particles too small to be seen.
2. These particles are in a constant motion.
3. The particles in a solid are closely packed and strongly attracted toward each other. Their motion consists of random vibrations.
4. Liquids consist of clusters of particles, but because of weak attraction, the individual particles may break away from one cluster and join a different one.
5. The particles in a gas are moving rapidly, have little attraction for each other, and are widely separated.

A visual representation of this model is shown in Figure 1.5.

Liquid Crystals

Recent investigations of the solid and liquid states have resulted in production of substances which appear to straddle the boundary, called liquid crystals. Some of these substances which change color

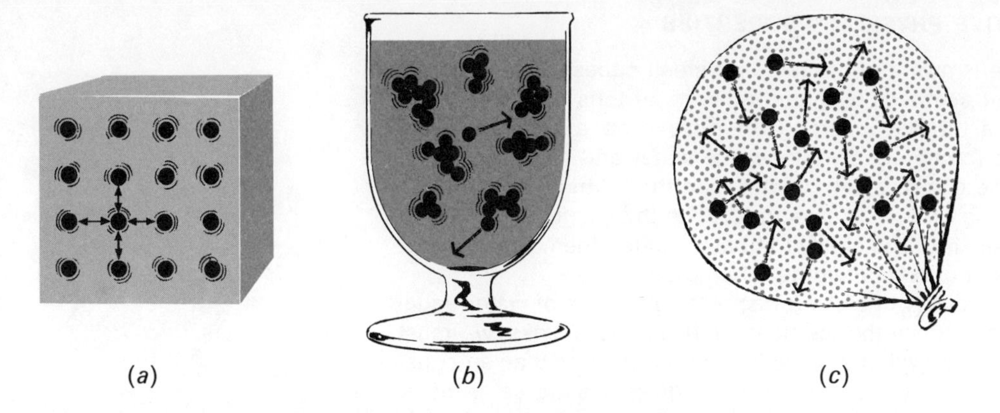

Figure 1.5
A schematic representation of the three physical states of matter. (a) The solid state. Because of strong attractions between particles, their motion is limited. (b) The liquid state. Particles are in clusters whose size is constantly changing because of weak attractive forces. (c) The gaseous state. Particles are widely separated because they have such high velocities, and attractions between them are very weak.

with extremely small temperature changes are being used as diagnostic agents in circulatory and metabolic disorders. When the crystals are applied to the skin, their color indicates the temperature in the underlying tissue. If the blood flow is restricted by blockage of a vessel, for example, heat accumulates in the tissue. The location of the blockage is indicated by the color response of the liquid crystal.

1.5 PHYSICAL AND CHEMICAL PROPERTIES OF MATTER

Water is one of the most abundant kinds of matter on earth. Although it exists predominantly as a liquid found in oceans, streams, rain, and in all living cells, it also occurs in the solid state as ice and snow, and in the atmosphere as water vapor.

Pure water is colorless except in deep layers. The problem of distinguishing between water and other colorless liquids is similar to that of differentiating between identical twins: the examination of a number of identifying features should ultimately reveal some way in which the two are different. The identifying features of matter are called *properties.* Those properties which pertain primarily to the appearance or physical state of matter, or which may be observed without causing the matter to change its identity, are termed *physical properties.* Physical properties of water include its lack of odor, its failure to be attracted to a magnet, and the fact that water will dissolve substances such as salt and sugar but not glass or wax. The lack of odor alone would enable us to distinguish water from anaesthetic ether, another colorless liquid.

In addition to its physical properties, water also possesses a number of behavioral properties; for example, it does not burn. These behavioral properties which represent the tendency of matter to undergo changes in identity are called *chemical properties.* Since anaesthetic ether is extremely flammable, this chemical property would also enable us to distinguish between it and water. Additional chemical properties will be discussed in Chapter 2.

1.6 QUANTITATIVE PHYSICAL PROPERTIES

Almost everyone is aware of the fact that small cubes of ice will float in water. Floating icebergs weighing millions of tons have also been observed. It is a bit disconcerting, then, to drop a piece of glass weighing only a fraction of an ounce into water and see it sink. The mass of the iceberg is so much greater than that of the glass, and yet the water is able to offer an opposing force to the ice great enough to prevent it from sinking. The ability to float on water, then, depends not on the total mass but on some other characteristic.

A solid in the process of sinking displaces a volume of water equal to its own volume. Since the ice floats and the glass sinks, we must assume that any given volume of ice has a smaller mass than an equal volume of water, while the same volume of glass has a greater mass. The mass of a substance which is contained in a unit of volume is another physical property of matter and is called *density.* A convenient way of expressing this idea is

$$\text{density} = \frac{\text{mass}}{\text{volume}}$$

and is read "density equals mass per unit of volume." The idea of dividing a mass by a volume may at first appear strange, but there are numerous familiar instances of such a procedure: the price of food, for example, is usually indicated in monetary units divided by units of weight, as in "cents per pound."

The calculation of the densities of glass and ice cannot be completed until we have some method of finding the mass and the volume. In addition to density, there are other physical properties which can be described only in terms of measurements of particular observations: for example, when liquid water is cooled by removal of heat energy, it eventually reaches the condition wherein the particles no longer have enough energy to remain in the liquid state, and the water becomes a solid (freezes). The energy possessed by the particles of water when this occurs can be measured by means of a thermometer and expressed in units of temperature called degrees. In the absence of unusual conditions, this change of state always occurs at the same temperature, called the *freezing point.*

When ice is heated, heat energy makes the particles move more rapidly: the ice loses its shape and melts. The temperature at which melting begins is the same as that at which freezing begins. The freezing point and the melting point are the same; if the temperature increases, the substance melts; if the temperature decreases, the substance freezes.

These examples illustrate the importance of measurement in chemistry. Were we to depend entirely upon such subjective properties as color, odor, and taste, our understanding of matter would be extremely limited. In order to determine any property of matter, such as density, we must first consider some of the tools which make measurement of observations possible.

1.7 MEASUREMENT AND MEASUREMENT SYSTEMS

Measurement is a procedure in which a numerical value is assigned to some aspect of matter such as its length, weight, or temperature. The magnitude of the measurement is determined by comparison with an accepted standard. Standards are *arbitrarily* selected. For example, a Biblical standard of length was the cubit, defined as the distance from the elbow to the tip of the middle finger. (Assuming that this distance averaged 18 inches, the dimensions of Noah's ark become roughly 450 feet long, 75 feet wide and 45 feet high.)

A more modern standard of length is the inch, which was derived from the combined lengths of three barleycorns taken from the middle of an ear of barley. The awkward process of laying out a trail of barleycorns was soon replaced by the use of a stick on which this distance was indicated by marks—the forerunner of the ruler.

The collection of measurement standards constitutes a *measurement system.* A measurement system need not be very complex: one which includes standards for length (or distance), weight, volume, temperature, heat energy, and time should be sufficient for ordinary purposes. But it is essential that the system include standards of varying sizes for measuring the same thing. It would be very awkward to measure the distance from New York to Los Angeles, using a stick with lines one inch apart!

1.8 BRITISH UNITS OF MEASUREMENT

The measurement units used in the United States today are essentially the same as those which were in use when this land was a British colony. However, this mismatched assortment of units is hardly a "system." British units of length include the inch, foot, yard, rod, chain, furlong, and mile, and there is no logical relationship between any two of them. Among the volume units are fluid ounces, pints, quarts, gills, gallons, and bushels.

Computations involving two different units in the British system can be extremely awkward as shown by the following problem: Calculate the number of square feet in an area 12 feet 2 inches long and 9 feet 6 inches wide. Each of the two measurements must first be expressed in terms of inches, their product obtained, and this result divided by 144. If either dimension includes fractional parts of an inch, the problem becomes even more complex. The awkwardness of such calculations is a persuasive argument for abandoning such a system.

1.9 THE METRIC SYSTEM

The metric system was introduced by France in 1790 and is now used by all major nations of the world except the United States, although our monetary system is constructed along metric lines. It has the advantages of simplicity and ease of calculations and is the universal measurement system for scientific work.

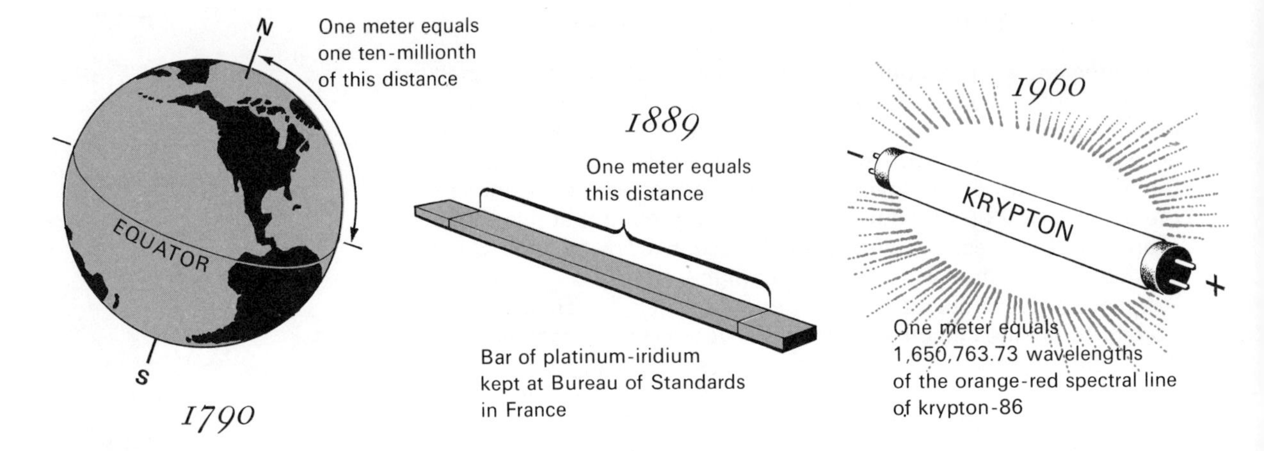

Figure 1.6
Historical changes in the definition of the meter.

Metric Units of Length

In the metric system, there is one basic unit for expressing a measurement such as length, and the name of this basic unit is modified by prefixes to produce the names of larger or smaller units. The basic unit of length in the metric system is the *meter* (m) which was initially defined as one ten-millionth of the distance from the North Pole to the equator. (Note the arbitrary nature of the standard). Its British equivalent is 39.37 inches. In the years since 1790 the meter has been redefined twice (see Figure 1.6) although its value remains essentially unchanged. Table 1.2 contains a partial listing of metric units of length and shows their sizes in terms of the basic unit—the meter. In addition to the units in Table 1.2, there are two useful units which do not contain the root word: they are the micron (μ), which is equal to 0.000,001 m (1×10^{-6} m), and the Angstrom unit (Å), which has a value of 0.000,000,000,1 m (1×10^{-10} m). The sizes of the fundamental units of matter are expressed in Angstrom units.

An examination of Table 1.2 shows that every unit is related to all other units by some factor of ten: for example, one meter is the same

TABLE 1.2 A Comparison of Some Metric Units of Length

		Magnitude		
Unit	Symbol	Decimal Form	Fractional Form	Exponential Form
kilometer	km	1000 m	1000 m	1×10^{3} m
hectometer	hm	100 m	100 m	1×10^{2} m
dekameter	dkm	10 m	10 m	1×10^{1} m
meter	m	(Basic unit equal to 39.37 inches)		
decimeter	dm	0.1 m	1/10 m	1×10^{-1} m
centimeter	cm	0.01 m	1/100 m	1×10^{-2} m
millimeter	mm	0.001 m	1/1000 m	1×10^{-3} m

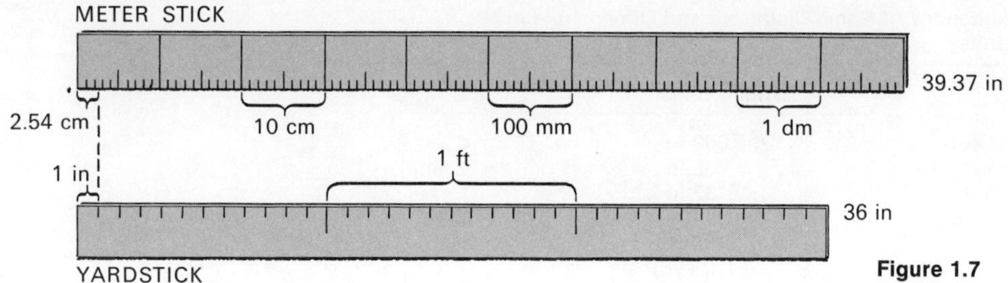

Figure 1.7
Comparison of meter stick and yardstick.

as 10 decimeters, or 100 centimeters, or 1000 millimeters. A one-meter length of wood or metal with divisions of 1 mm is called a meter stick, and is used in the same way that one uses a yardstick. Figure 1.7 shows a comparison of the two measuring instruments.

The decimal aspect of the metric system makes calculations much simpler than corresponding calculations involving British units, as shown in Figure 1.8. Detailed examples of metric calculations are given in Appendix A.

In order to provide some reference point for this new system, approximations of some distances and the dimensions of familiar objects are given in Table 1.3.

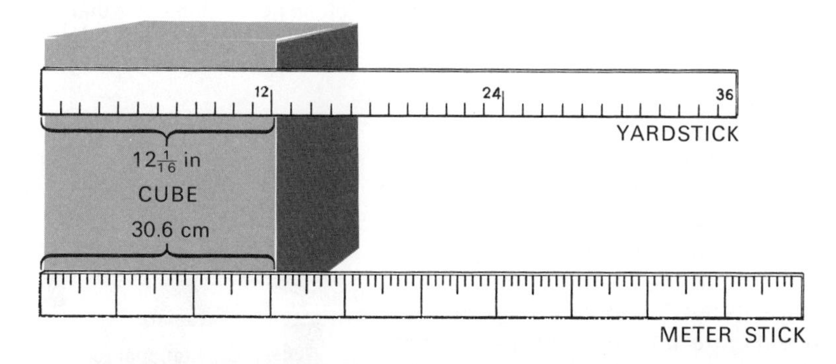

	British Measurements	Metric Measurements
Step 1	$12\frac{1}{16}$ in × $12\frac{1}{16}$ in × $12\frac{1}{16}$ in	30.6 cm × 30.6 cm × 30.6 cm
Step 2	$\frac{193}{16}$ in × $\frac{193}{16}$ in × $\frac{193}{16}$ in	Answer: 28,652 cubic centimeters (cc or cm^3)
Step 3	$\frac{7{,}199{,}057 \text{ in} \times \text{in} \times \text{in}}{4096}$	
Answer:	$1757\frac{2385}{4096}$ cubic inches (in^3)	

Figure 1.8
Comparison of a volume calculation using metric and British units of length.

TABLE 1.3 Approximations of Some Distances and Dimensions in British and Metric Units

	British Units	Metric Units
Diameter of bacterial cell	1/25,000 in	1 μ
Thickness of new dime	1/32 in	1 mm
Diameter of lead pencil	13/32 in	1 cm
Diameter of human eyeball	1 1/2 in	4 cm
Length of football field	300 ft (100 yd)	91 m
One mile	5280 ft	1600 m (1.60 km)

Metric Units of Mass

As noted earlier, the force of attraction between two objects is directly related to their masses. Because the mass of the earth is so enormous, it is possible to use the force of attraction between the earth and two smaller objects to compare the masses of the smaller objects. Suppose that we have two small spheres which appear to be identical in all respects, and we wish to know which of the two has the greater mass. For this purpose we use an instrument called a "balance," which consists of pans attached to the ends of a beam supported on a sharp edge (Figure 1.9). A sphere is placed in each pan. The sphere having the greater mass will experience the greater gravitational attraction and will move closer to the earth (Figure 1.10).

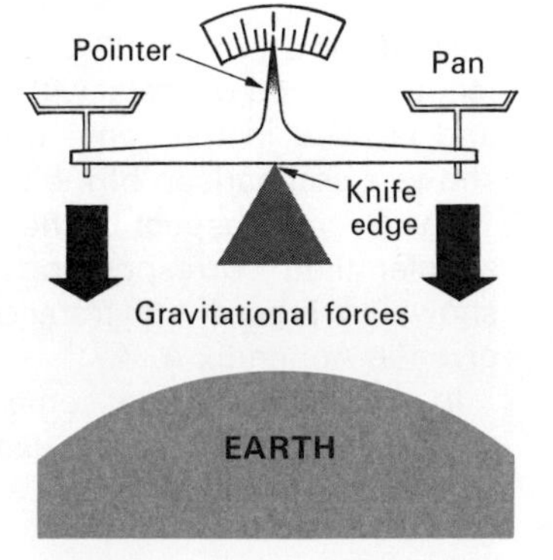

Figure 1.9
A simple balance. The position of the pointer indicates that the gravitational forces acting on the two pans are the same.

The actual mass of an object may be obtained by comparison with other objects whose masses are known. This procedure, commonly—but incorrectly—called "weighing," is shown in Figure 1.11.

The distinction between mass and weight is an important one. The *weight* of an object results from the gravitational force working on it, and is directly proportional to its mass. The weight of the body decreases as its distance from earth increases—a fact well known to space explorers who often float in their vehicles in a weightless condition. Note that, although the weight of the object varies with gravitational attraction, the quantity of matter (and therefore the mass) does not. The familiar instrument called a "scale," shown in Figure 1.12 measures weight, whereas a balance (Figure 1.11) compares an object of known mass with an object of unknown mass. Since we can assume that the location of our laboratory on the earth will not change, we will use the terms mass and weight interchangeably.

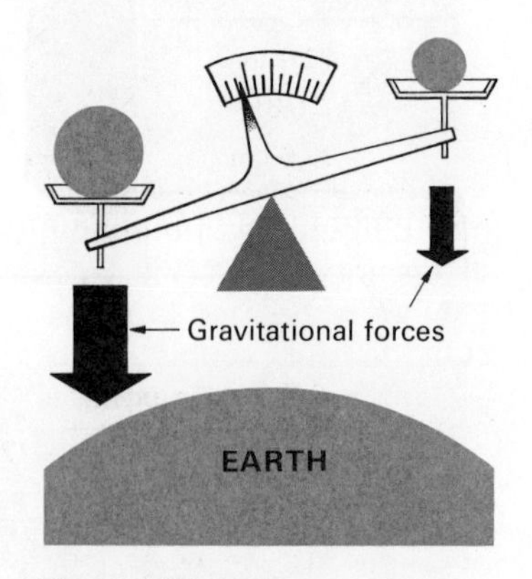

Figure 1.10
Comparison of two masses. The position of the pointer indicates that the sphere on the left has the greater mass.

The basic unit of mass in the metric system is the *gram* (g). The gram is the mass arbitrarily assigned to one cubic centimeter of water at the temperature at which water has its greatest density. The value of the basic unit is modified by the use of the same prefixes which were used for the meter, with the addition of the microgram (μg), which equals 0.000,001 g (1×10^{-6} g). A comparison of common British and metric units of weight is given in Table 1.4.

Astronaut Edward H. White II floating in a weightless condition outside the Gemini-4 spacecraft. (Courtesy NASA)

Astronauts aboard space vehicles experience weightlessness for prolonged periods of time. Weightlessness can result in weakened muscles and loss of minerals from the bone. Some individuals experience symptoms of motion sickness in a zero gravity environment. Other biological effects are still being studied. To solve the problems of weightlessness, space stations may be designed to rotate so that an artificial gravitational field is produced.

TABLE 1.4 Equivalence of Some British and Metric Units of Weight

1 ounce (oz)	=	28.35 grams (g)
1 pound (lb)	=	453.6 g
2.20 lb	=	1 kilogram (kg)

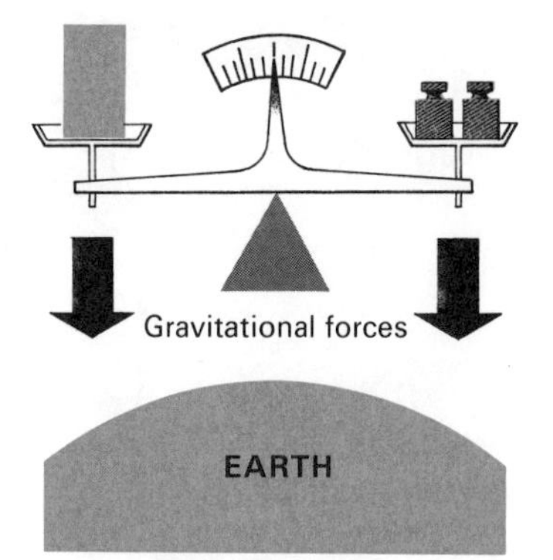

Figure 1.11
"Weighing" with a simple beam balance. The position of the pointer shows that the gravitational forces on both pans are the same, thus the mass of the object on the left equals the combined known masses on the right.

Metric Units of Volume

The volume of many regular shapes (cubes, cones, cylinders, etc.) may be obtained from multiplications involving several of the linear dimensions. Since the volumes are derived from linear dimensions, they are expressed in terms of cubic units of length, as for example, cubic feet (ft^3), cubic inches (in^3), and cubic yards (yd^3). The British system includes a number of volume equivalents such as the gallon (231 in^3), the bushel (32 qt), and the peck (1/4 bushel), but no basic unit of volume.

Volumes in the metric system may also be expressed in cubic units of length such as the cubic centimeter (cc or cm^3), or they may be given in terms of the basic unit of volume, the *liter.* The liter was originally defined as the volume occupied by 1000 g of pure water at its greatest density. Since 1 cc of water at this temperature weighs 1 g, then 1000 g of water also occupies a volume of 1000 cc. This makes a liter equal to 1000 cc and makes one milliliter (ml) equal to 1 cc. Because of this equivalence, measuring devices such as hypodermic syringes may be graduated either in milliliters or cubic centimeters. Several pieces of common volumetric apparatus are shown in Figure 1.13.

As with units of length and mass, the basic unit of volume is modified by prefixes to produce larger or smaller units: thus one liter = 10 deciliters (dl) = 100 centiliters (cl) = 1000 milliliters (ml). The relationship between a liter and the British quart is among those shown in Table 1.5.

The interconversion of metric and British units of volume is explained in Appendix A.

TABLE 1.5 Relationships Between Some Metric and British Units of Volume

1 liter (1000 ml)	= 1.057 qt
946 ml	= 1.0 qt
30 ml	= 1 fluid ounce (fl. oz)
about 5 ml	= 1 teaspoon (tsp)

Figure 1.12
A spring scale. Because of the difference in gravitational attractions at different locations, the object on the scale would have a slightly lower weight on a mountain peak than at sea level.

A triple beam balance. (Courtesy Ohaus Scale Corporation, Florham Park, New Jersey)

A single-pan, top-loading balance. (Courtesy Mettler Instrument Corporation, Princeton, New Jersey)

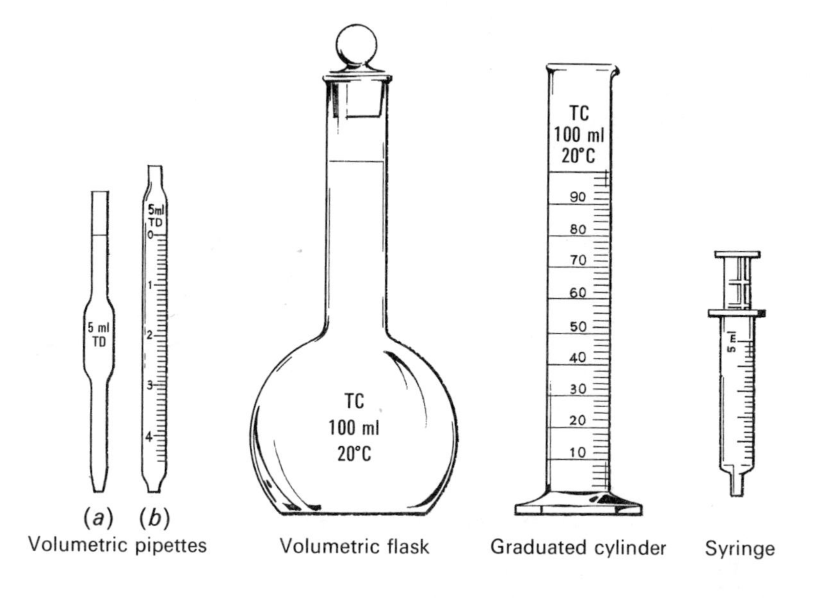

Figure 1.13
Common volumetric apparatus used in chemistry, biology, and medicine.

1.10 HEAT ENERGY AND TEMPERATURE

One of the most elusive, yet most important concepts in science is that of energy. Energy-storing and energy-releasing mechanisms are responsible for all changes which take place in our universe, whether they be the formation of stars, the melting of snow, or the growth processes of a living creature. Matter stores energy in a variety of forms. We have all witnessed the release of energy in the form of heat and light when a fuel (matter) such as coal or wood is burned. What is not generally appreciated is that energy was stored in the inanimate log or lump of coal by chemical processes which trapped some of the energy being released by the sun.

Perhaps what is so puzzling is that energy exists in so many forms: electrical, mechanical, chemical, atomic, and radiant. However, there is one property common to all forms of energy—the ability to do work—which strongly suggests that any form of energy is potentially convertible into any other form—a hypothesis which has been shown to be correct. We know, for example, that the energy of falling water can be transformed into electrical energy, which in turn can produce radiant energy (heat and light) (Figure 1.14); and that a living organism changes the energy stored in food into heat and motion.

Since the average person is probably more familiar with the characteristics of heat energy than with the other energy forms, the emphasis in this discussion will be on heat energy, and other forms of energy will be considered in later chapters.

There is a difference between heat and temperature: *heat* is a form of energy which, when transmitted to a system, increases the average kinetic energy of the particles in the system, while *temperature* is a measure of the energy level of the system. A number of instruments for measuring temperature have been devised, but one of the simplest and most familiar is the mercury thermometer—a slender glass tube sealed at one end and having at the other a small bulb containing

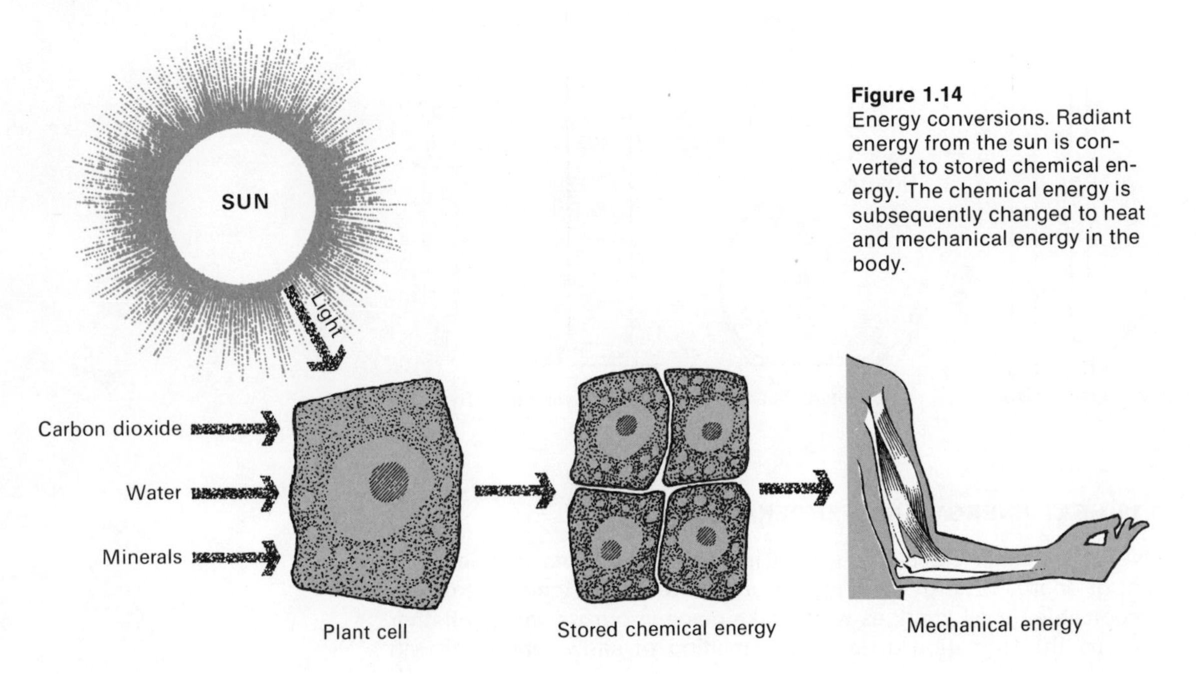

Figure 1.14
Energy conversions. Radiant energy from the sun is converted to stored chemical energy. The chemical energy is subsequently changed to heat and mechanical energy in the body.

mercury. Mercury is a silvery liquid which was chosen because it expands uniformly when heated.

The use of different reference points in calibrating a thermometer produces different temperature scales, the most important being the Fahrenheit, Celsius (or centigrade), and Kelvin (absolute). The original reference points for the Fahrenheit scale were the temperature of a mixture of salt and snow, and the armpit temperature of a human being. The thermometer bulb was first immersed in the snow-salt mixture and a mark was made on the glass at the level of the mercury column. The mark was labeled 0°F. The thermometer was then placed between the arm and the body and the level of the mercury column was marked once again. This mark was labeled 100°F. The distance between these points was divided into 100 equal degrees. The extension of this scale above the upper reference point produced a boiling point of 212°F for water. The subject whose body temperature was used in this calibration must have had a slight fever: at any rate the scale was later adjusted so that it is now 98.6°F for normal body temperature.

The reference points for the Celsius (C) scale were the temperature of an ice-water mixture (called 0°C) and the temperature of boiling water (100°C). Normal body temperature on the Celsius scale is 37°C. Temperatures on the Fahrenheit and Celsius scales have the following relationship:

$$°C = 5/9(°F - 32°)$$

Although the interconversion of Fahrenheit and Celsius temperatures is discussed in Appendix A, one example may help to clarify the relationship between the Fahrenheit and Celsius scales.

Sample problem 1.1 An American is planning a trip to Paris and wants to make sure that his wardrobe will be suitable for the climate. From his guidebook he learns that the daytime temperature in Paris at that time of year averages 15°C. What Fahrenheit temperature is this equivalent to?

Solution $15° = \frac{5(°F - 32°)}{9}$

$$9 \times 15° = 5F - 160°$$

$$135° = 5F - 160°$$

$$135° + 160° = 5F$$

$$\frac{295°}{5} = F$$

Answer $F = 59°$

It is logical to assume that if all of the heat were removed from a body, it would then be at its lowest possible temperature. Experiments have indicated that this temperature, called *absolute zero,* would be

−459.6°F or −273.16°C. If the Celsius scale is renumbered so that the absolute zero temperature becomes 0°, the Kelvin (K) [or Absolute (A)] temperature scale is produced. On the Kelvin scale the boiling point of water is 373.16°K and normal body temperature is 310.16°K. The three temperature scales are shown in Figure 1.15. We can ignore the fractional parts of a degree, and the relationship between Kelvin and Celsius temperatures becomes: °K = °C + 273°. The behavior of gases and other energy-related phenomena are discussed in terms of Kelvin temperatures.

The study of the behavior of matter at temperatures near absolute zero is called *cryogenics.* Among the useful by-products of cryogenic studies is the discovery that removal of heat drastically alters the physical and chemical properties of matter. An electric current induced in a coil of wire at temperatures near absolute zero continues to circulate through the coil indefinitely and is the nearest thing to perpetual motion yet discovered. The life processes of an organism may be stopped, either temporarily or permanently, by supercold.

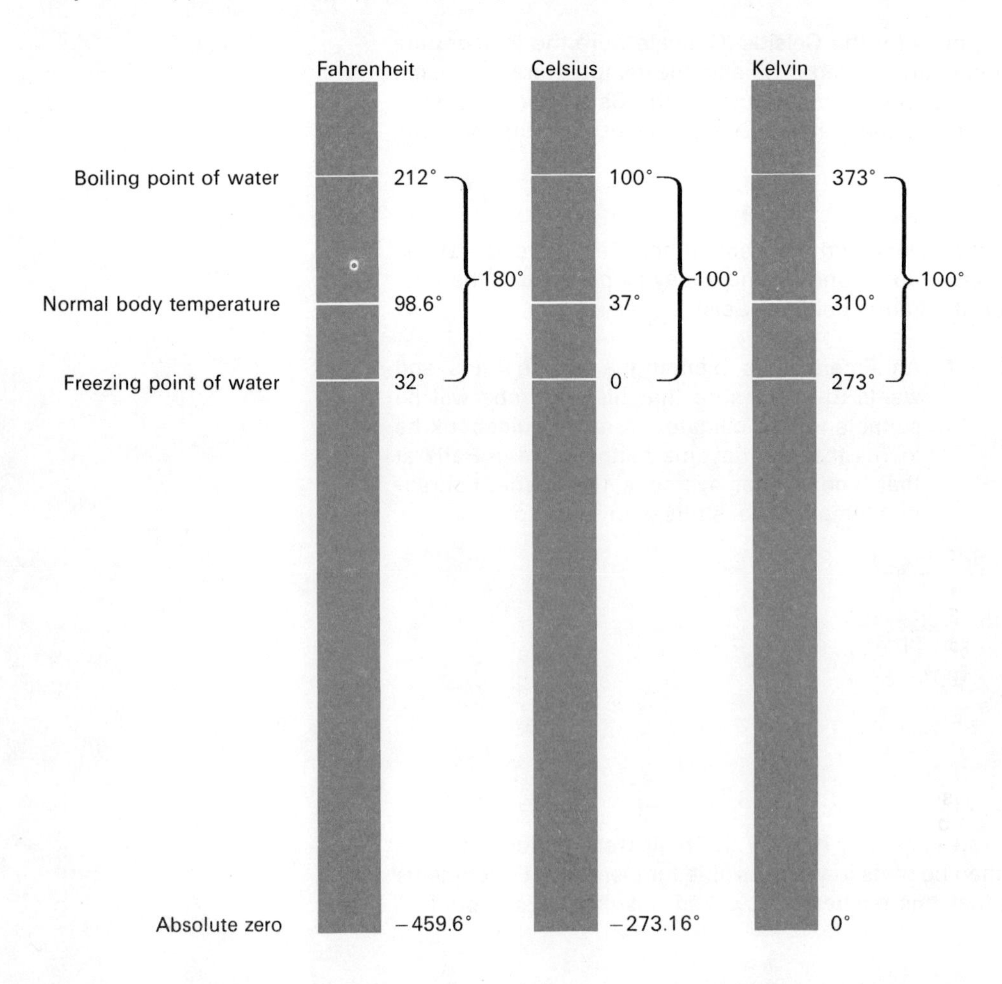

Figure 1.15
The Fahrenheit, Celsius, and Kelvin (absolute) temperature scales.

Human tissues can be preserved for extended periods by this method, permitting the establishment of organ and tissue banks. Food is also preserved from spoilage by low temperatures. But perhaps the most intriguing application of cryogenics is in the area of surgery. Instead of a scalpel the surgeon uses a needlelike probe cooled to 77°K by liquid nitrogen. Bleeding and trauma are less severe than with knife surgery. Certain types of stomach ulcers may be treated by inserting an insulated tube into the stomach through the esophagus and applying liquid nitrogen to the diseased tissue while monitoring the operation through a fluoroscope. There is every reason to believe that the benefits of cryogenic studies in biology and medicine are only beginning to be realized.

The concept of energy is a quantitative one: a large amount of change implies the action of a large amount of energy. The amount of heat energy is expressed in units called calories. A *calorie* (cal) is the amount of heat energy which will raise the temperature of 1 g of water 1°C. The dietary calorie (Cal) is equal to 1000 calories or one kilocalorie (kcal).

The apparatus used to determine the caloric value of foods is called a calorimeter (see Figure 1.16). A weighed sample of dehydrated food is burned completely in an insulated container. The energy released by the burning food is absorbed by the container and by a known quantity of water which surrounds the combustion chamber. The change in temperature of the water is directly related to the number of calories released. Because we know the amount of energy used by the body for various activities, we can adjust the intake of food so that we receive only enough calories to sustain our activities. In this way our body weight can be controlled.

Future Life for a Body Deep-Frozen?

By Walker A. Tompkins
Santa Barbara News-Press Staff Writer Copyright 1967, News-Press Publishing Co.

The body of a 74-year-old Santa Barbara feminist and peace crusader was deposited today in a hermetically sealed iron capsule in Phoenix, refrigerated by liquid nitrogen to minus 196 degrees centigrade.

At a prepaid cost of $10,000, she will be kept in "cryogenic storage" until such time as medical science develops the technology to revive her, perhaps 1,000 years hence.

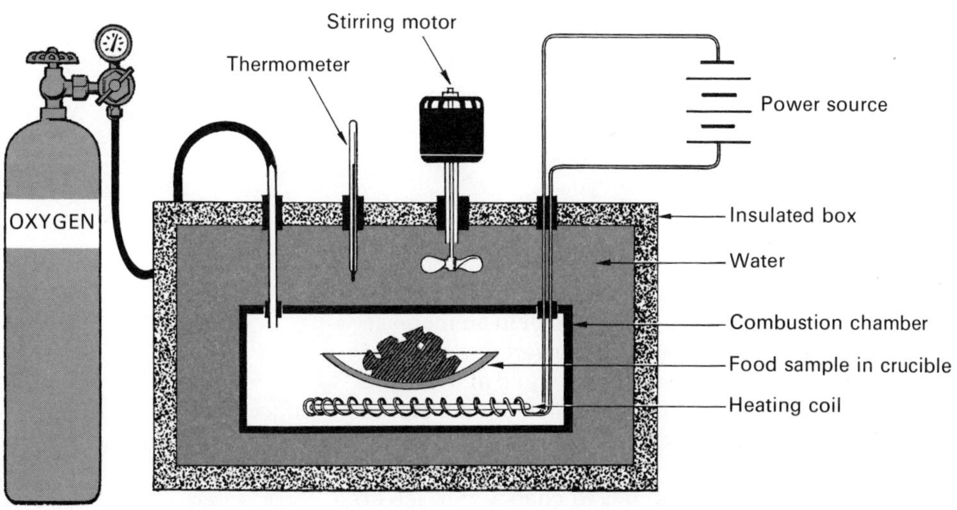

Figure 1.16
A calorimeter. Heat released by the combustion of a dried, weighed sample of food is absorbed by the calorimeter and raises its temperature. The change in temperature is directly proportional to the heat energy contained in the food.

1.11 SPECIFIC HEAT

Students in chemistry labs often make the painful discovery that while water seems to take forever to boil, the iron ring stand on which the container of water is heated quickly reaches the temperature of a branding iron. This is because different kinds of matter will store the same amount of heat with different changes in temperature. The absorption of 1 calorie will change the temperature of a gram of water by 1°C, but it takes only 0.12 calorie to raise the temperature of a gram of iron by the same amount. The number of calories needed to cause a change of 1°C in the temperature of 1 gram of a substance is called the *specific heat* of the substance. Specific heats of a number of substances are given in Table 1.6.

The relatively high specific heat of water is of extreme biological and climatic importance. Living creatures are mostly water, and the ability of water to absorb large amounts of heat without appreciable change in temperature enables an organism to maintain a fairly stable temperature even during exposure to direct sunlight. Large bodies of water may greatly influence the climate of a region. Water has a higher specific heat than either rocks or soil. During the winter months, a body of water slowly releases heat to the air long after the land has cooled, resulting in a more moderate climate.

Caloric Values of Selected Foods

Food	Calories
Apple, fresh (large)	117
Beef, hamburger (large patty)	300
Beer (8 oz)	115
Bread, white (slice)	60
Chocolate malted milk (8 oz milk)	502
Cola beverage (8 oz)	105
Egg, boiled or poached	75
Grapefruit ($\frac{1}{2}$ medium)	70
Lettuce (6 large leaves)	18
Mayonnaise (1 tablespoon)	109
Milk, whole (8 oz)	166
Peanut butter (2 tablespoons)	180
Tuna fish, canned in oil ($\frac{1}{2}$ cup)	300

TABLE 1.6 Specific Heats of Some Pure Substances

Substance	Specific Heat (cal/g/°C)	Substance	Specific Heat (cal/g/°C)
Water (liquid)	1.0	Sugar	0.27
Paraffin	0.69	Calcium carbonate (limestone)	0.20
Water (ice)	0.5	Glass	0.20
Rubber	0.45	Carbon (diamond)	0.12
Wood	0.42	Steel	0.11
Leather	0.36	Silver	0.056
Sulfuric acid (liquid)	0.34	Gold, lead, platinum	0.03

1.12 HEAT OF FUSION AND HEAT OF VAPORIZATION

When a pure substance undergoes a change of state from solid to liquid, it is the result of the absorption of a rather large amount of energy. This additional energy is needed to counteract the attractive forces which hold the solid together and does not result in an increase in temperature. The amount of heat necessary to convert one gram of a solid substance at its melting point to one gram of liquid at the same temperature is called its *heat of fusion.* The heat of fusion of water is 80 calories per gram. This quantity of energy will convert one gram of ice at 0°C to a gram of water also at 0°C. This same amount of heat must be removed from one gram of water at 0°C to change it to ice at the same temperature (Figure 1.17).

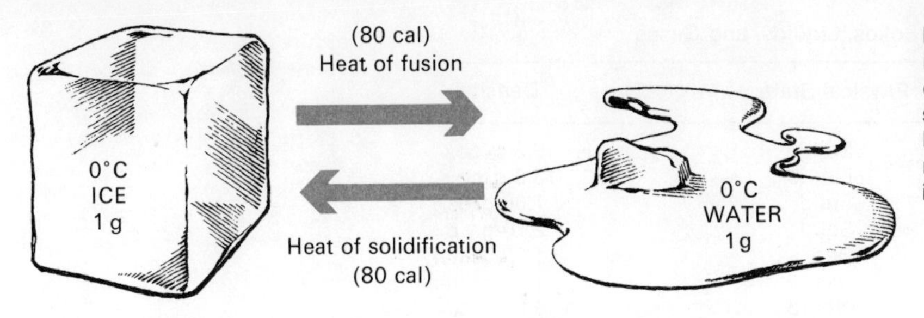

Figure 1.17
Energy changes relating to a change of state. A solid absorbs energy in changing to a liquid, and this energy must be removed from the liquid before it can solidify.

The number of calories absorbed when one gram of a liquid at its boiling point becomes one gram of vapor at the same temperature is called the *heat of vaporization*. The heat of vaporization of water is 540 calories per gram. One important aspect of the heat of vaporization of water is that steam at 100°C has a much higher energy content than water at the same temperature. When surgical instruments are being sterilized, they are placed in a tray in an atmosphere of steam rather than in the boiling water itself. Burns caused by steam tend to be more severe than those caused by hot water as one gram of steam condensing on the skin releases 540 calories of heat.

1.13 DENSITY AND SPECIFIC GRAVITY

Recall from Section 1.6 that the density of a substance is one of its physical properties and is the amount of mass contained in a given volume of the material. Expressed another way it is the number of mass units divided by the number of volume units. Now that we have units for expressing both mass and volume, it is possible to complete the calculation of the density of ice which we began earlier.

At a temperature one or two degrees below its freezing point a cube of ice measuring 2 feet on a side would have a mass of approximately 459 pounds. The total volume of the cube is 2 ft $\times$ 2 ft $\times$ 2 ft or 8 ft^3. The density of the ice is 459 lb/8 ft^3 or roughly 57 lb per cubic foot. Since pure water has a density of about 62 lb per cubic foot, it is easy to see that ice will float on water.

Note that a correct density expression includes a number together with a mass unit and a volume unit. In the metric system the densities of solids and liquids are usually given in grams per cc and the densities of gases in grams per liter.

Most substances tend to expand when heated and contract when cooled; thus the density is dependent upon the temperature. It is a common practice to include in the density statement a superscript number which indicates the temperature at which the measurement was made, e.g.,

$$D = \frac{4.56\ g^{4^\circ c}}{ml}$$

In the absence of a stated temperature it is assumed that the measurement was performed at room temperature, which is usually taken as 20°C. Table 1.7 contains the densities of a number of substances.

TABLE 1.7 Densities of Some Solids, Liquids, and Gases

Substance	Physical State	Density[1]
Gold	Solid	19.3 g/cc
Mercury	Liquid	13.6 g/cc
Iron	Solid	7.86 g/cc
Aluminum	Solid	2.70 g/cc
Carbon dioxide	Gas	1.92 g/liter
Oxygen	Gas	1.43 g/liter
Water	Liquid	0.99 g/cc
Alcohol, methyl	Liquid	0.79 g/cc
Hydrogen	Gas	0.0899 g/liter

[1] Densities shown for gases are at 1 atmosphere of pressure and 0°C. Densities for solids and liquids are at 20°C.

The density of pure water at 4°C is 1 g/cc. Iron has a density of 7.86 g/cc. A comparison of the two densities shows that a cubic centimeter of iron weighs 7.86 times as much as the same volume of water. In obtaining this information we divided the density of iron by the density of water. Note that in this operation the mass and the volume units cancel, leaving only the numerical value:

$$\frac{\frac{7.86\text{ g}}{\text{cc}}}{\frac{1.0\text{ g}}{\text{cc}}} = \frac{7.86\text{ g}\!\!\!/}{\text{cc}\!\!\!/} \times \frac{\text{cc}\!\!\!/}{1.0\text{ g}\!\!\!/} = 7.86$$

The density of a substance relative to the density of water is called the *specific gravity* of that substance. In the example just given, it is as meaningful to say that the specific gravity of iron is 7.86 as it is to say that the density of iron is 7.86 g/cc. Despite the reference to gravity, the specific gravity is a comparison of densities and has nothing directly to do with gravity.

While the determination of the specific gravity of a solid or gas usually requires both a weight and a volume measurement, it is possible to measure the specific gravity of a liquid directly by means of a *hydrometer.* A hydrometer is a sealed glass tube, weighted on one end with small metal shot, and bearing a graduated scale. It is calibrated by immersion in liquids of known densities. A hydrometer with calibrations from 0.500 to 1.000 is generally used for liquids less dense than water, and one with readings from 1.000 to 2.000, or higher, is used for denser liquids. Figure 1.18 shows hydrometers being used to determine the specific gravities of urine and battery acid. A hydrometer is also useful in wine making as the specific gravity bears a direct relationship to the alcohol content.

1.14 PRESSURE

Pressure is defined as the number of units of force on a unit of area. Imagine that you have a cube of gold each face of which measures

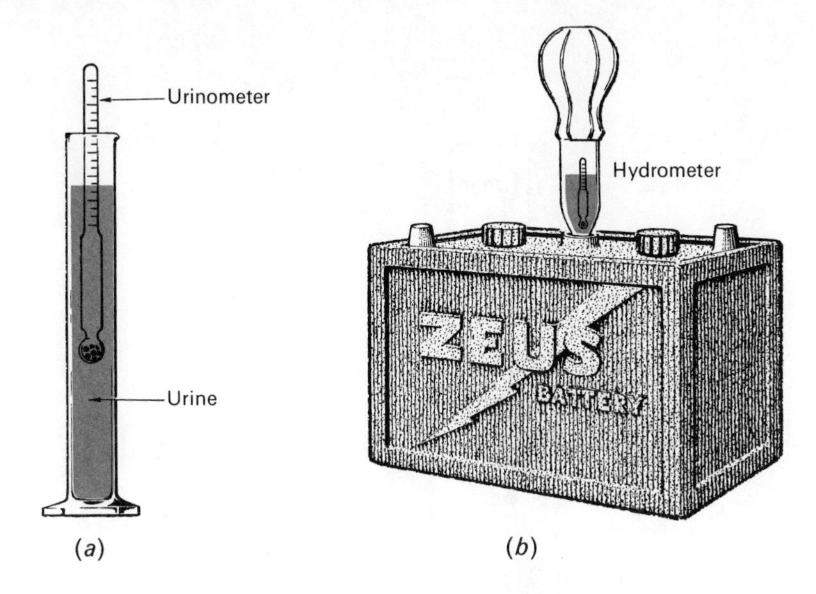

Figure 1.18
Use of hydrometers in specific gravity measurements. (a) Determination of the specific gravity of urine. The normal range is about 1.008 to 1.120: variations may indicate pathological conditions such as diabetes. (b) Measurement of the condition of a storage battery by means of a hydrometer. Specific gravity values below 1.225 mean that the battery needs to be recharged. The specific gravity of a fully charged battery is about 1.260.

2 cm by 2 cm, and that the weight of the gold is 154 g. When the cube is placed on a table, its weight is uniformly distributed over the surface on which it rests, the resulting pressure being 38.6 g per square centimeter. Since the pressure on the table can be varied by increasing or decreasing the quantity of gold, pressure cannot be used as a physical property of a material. An understanding of the concept of pressure, however, is a prerequisite to understanding the behavior of gases. And since pressure, like density, is expressed in compound units, it is logically considered in this section on measurements.

The air which makes up the earth's atmosphere is acted upon by gravitational force causing it to exert pressure on the earth. The gravitational force on the air closest to the earth is greatest, causing the air in this region to be compressed. This compression increases the density and also increases the pressure of the air—thus air density and air pressure decrease as the altitude increases.

Atmospheric pressure may be measured by an instrument called a *barometer,* the idea for which was suggested in the seventeenth century by an Italian physicist named Torricelli. A barometer may be constructed from some mercury, a glass tube about a meter in length and sealed on one end, and an open dish. The tube is filled with mercury, the open end is closed by holding something over it, and the tube is inverted in the open dish which is about half-filled with mercury. When the closure is removed, mercury drains out of the tube until the pressure exerted by the column of mercury counterbalances that exerted by the atmosphere. The height of the mercury column at this equilibrium condition will be about 29.92 inches or 760 mm at sea level (see Figure 1.19). A column of mercury 1 mm high is said to exert 1 *torr* of pressure (named after Torricelli) and so *standard atmospheric pressure* is considered to be 760 torr.

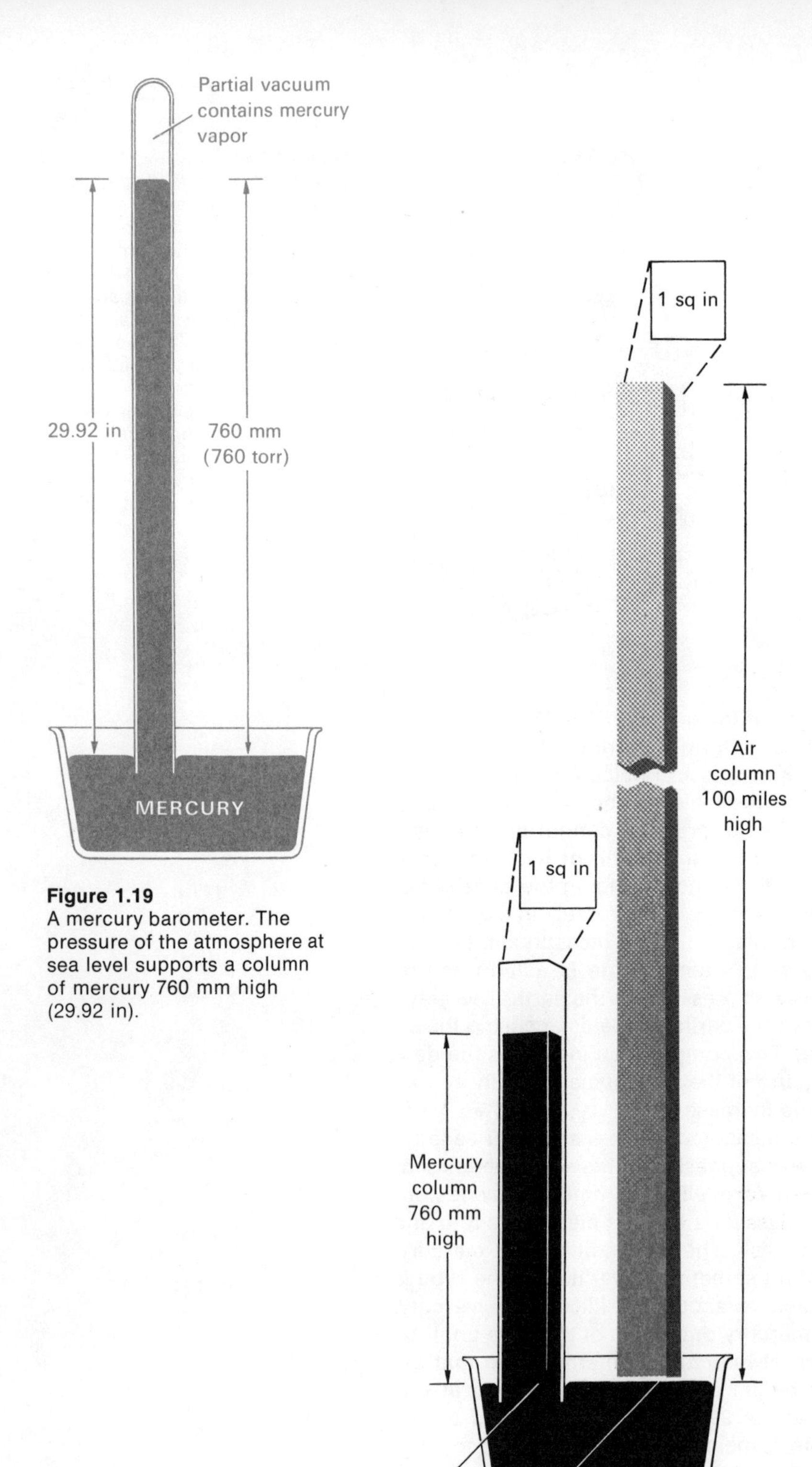

Figure 1.19
A mercury barometer. The pressure of the atmosphere at sea level supports a column of mercury 760 mm high (29.92 in).

Figure 1.20
Equilibrium between atmospheric and mercury pressure in a barometer. The pressure under the mercury column equals that under the column of air. Drawing is not to scale!

A barometer could also be constructed using a square tube with a cross-sectional area of exactly one square inch. The height of the mercury column would still be 760 mm and the column of mercury would weigh 14.7 pounds: this would create a pressure of 14.7 lb/in^2 under the column. This must be the same as the pressure under a column of air having the same cross-sectional area but extending to a height of approximately 100 miles (see Figure 1.20). Because of this equality one atmosphere of pressure may also be expressed as 14.7 psi (*p*ounds per *s*quare *i*nch).

Because mercury is 13.6 times as dense as water, a water barometer would have to be 13.6 times as high as a mercury barometer—or 10.4 meters in height.

Another instrument that measures pressure is a sphygmomanometer. This is the device used to measure the arterial pressure of the blood. The upper arm is encircled by a hollow cuff that is then inflated with air until the pressure that it exerts on the arm is greater than that which pushes the blood through the arteries, and the arteries collapse. By listening with a stethescope as the pressure in the inflated cuff is reduced, it is possible to detect the reopening of the vessels. (The pressure at which the vessels reopen is the same as the pressure exerted by the heart at the peak of its pumping cycle.) When he hears the vessels opening, the operator notes the pressure shown by the gauge attached to the cuff. The pressure varies with age, sex, physical condition, etc., but a pressure of about 120-140 mm would be normal for a 20-year-old male.

APPLICATION OF PRINCIPLES

Before you attempt to answer these questions, you should read Appendix A. You will find there a number of similar problems which have been worked out in stepwise fashion, together with some techniques for handling calculations that you may find extremely useful. The correct answers to some of the questions below are contained in Appendix B.

1. Make the following conversions:
   (a) 0.050 liter to ml (e) 5.0 in to mm
   (b) 1.42 g to cg (f) 8 oz to g
   (c) 0.40 m to mm (g) 2 pints to ml
   (d) 0.70 kg to g (h) 100 cc to ml
2. A sign on the road from Tijuana to Ensenada reads, "Ensenada 19 km." What is this distance in miles?
3. In cafes in Europe the wine is sold in glasses that contain either 1/4 liter or 1/8 liter. How many milliliters is 1/8 liter? How many fluid ounces is this?
4. What is the volume in cc of a box whose dimensions are 20 mm by 0.30 m by 10 cm?
5. The melting point of mercury is −39°C. What is this temperature on the Fahrenheit and Kelvin scales?

6. The density of concentrated sulfuric acid is 1.84 g/ml. What volume would be occupied by 1 kilogram of the acid?
7. The temperature of Dry Ice® is −108.4°F. What is this temperature on the Kelvin and Celsius scales?
8. A hospital birth record shows the following data: length—48 cm; weight—3512 g. What are these dimensions in the British system?
9. A rock weighing 35 g is placed in a graduated cylinder containing 20 ml of water. After the addition of the rock, the level of the water in the cylinder stands at 30 ml. What is the density of the rock? What is its specific gravity?
10. Oxygen gas is stored in metal cylinders at a pressure of about 2500 psi. How many atmospheres of pressure is this? How many torrs?
11. An object is weighed on a pan balance and then on a spring scale. Both weighings are carried out on the upper slopes of Mt. Everest. On which of the two instruments will the object have the greater weight? Why?
12. The average diameter of a red blood cell is about 8 μ. If the image of the cell is enlarged 1000 times by a microscope, what will the diameter of the image be?
13. How many calories of heat are given off by 20 g of water in cooling from 50°C to 20°C?
14. The liquid in the cooling system of an automobile is supposed to carry away excess heat generated by the engine. For this purpose would you recommend a liquid with a high specific heat, or one whose specific heat is low? Why?
15. What assumption would you add to the model of the three physical states that would explain why mercury rises in a thermometer?
16. Is 100 ml less than, equal to, or more than 100 cc? Explain your answer.

An alchemist's laboratory. (Courtesy Science Museum, London)

2

THE BEHAVIOR AND CLASSIFICATION OF MATTER

LEARNING OBJECTIVES

1. Distinguish between a physical change and a chemical change.
2. Given the complete description of a matter change, classify the change as either physical or chemical.
3. Compare the properties of compounds with those of mixtures.
4. Distinguish between an element and a compound.
5. Explain what brings about the separation of the various components of a mixture in each of the following processes: chromatography, distillation, electrophoresis, evaporation, and filtration.
6. Apply the principles explained in Objective 5 to predict how a given mixture might be separated.
7. Describe a classification scheme that applies to the different categories of matter.

KEY TERMS AND CONCEPTS

adsorption
chemical change
chemical property
chromatography
compound
distillate
distillation
electrolysis
electrophoresis
element
evaporation
filtration
fractional distillation
heterogeneous
homogeneous
mixture
physical change
solute
solution
solvent
symbol

If you were to open the bottles, jars, and boxes that fill the shelves of your local pharmacy, you would find that many of them contain white solids. A number of these white solids are water soluble, and some of them form solutions that are tasteless and odorless. Despite these similarities, each solution could have a profoundly different effect if they were all taken internally. Since the effect of a chemical on a living organism is directly related to its chemical properties, it is important to understand what chemical properties are, and how they differ from physical properties.

2.1 A CLOSER LOOK AT CHEMICAL PROPERTIES

Chemical properties are those that represent the tendency of matter to undergo changes in identity. The word "tendency" implies not only the inclination toward some activity, but also the inclination to refrain from action. We can say, for example, that buried treasure has a tendency to stay buried: meaning that the activities leading to its discovery are not likely to take place. If chemical properties are measurements of tendencies, then we might consider the failure to undergo a change under a given set of conditions to be a tendency toward stability, as shown by the following example.

At a temperature of 100°C water changes to steam. If the steam is heated to temperatures as high as 3000°C, its identity remains unchanged, but at temperatures beyond 3000°C water changes into two gaseous substances—hydrogen and oxygen. The behavior of water at this temperature is one of its chemical properties. To the chemist the stability of the steam at extremely high temperatures is as significant as its eventual decomposition, because it implies that the hydrogen and oxygen are held together by an attractive force of considerable size. We can emphasize one or the other aspect of this chemical property, saying either that water has high thermal stability, or that it is decomposed by high temperatures.

Water may also be decomposed into hydrogen and oxygen by direct electric current following the addition of dilute acid, as shown in Figure 2.1. Whenever water is decomposed into hydrogen and oxygen, the volume of hydrogen is always twice that of oxygen. This procedure is called *electrolysis* [< Gr. lysis, a loosening].

Among the other chemical properties of water are its varying reactions with metals: some metals, such as calcium, react with water at room temperature, decomposing the water and liberating hydrogen, while magnesium reacts in the same way with steam, and lead under these conditions is inactive.

Figure 2.1
Electrolysis of water using the Hoffman apparatus. Water is decomposed by this procedure into hydrogen gas and oxygen gas.

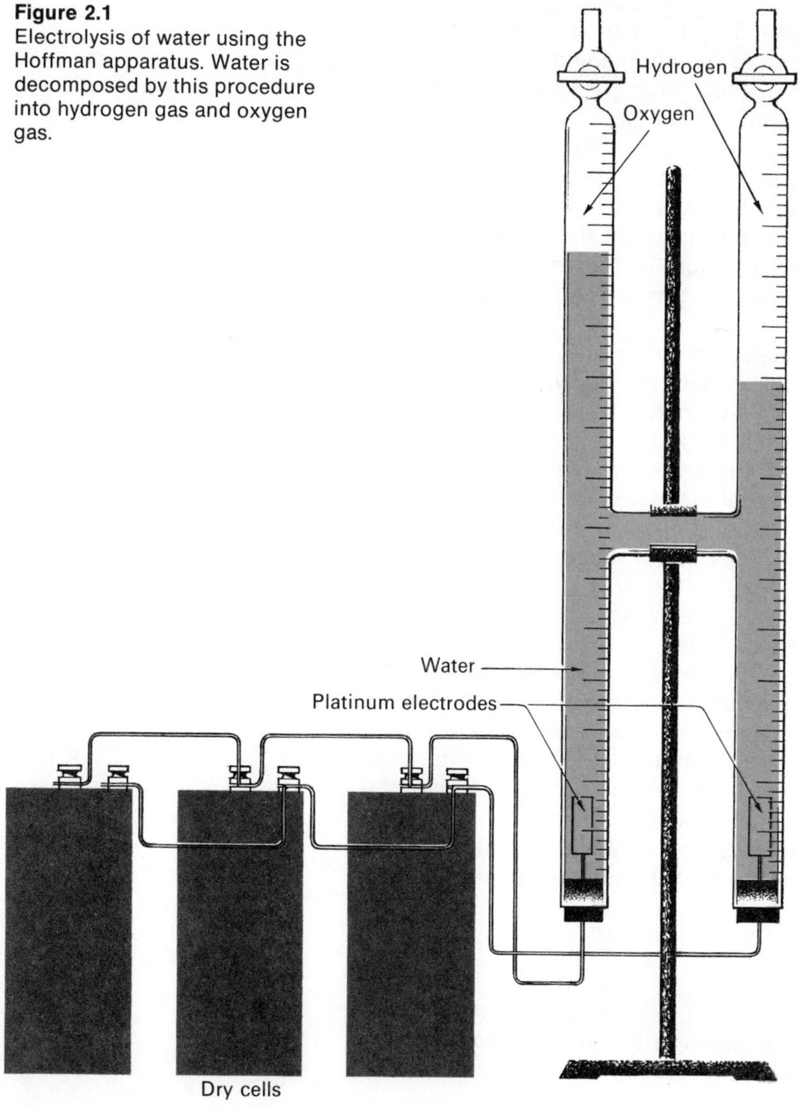

2.2 PHYSICAL AND CHEMICAL CHANGES

Many of the physical and chemical properties of water are associated with changes in the water, such as the change from liquid to steam, or its decomposition. The changes themselves can also be classified as either physical or chemical. The distinction between them is that a *chemical change* produces a new substance having a new set of identifying properties, while a *physical change* does not. For example, liquid water and steam are identical substances, while water before electrolysis and the hydrogen and oxygen obtained after hydrolysis are not. Against this background most changes are rather easily classified. For example, a piece of wood sawed into smaller lengths is still wood; but when the same wood is burned, it is changed into a mixture of gases and a residue called ash. The application of the definitions given above will show that cutting, grinding, melting, and boiling are physical changes, while decomposition and burning, because they produce new substances, are chemical changes.

So many changes are accompanied by the evolution of heat that it is easy to fall into the trap of assuming that *all* changes in which heat is released are chemical. Two very familiar exceptions to this misleading generalization are the changes produced by electricity in an electric heater and in a light bulb. Both devices include metal wires, or filaments, which resist the passage of electric current, causing some of the energy to be transformed into heat and light. The production of heat and light is, in both instances, the result of a physical change. Although an electric heater usually lasts for many years, there invariably comes a point in time (usually inconvenient) when a light bulb fails. The bulb fails because it operates at a higher temperature than the heater, and a gradual chemical reaction between traces of water vapor and the thin metal filament eventually weakens the metal until it breaks. This slow erosion has nothing to do with the production of heat and light—a process which does not produce a new substance and is, therefore, a physical change.

Another pitfall for the unwary is the process called dissolving, exemplified by the addition of salt to water. That a change does take place is obvious: the white crystals disappear and the water acquires a different taste. The salt can be recovered, however, by simply allowing the mixture (called a *solution*) to stand until the water has evaporated. Tests on the recovered salt will show that it still retains all of its original properties, including that of dissolving in water. Although the process called dissolving is sometimes accompanied by chemical changes, it is generally considered to be a physical change. Some of the exceptions to this generalization will be pointed out in later chapters.

2.3 THE ELEMENTS

The electrolysis of water (refer to Section 2.1) always produces the two gases hydrogen and oxygen. Yet all attempts to decompose either hydrogen or oxygen into simpler substances by chemical techniques fail. This fact, together with the observation that other substances besides water contain either hydrogen or oxygen, or both, suggests that these two gases are basic types of matter. The fundamental types of matter are called *elements,* and are defined as substances that cannot be decomposed into simpler substances by ordinary chemical means.

At the time of this writing 105 elements have been either discovered or synthesized. The majority of these are solids, about a dozen are gases, and only a few are liquids at room temperature. Only fifteen of the elements are synthetic: they have not been detected in nature, but the chemist has been able to produce these unique elements in the laboratory. The remaining 90 elements which are found in nature might be considered the primary building blocks of the universe, producing by their various combinations such diverse materials as plastic spoons and chromosomes. The study of chemistry is simplified, however, by the observation that, singly or in combinations, only about two dozen elements account for about 99% of all matter in the universe! Table 2.1 shows the approximate composition of the earth's crust, and Table 2.2 lists the elements found in the human body.

It is quite likely that the names, if not the properties, of many elements are already familiar to you. One element, however, deserves special mention—not only because it has some unusual properties, but also because of its role in modifying world history. The element is gold. For many years the "chemists" all over the world were involved in a search for some method of turning other metals to gold. This branch of endeavor, called alchemy, was the precursor to the later development of the science of chemistry. Wars have been fought and new continents discovered in man's search for a source of more gold.

Alchemy is likened by Francis Bacon to . . . "the man who told his sons that he had left them gold buried somewhere in his vineyard; where by digging they found no gold, but turning up the earth about the roots of the vine produced a plentiful harvest."

In its pure form gold is a lustrous solid with a tawny yellow color. With a specific gravity of 19.3 it ranks as one of the densest materials known to man. It is one of the best conductors of electric current—a fact of considerable importance to the electronics industry. Gold melts at 1063°C and boils at 2660°C. It can be drawn into fine wires (a property called *ductility*) and hammered into sheets only a few microns in

TABLE 2.1 Approximate Composition of the Earth's Crust[1]

Element	Percentage by Weight	Element	Percentage by Weight
Oxygen	45.60	Titanium	0.63
Silicon	27.30	Phosphorus	0.11
Aluminum	8.36	Fluorine	0.054
Iron	6.22	Barium	0.039
Calcium	4.66	Strontium	0.038
Magnesium	2.76	Sulfur	0.034
Sodium	2.27	Carbon	0.018
Potassium	1.84	Chlorine	0.013
		Miscellaneous	0.05

[1] From *The Encyclopedia of Geochemistry and Environmental Sciences,* Vol. 4A, edited by Rhodes Fairbridge. © 1972. Reprinted by permission of Van Nostrand Reinhold Company.

TABLE 2.2 Elemental Composition of the Human Body

Element	Approximate Percentage
Oxygen	65.0
Carbon	18.0
Hydrogen	10.0
Nitrogen	3.0
Calcium	2.0
Phosphorus	1.1
Potassium	0.35
Sulfur	0.25
Sodium	0.15
Chlorine	0.15
Magnesium	0.05
Iron	0.004
Iodine	0.0004
Copper	0.00015
Manganese	0.00013
Cobalt	Trace
Fluorine	Trace
Zinc	Trace

thickness (*malleability*). Gold is often found free in nature, but usually as a mixture with silver. It does not react with acids except aqua regia (a mixture of concentrated nitric and hydrochloric acids), and shows little reactivity toward most other substances, making it almost ideal for coins, jewelry, and dental fillings. Its softness is counteracted by adding varying amounts of another metal, usually silver, to liquid gold. The resulting mixtures are called *alloys*. In one of its more exotic applications, a thin film of gold is deposited at an angle on a specimen to be viewed through the electron microscope. The thickness of various features of the specimen can then be determined in the same manner as the height of a tree can be obtained from the length of its shadow. The high density of gold makes it particularly suitable for shadowing. The contrast between the specimen and the background results from differences in scattering of the electron beam, and a dense substance like gold or tungsten scatters the beam much more than the less dense specimen.

2.4 CHEMICAL SYMBOLS

It was mentioned earlier that one of the goals of the alchemists was the conversion of less valuable metals into gold. You can imagine that the recipe and procedure for each trial was carefully recorded, for what could be more heartbreaking than to succeed in the transformation and then be unable to recall exactly how you had done it. Alchemists also lived in fear that their secrets might be seen and used by others, and so they often used symbols in their records to indicate various substances. Many of these symbols were borrowed from astronomy. Gold, for example, was often represented by a modified astronomical symbol for the sun.

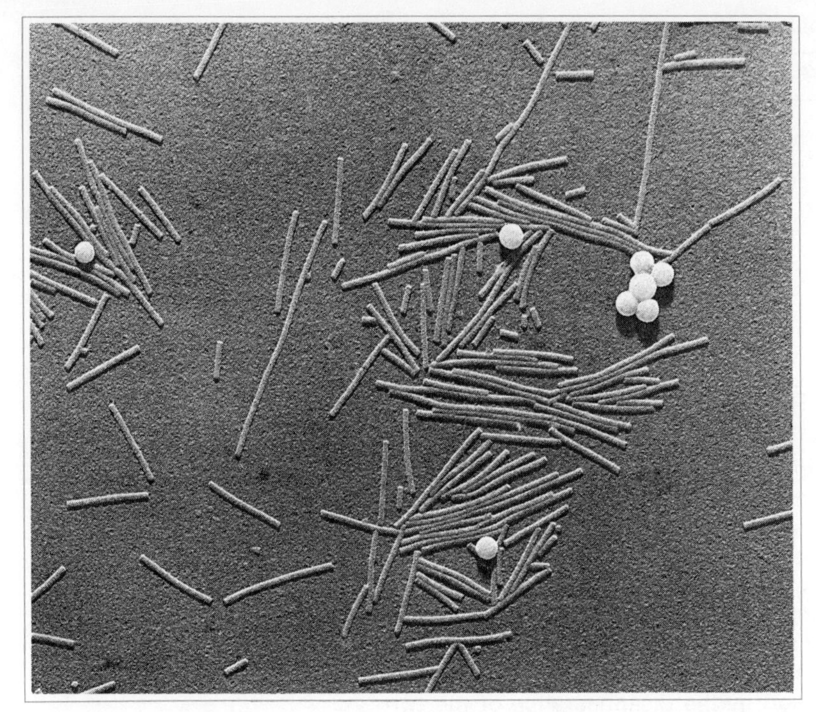

Chromium-shadowed tobacco mosaic virus in the presence of polystyrene spheres of known diameter. The thickness of a virus particle can be determined by comparing the length of its shadow with that of a sphere. (Courtesy Dr. R. B. Park, University of California, Berkeley)

After years of failure the alchemists abandoned their quest, but men continued to experiment, intrigued by the properties of matter that they observed. Their observations were shared with others, rather than hidden, but the use of nonuniform names and symbols for chemicals made communication difficult. In the early nineteenth century Jons Berzelius, a Swedish physician-chemist, proposed a system of letter symbols to replace the signs and drawings that were used to represent elements. Under his system each element is represented by a symbol consisting of either one or two letters. If a symbol consists of a single letter, it is capitalized. If it consists of two letters, only the first one is a capital. Oxygen, for example, is represented by the symbol O, and neon by Ne. A few of the symbols are derived from Latin or Greek names that are no longer used. For example, gold is represented by the symbol Au because the Latin name for gold was aurum, meaning shining dawn. Since symbols for the elements are used as frequently as their names, you will do well to learn them both. The names and symbols of some of the most important elements are given in Table 2.3. A complete list of the elements is found on the inside of the front cover.

The evolution of the atomic symbol for lead

1500	1600	1700	1808	1814
			Ⓛ	Pb

2.5 COMPOUNDS

Water has been used as an example in much of the introductory material not only because it is so familiar, but also because it is representative of an almost limitless group of materials, each having character-

TABLE 2.3 Names and Symbols of Some Important Elements

Name	Symbol	Name	Symbol
Aluminum	Al	Magnesium	Mg
Bromine	Br	Mercury	Hg
Calcium	Ca	Neon	Ne
Carbon	C	Nitrogen	N
Chlorine	Cl	Oxygen	O
Fluorine	F	Phosphorus	P
Gold	Au	Potassium	K
Hydrogen	H	Sodium	Na
Iodine	I	Silver	Ag
Iron	Fe	Sulfur	S
Lead	Pb	Zinc	Zn

istic properties and a definite composition. These substances are called compounds. A *compound* is a substance composed of two or more elements held together in a specific way by chemical bonds.

It was pointed out earlier that electrolysis of water always produces the same two elements, hydrogen and oxygen, and always in a fixed ratio. This observation is taken as evidence that water is a compound of hydrogen and oxygen. Confirmation that this hypothesis is correct is obtained by mixing two volumes of hydrogen with one of oxygen in a sealed container and passing an electric spark through the mixture: there is an explosion accompanied by the disappearance of the two gases and the appearance of water. Table 2.4 lists a number of other compounds which contain hydrogen and oxygen.

Despite their diversity, all compounds have the following characteristics in common:

1. They have a definite composition. A given compound always consists of the same elements in unvarying proportions.
2. They cannot be separated into their component elements without a change of chemical properties.
3. They have properties different from those of their constituent elements.
4. They are homogeneous. (All samples of the same compound are uniform and have the same properties throughout.)

TABLE 2.4 Composition of Some Common Compounds

Common Name	Chemical Name	Elements Present
Baking soda	Sodium bicarbonate	Carbon, hydrogen, oxygen, sodium
Table sugar	Sucrose	Carbon, hydrogen, oxygen
Milk of magnesia	Magnesium hydroxide	Hydrogen, magnesium, oxygen
Grain alcohol	Ethanol	Carbon, hydrogen, oxygen
Anaesthetic ether	Diethyl ether	Carbon, hydrogen, oxygen

2.6 MIXTURES

One of television's visual clinches is the image of a man in a long white coat against a background of fuming liquids which bubble through endless columns and spirals of glass. Laboratory manuals often include an introductory exercise in which the student uses pieces of glass tubing in an attempt to produce a slightly less exotic piece of apparatus called a washbottle. During the course of this project, the student usually makes the following observations: (1) glass which is hot enough to cause burns looks the same as glass at room temperature, and (2) progressively higher temperatures cause glass to become increasingly pliable. While the first observation is meant to be a kind of ironic humor, the behavior of glass as it is heated *is* significant.

A pure compound usually melts completely at a sharply defined temperature: the presence of any other compound has the effect of spreading the melting process over a number of degrees. Water, for example, melts at 0°C and table salt or sodium chloride at 801°C. Glass, on the other hand, does not liquefy at a specific temperature. In its initial stage, it resembles stiff taffy, becoming softer only in response to an increase in temperature. This behavior suggests that glass is not a compound but rather a mixture. A *mixture* is a combination of two or more substances, each of which retains its own identity and properties. Another way to define a mixture is to compare it with a compound as is done in Table 2.5.

TABLE 2.5 A Comparison of Compounds and Mixtures

Compounds	Mixtures
1. Have a definite composition: are composed of two or more elements in unvarying proportions.	1. May have a variable composition: both the components and their proportions may vary.
2. A compound has different properties from those of its constituent elements.	2. The components of a mixture retain their identities and properties.
3. The elements in a compound cannot be separated by physical methods.	3. The components of a mixture can be separated by physical methods.
4. Are homogeneous.	4. May be either homogeneous or heterogeneous.

Glass is rather a unique mixture because it is homogeneous—a characteristic which it shares with all compounds. Some insight into this fact may result from considering glass as a solution of compounds of sodium dissolved in oxides of silicon. As with any solution, the composition of glass can vary, resulting in slightly different characteristics. The addition of boron compounds, for example, creates glass which can undergo rapid temperature changes without cracking. Glass having this property is usually called "Pyrex," ®

although this is actually a trade name for glass products made by a particular company.

A large number of natural materials are mixtures: among them are gasoline, cheese, wood, urine, blood, dirt, tar, and air. The average composition of dry air is given in Table 2.6. The composition of air varies with the relative humidity and other local factors.

TABLE 2.6 Composition of Clean Dry Air at Sea Level

Component	Symbol	% by Volume[1]	% by Weight[2]
Nitrogen	(N_2)	78.09	75.6
Oxygen	(O_2)	20.94	23.1
Argon	(Ar)	0.93	1.28
Carbon dioxide	(CO_2)	0.0318	–
Neon	(Ne)	0.0018	0.0013
Helium	(He)	0.00052	0.000072
Krypton	(Kr)	0.0001	0.00029
Xenon	(Xe)	0.000008	0.000036
Nitrous oxide	(N_2O)	0.000025	–
Hydrogen	(H_2)	0.00005	0.0000035
Methane	(CH_4)	0.00015	–

[1] Reprinted from "Cleaning Our Environment—The Chemical Basis for Action," a Report by the Subcommittee on Environmental Improvement, Committee on Chemistry and Public Affairs, American Chemical Society, 1969, p. 24. Reprinted by permission of the copyright owner.
[2] "Composition of Clean Dry Air at Sea Level by Weight." *J. Chem. Ed.*, Oct. 1969, p. 626.

2.7 SEPARATION OF MIXTURES

An appreciation of the importance of mixtures may result from the realization that gasoline, motor oil, kerosene, diesel fuel, tar, paraffin, mineral oil, and hundreds of other products are obtained from a mixture called petroleum. This bounty is made possible by the development of techniques for separating mixtures into their component parts. Different mixtures require many different methods of separation. For example, the procedure that works so well on crude oil would turn blood into a charred black solid. The following brief survey of separation techniques is intended only to illustrate the variety of problems encountered with mixtures.

Iron in a mixture can be separated from the other components by a magnet, whereas iron in a compound is not attracted by a magnet.

Filtration

The use of filtration is by no means restricted to chemistry: it is involved in the production of beer, wines, sugar, and fruit juices, and in the usual method of preparation of our most popular beverage—coffee. A filter is essentially a fine screen which allows the passage of liquids but retains solid matter. The filters used in chemistry are normally made of special paper, but porous clay, pads of asbestos, fritted glass, and plastic membranes are all used in biology and medicine. Extremely fine filters made of tiny bits of glass that have been fused together at the edges are used to remove bacteria from liquids. An even more unique type of filter called a molecular sieve consists of an array of regularly spaced large molecules. Smaller molecules are absorbed into the spaces between the larger ones.

The use of a filter is exemplified by the brewing of coffee. When boiling water comes into contact with ground coffee beans, only those components which are water-soluble dissolve, passing in solution through the filter and leaving the insoluble solids behind.

A simple filtration apparatus consists of a circle of paper which has been folded into a cone and is supported by a glass funnel, as shown in Figure 2.2.

Evaporation

Probably the oldest known method of separation is evaporation, in which a dissolved material (called the *solute*) is separated from its dissolving liquid, or *solvent*. This technique relies upon the fact that the solvent usually changes to a vapor at a lower temperature than the

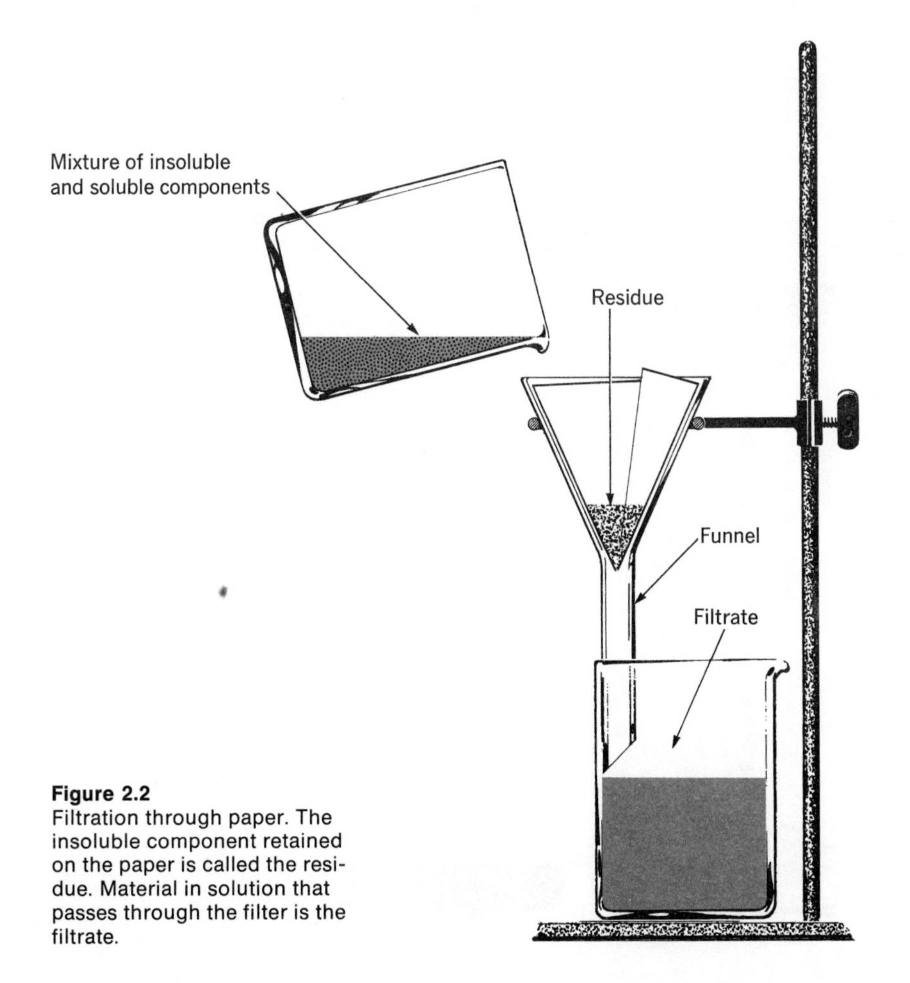

Figure 2.2
Filtration through paper. The insoluble component retained on the paper is called the residue. Material in solution that passes through the filter is the filtrate.

solute. One of its best-known applications is in the recovery of salt from sea water: this is accomplished by flooding lagoons to a depth of about 6 inches and letting the sun's energy vaporize the water.

Distillation

This process, like evaporation, makes use of the observation that components of a mixture usually have different boiling points. The temperature of the mixture is raised gradually, and as each component vaporizes, it is conducted to a different container where it is cooled and resumes its liquid state (see Figure 2.3). The condensed material is called the *distillate*. In some cases the boiling points are so close together that the distillate will not be a pure material. However, repeated distillation of the impure fraction will eventually yield relatively pure substances. This procedure is called *fractional distillation,* and is employed in the separation of crude petroleum (Figure 2.4).

The distillation of air is also of considerable importance. It is by this method that we obtain pure oxygen, nitrogen, neon, and carbon dioxide. The air is compressed and cooled. At 216°K and a pressure of 5 atmospheres the carbon dioxide becomes a solid and is removed.

Useful Products Obtained from Crude Oil by Fractional Distillation

Boiling Point Range	Product
below 20°C	Natural gas
20–60°C	Petroleum ether
60–100°C	Naphtha, ligroin
40–200°C	Gasoline
175–325°C	Kerosene, jet fuel
250–400°C	Fuel oil, diesel oil
Nonvolatile liquid	Mineral oil, grease, lube oil
Nonvolatile solid	Asphalt, paraffin wax

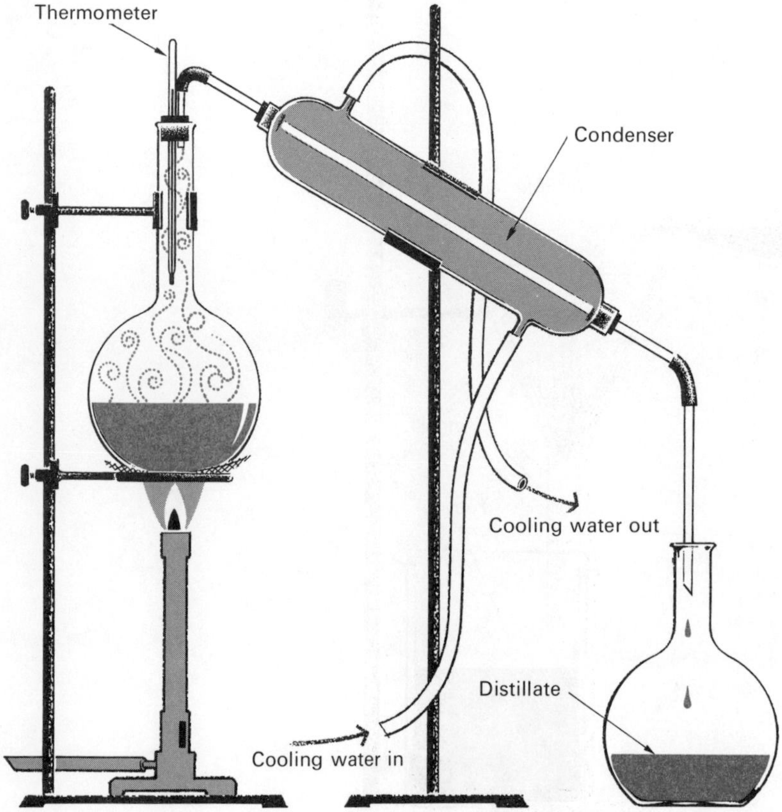

Figure 2.3
Distillation. Cold water passing through the outer jacket of the condenser causes the vapor to liquefy. The condensed vapor is called the distillate.

The cooling is continued to about 75°K, at which point all of the major components except neon have liquified. The neon is separated from the mixture and the temperature is slowly raised. The remaining gases boil off in the following order: nitrogen (77°K), argon (87°K) and oxygen (90°K).

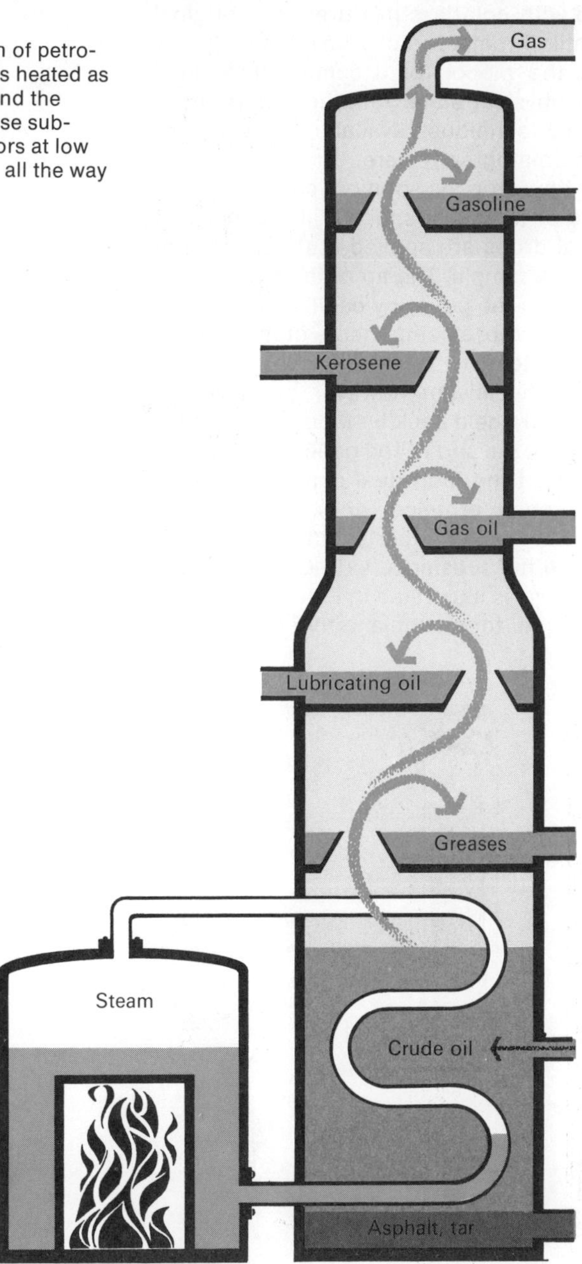

Figure 2.4
Fractional distillation of petroleum. The crude oil is heated as it enters the tower, and the vapors rise. Only those substances that are vapors at low temperatures will go all the way to the top.

Chromatography

The name of this procedure means "color graphing" and is somewhat misleading as it implies an involvement only with colored matter. The name resulted from the initial use of the technique for the separation of plant pigments, and has persisted despite the fact that most of its current applications have to do with colorless mixtures. The original concept has been expanded until chromatography now includes any separation technique involving the motion of a sample through a material which is in a different physical state. Although there are a large number of chromatographic techniques available, we will consider only paper and column chromatography here.

In paper chromatography, a tiny drop of solution containing the mixture is applied near one end of a strip of special filter paper and allowed to dry. Several additional drops are applied at the same spot to increase the concentration of the sample. The tip of the paper is immersed in a solvent which travels up the paper by capillary action. As the solvent passes the dried sample, those components of the mixture which have the greatest solubility tend to be carried along with it. Their movement is not as rapid as the solvent, however, because they are generally larger molecules and are held back by their attraction for the paper. When the solvent reaches the end of the paper, the process is complete. The location of each of the individual components may be determined by a variety of methods: proteins, for example, appear purple under ultraviolet light. The section of paper containing a particular substance may be cut out and the substance extracted into fresh solvent for further study. The process is illustrated in Figure 2.5. There are many variations of this technique: for example, a thin film of solid

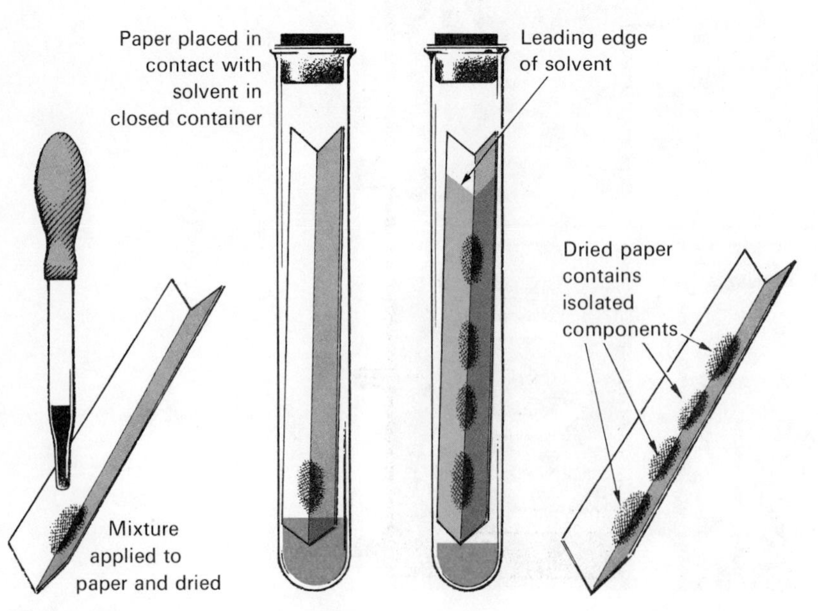

Figure 2.5
Paper chromatography. Because the paper acts like a blotter, the solvent is carried upward against the pull of gravity, carrying soluble components of the mixture with it.

deposited on a layer of glass may replace the paper. In any case, the process is the same.

In column chromatography, the sample is added to the top of a long tube packed with an adsorbent powder such as aluminum oxide. (*Adsorption* is a process in which one substance becomes attached to the surface of another.) The sample is followed by fresh solvent which flushes the mixture slowly down the column. As in paper chromatography, the progress of a given substance is the result of a tug-of-war between the solvent and the adsorbent. The individual components are collected as they emerge from the bottom of the column (Figure 2.6). By the correct choice of solvents and adsorbents this simple technique can be used to separate an astounding variety of mixtures.

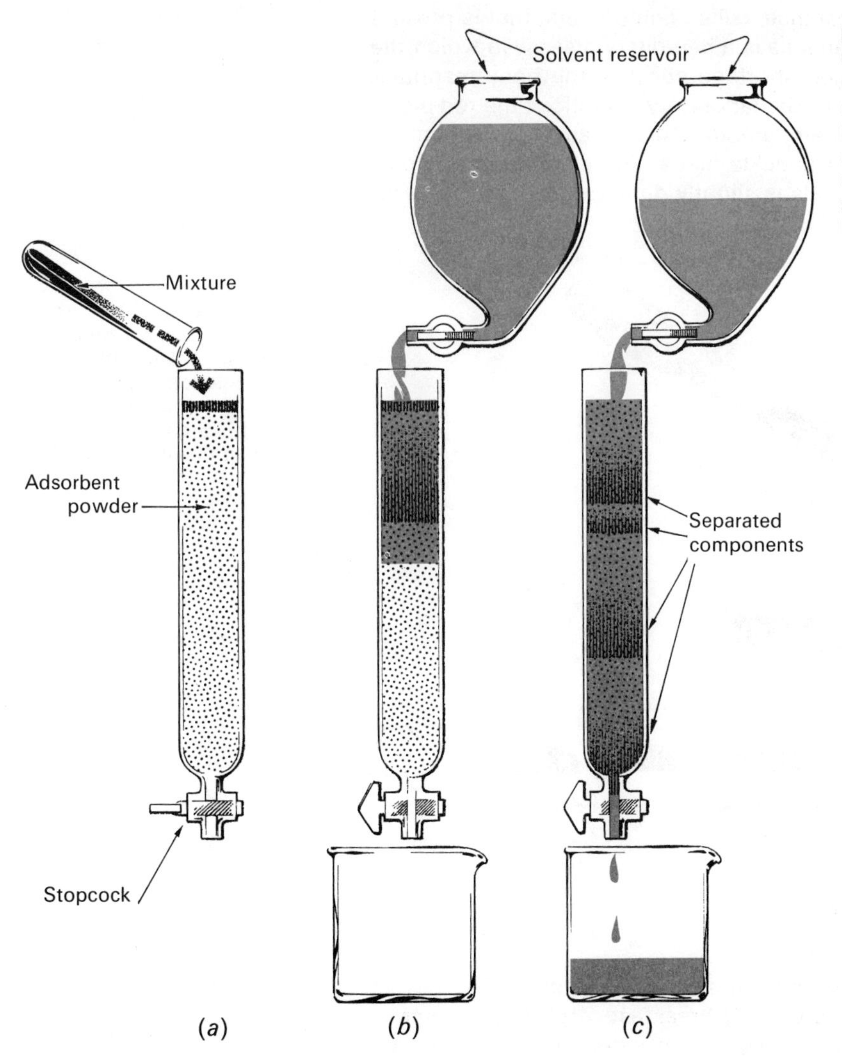

Figure 2.6
Column chromatography. (a) Mixture is introduced to column. (b) Solvent carries soluble components with it as it passes through column. (c) Components of mixture have separated and the first one is being collected.

Electrophoresis

This procedure is based on the fact that a charged substance will move in an electric field toward that electrode which has the opposite charge. Many compounds—for example, the proteins found in blood—have a slight electrical charge under the proper conditions. A sample of blood serum is placed in the middle of a horizontal paper strip which is saturated with a salt solution. The ends of the strip are connected to the anode and cathode of a direct-current power source. After electricity has coursed through the paper for some time, the strip is treated as in paper chromatography (see Figure 2.7). One of the applications of this method is the detection of abnormal chemicals in the blood, such as those present in sickle cell anemia.

Oxygen is transported to the tissues in a normal individual by attachment to a chemical substance, called hemoglobin, that is present in the red cells. Sickle cell anemia is a hereditary disease in which the tissues suffer from an oxygen shortage because the body manufactures defective hemoglobin. Under laboratory conditions the red blood cells containing abnormal hemoglobin can be caused to assume a shape somewhat like that of a sickle blade. The electrophoretic pattern of normal blood chemicals is slightly different from that of individuals with sickle cell anemia.

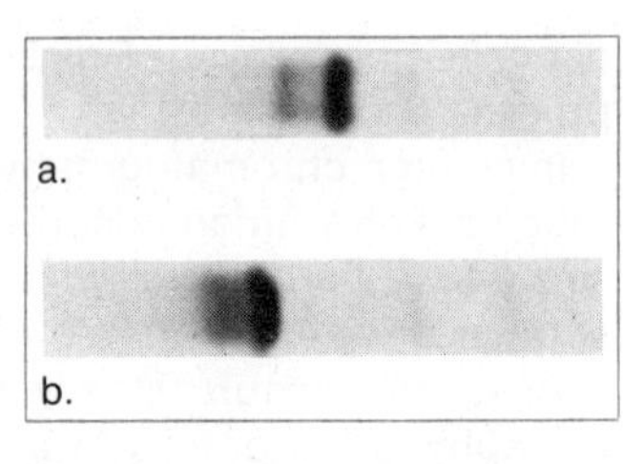

Electrophoretic patterns of hemoglobin from (a) a normal individual, and (b) a person with sickle cell anemia. (Courtesy Pathology Department, Santa Barbara Cottage Hospital)

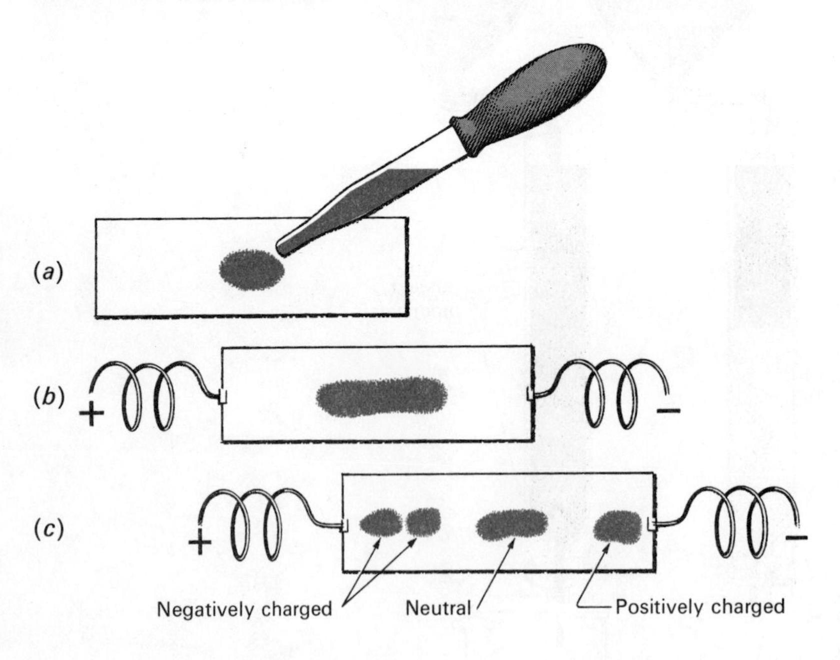

Figure 2.7
Electrophoresis. (a) Mixture is applied to a strip of porous material saturated with conductive solution. (b) Electric current passes through the strip. (c) Charged components migrate in the electrical field.

2.8 A PROPOSAL FOR THE CLASSIFICATION OF MATTER

All classification schemes, like measurement standards, are somewhat arbitrary. A person wishing to classify breakfast cereals, for example, could start by dividing them into two major categories—those which

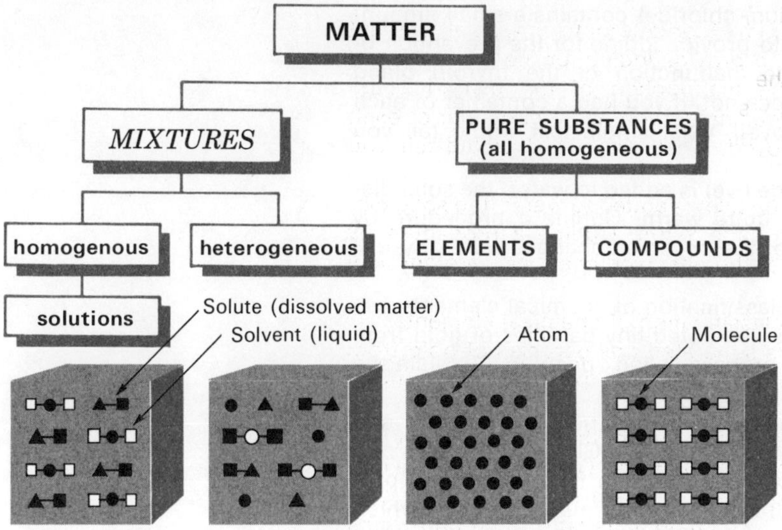

Figure 2.8
A diagrammatic classification of matter.

are served hot and those which are eaten cold. An equally logical division would be between those cereals which are presweetened and those which are not. The diagrammatic classification scheme for matter shown in Figure 2.8 is only one way of showing the differences in types of matter.

According to this proposal, every sample of matter belongs to one of two major categories: it is either a mixture or a pure substance. A pure substance is homogeneous; that is, all parts of a given sample are identical in composition. Mixtures may be either heterogeneous or homogeneous. Homogenized milk, sand, and protoplasm are heterogeneous mixtures, while solutions are homogeneous.

APPLICATION OF PRINCIPLES

1. Most biological materials (e.g., plant extracts, blood, egg white) are mixtures. Which of the separation techniques described in Chapter 2 would be *least* appropriate for such mixtures? Why?
2. Classify each of the following events as either a physical or chemical change. (You may need to do a little reading on these.)
   (a) souring of milk
   (b) drying of paint
   (c) hardening of concrete
   (d) fizzing of champagne when it is opened
   (e) thickening of cream as it is whipped
   (f) solidification of a gelatine dessert
   (g) production of heat when a wire is bent rapidly back and forth
3. "The membrane surrounding a living cell is a filter." Do you feel that this is a correct statement? Support your answer.

4. Commercial table salt (sodium chloride) contains a small amount of added potassium iodide to provide iodine for the prevention of a condition caused by the malfunction of the thyroid gland (goiter), but the pure salt does not. If you had a container of each and the labels were destroyed, what single test would tell you which was which?
5. When solid sodium hydroxide (lye) is added to water, the solid dissolves and the water gets quite warm. Outline a procedure by which you could tell whether the observed changes were physical or chemical.
6. Propose a scheme for the classification of chemical elements.
7. Early California gold miners separated tiny particles of gold from large quantities of dirt by a process called "panning." Explain the theory behind this process.
8. You are given two colorless liquids. One of them is pure water and the other is a solution of glycerine in water. Glycerine is also a colorless liquid at room temperature. Explain how you would proceed in the laboratory to tell which liquid is pure water and which is the glycerine solution?
9. Classify each of the following materials as an element, compound, or mixture:
   (a) beer
   (b) wood
   (c) toothpaste
   (d) baking soda
   (e) milk of magnesia
   (f) soap
   (g) paint
   (h) calcium
   (i) rubbing alcohol
10. The Food and Drug Administration has cautioned against eating certain kinds of fish because they contain large amounts of mercury. Since mercury is completely insoluble in water, how can this be so?

Night scene in Las Vegas. The different colors of the so-called neon signs are generated by excited electrons in atoms of various elements. (Courtesy Las Vegas News Bureau)

3

ATOMS: THE EVOLUTION OF AN IDEA

LEARNING OBJECTIVES

1. Summarize Dalton's Atomic Theory and describe how it has been modified by subsequent discoveries.
2. Summarize the behavior of charged particles as described by Coulomb's Law.
3. Explain the origin and significance of the lines in the emission spectrum of hydrogen.
4. Summarize the main features of the Thomson, Rutherford, and Bohr models of the atom.
5. Describe the electromagnetic radiation spectrum and the relationship between the wave length, velocity, and frequency of electromagnetic radiation.
6. Describe the part played by the Heisenberg Uncertainty Principle in the development of modern atomic theory.
7. Describe the quantum mechanical model of the atom.
8. Apply Hund's Rule to determine the orbital notation of an element having an atomic number between 1 and 20.
9. Apply the "$n + l$" Rule to determine the order in which orbitals are filled.
10. Apply the rules for the construction of Lewis diagrams.

KEY TERMS AND CONCEPTS

alpha particle	Lewis diagram
atom	mass number
atomic number	"$n + l$" Rule
Aufbau Principle	neutron
beta particle	nucleus
Coulomb's Law	Octet Rule
electromagnetic radiation	orbital
electron	orbital notation
emission spectrum	Pauli Exclusion Principle
energy level	proton
gamma radiation	quantum number
Heisenberg Uncertainty Principle	radiation
	radioactivity
Hund's Rule	spectrum
isotope	valence

Few topics have intrigued man as much as the questions of his origin and essence. In the absence of conclusive evidence man frequently attempted to answer these questions by proposing imaginative theories concerning the nature of the universe. Since these theories were philosophical rather than scientific, it is not surprising that early concepts of matter also had a strong philosophical flavor.

3.1 THE "ATOMOS" OF THE GREEKS—MORE PHILOSOPHY THAN REALITY

The idea of the atom may have evolved indirectly from the ancient Chinese view of matter which was contained in the Shu Ching (about 1700 B.C.), wherein it was proposed that all matter consisted of combinations of five elements: earth, metal, air, fire, and water. The existence of elements was restated in 444 B.C. by Empedocles, who rejected metal but adopted the rest (Figure 3.1). In addition to being a philosopher, Empedocles was something of a scientist, and performed an experiment by placing his finger over the smaller opening of a funnel and submerging the larger end in water. The air trapped in the funnel prevented water from entering until he removed his finger, and so he concluded that air was a form of matter.

In 430 B.C. Leucippus suggested that any primary substance existed in the form of minute particles which were identical and which were separated from each other by a void. His pupil, Democritus, conceived of these particles as hard spheres of varying sizes and applied the word "atomos" [< Gr., uncut] to them, meaning that they were the fundamental units of matter and could not be further subdivided.

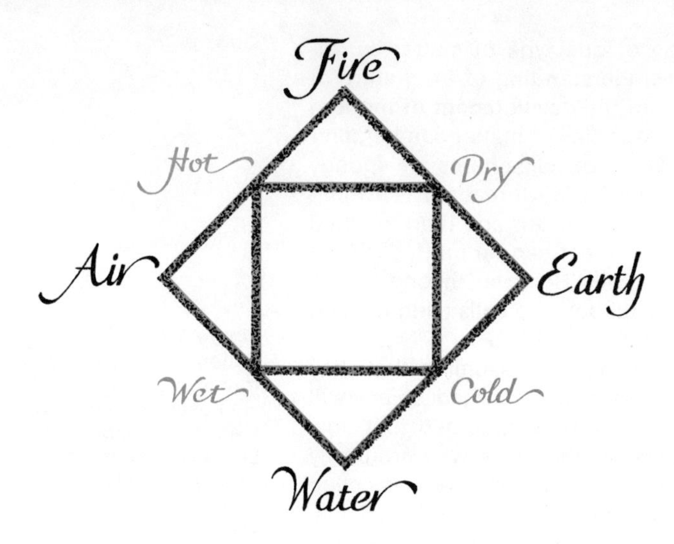

Figure 3.1
An ancient symbol showing how all matter was thought to be composed of four elements.

Today the word "atom" still refers to the particles of matter, but they are no longer considered to be indivisible.

For almost twenty centuries the concept of the atom was embroidered by almost everyone who considered himself to be either philosopher or scientist. Among the theories was one that declared atoms to have the properties of the substances they composed. For example, atoms of skin were soft and smooth while those of rock were hard and rough. The discovery of bacteria by Anton van Leeuwenhoek resulted from his microscopic examination of pepper in an attempt to learn whether its atoms produced a sharp sensation on the tongue because they were pointed and sharp.

Although microscopes thousands of times more powerful than Leeuwenhoek's have been perfected, man is still waiting for his first glimpse of an atom. How, then, do we know that atoms exist? Although absolute proof is lacking, the evidence is overwhelming. Surprisingly, the first clues were obtained from experiments involving static electricity.

3.2 THE ELECTRICAL NATURE OF MATTER

In the early part of the seventeenth century William Gilbert observed that when amber was rubbed against fur, the amber then attracted feathers and other lightweight objects. He coined the work "electricity" to describe this property—the Greek word for amber being elektron. It was later discovered that this was not an exclusive property of amber, and that the same effect was produced by rubbing a glass rod with silk, or hard rubber with fur. Most of us have probably observed the way in which different fabrics cling to each other when clothes are removed from a dryer, or that in dry weather a comb attracts bits of tissue after use.

Experiments involving the attraction of one type of matter by another have contributed heavily to our understanding of the nature of matter and have had a prominent role in the development of modern atomic theory. Some of the electrical properties of matter can be demonstrated by the use of two small foil-covered spheres of spongy wood called pith balls, which are suspended by threads a few millimeters apart. When a rubber rod is stroked with fur and then touched to both balls, they move apart. The same effect can be produced by a glass rod which has been rubbed with silk. However, if one ball is touched by a glass rod and the other by rubber, the balls cling to each other. These experiments are illustrated in Figure 3.2.

The behavior of the pith balls is explained by assuming that they acquire a charge from the rods, and, since any charged object will either attract or repel a charged ball, it is further assumed that only two types of charges exist. The charge on the glass was arbitrarily designated positive (+) while that on the rubber and amber was called negative (−). It may be concluded from these experiments that like charges repel and unlike charges attract. Further studies showed the force of repulsion or attraction to increase as the charge on either ball increased, and to diminish rapidly as the balls were moved apart. This

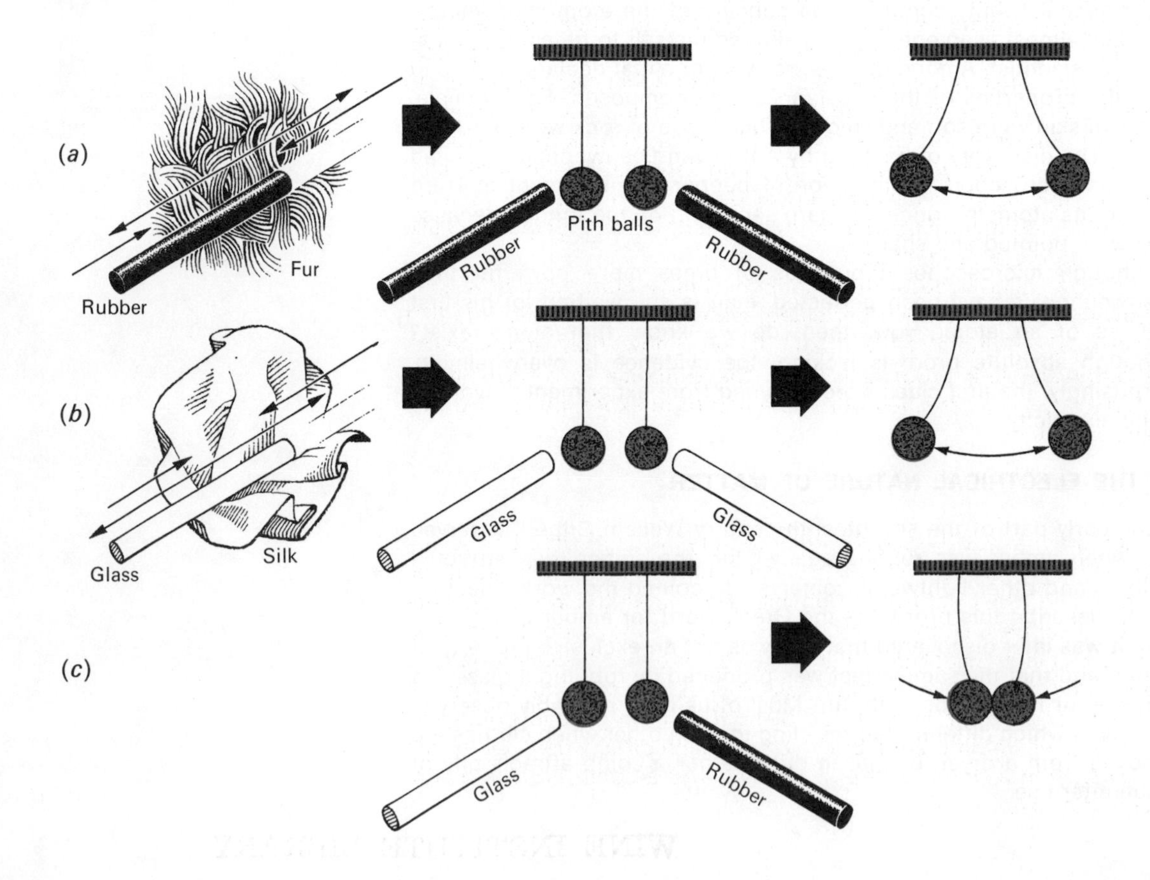

Figure 3.2
Experiments demonstrating the forces of attraction and repulsion between various types of matter. (a) Rubber rods stroked by fur are touched to pith balls, causing them to repel each other. (b) Glass rods stroked with silk produce a similar effect. (c) Pith balls touched by different rods attract each other.

behavior is summarized in *Coulomb's Law,* which says that the force is directly proportional to the product of the separate charges, and inversely proportional to the square of the distance between them. The symbolic statement of Coulomb's Law is

$$F \propto \frac{q_1 q_2}{d^2}$$

where q_1 and q_2 are the respective charges, d is the distance between their centers, and $\propto$ means "is proportional to." This is an important concept and will be used in connection with the structure of the atom.

In 1647 Otto von Guericke invented a machine to produce static electricity using a moving belt which rubbed against a ball of sulfur; this was the first of many such devices which mimicked the action of the amber on fur. In 1745 E. von Kleist showed that electricity could be stored in a container which is called a Leyden jar (Figure 3.3). It was shortly after this that Benjamin Franklin performed the experiment which demonstrated that lightning and static electricity are essentially the same. Finally, as the eighteenth century drew to a close, Alessandro Volta produced electricity in a way which involved no moving parts, using instead a number of metal plates immersed in chemical solutions, and provided additional evidence that matter could be separated into positive and negative components.

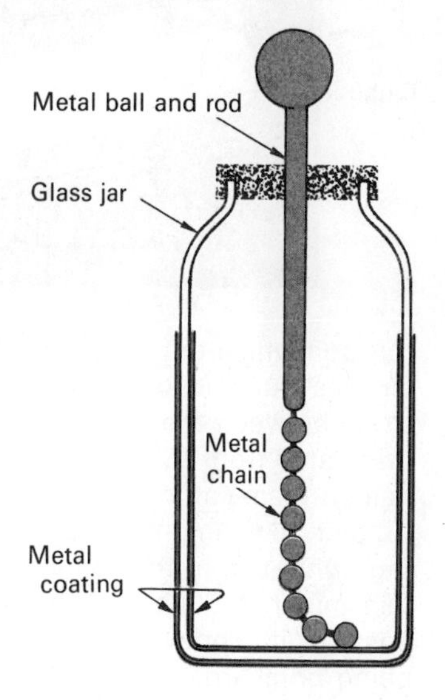

Figure 3.3
A Leyden jar for storing static electricity.

3.3 THE DALTONIAN ATOM

In 1803 an English schoolmaster named John Dalton revived the idea that all substances are composed of atoms and published a paper in which he used the atomic concept to explain certain aspects of chemical reactions. The major points contained in his paper have come to be known as the *atomic theory* and may be summarized as follows:

1. All matter is composed of tiny particles called atoms.
2. Atoms cannot be created, destroyed, subdivided, or interconverted.
3. Atoms of any particular element are identical in all properties and are different from atoms of other elements in these properties.
4. Chemical change is a union, separation, or rearrangement of atoms.
5. Atoms combine in simple whole-number ratios.

Subsequent discoveries have required only slight modification of this theory; it remains as one of the foundation stones of modern chemistry.

3.4 SUBATOMIC PARTICLES

One of the experiments which was to lead eventually to a change in the atomic theory was performed by William Crookes in 1879. His apparatus was a sausage-shaped glass tube having a metal disc, called an electrode, sealed in either end and connected to a source of electricity (Figure 3.4). In the center of the tube was an opening

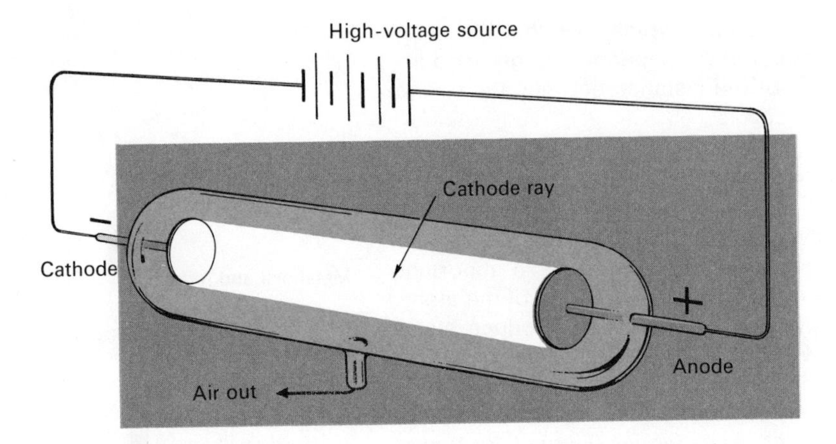

Figure 3.4
A Crookes cathode ray tube. As air is removed from the tube, the cathode ray appears between the electrodes.

through which the air could be removed. Crookes noticed that as the air was pumped out, a flickering ray of greenish light appeared to travel between the two electrodes. He was able to show that the rays originated from the negatively charged electrode, or *cathode,* and so named them cathode rays.

Later experiments showed that cathode rays could affect photographic film in the same manner as sunlight, and could cause certain minerals such as sphalerite (zinc sulfide) to glow. Glass plates coated with sphalerite were observed under high magnification while being bombarded with cathode rays. The light given off by the plate was observed to consist of multiple pinpoint flashes, suggesting that the cathode rays were actually a stream of extremely small particles.

Much of the credit for solving the riddle of the cathode rays must go to J. J. Thomson. By varying the material of the cathode he showed that cathode ray production was a general property of matter. He also demonstrated that the stream could be deflected by using charged plates or a magnetic field. The cathode ray was repelled by the north pole (negative plate) and attracted to the south pole (positive plate) (Figure 3.5). Since light did not have this property, he concluded that cathode rays were streams of negatively charged particles, which he called *electrons*. His conclusion was shown to be correct several years later when the weight of an electron was indirectly determined by Robert Millikan.

At almost the same time Eugene Goldstein used a Crooke's tube with a perforated cathode and noticed colored rays passing through the holes in the direction away from the anode. Because of their appearance, he called them *canal rays,* and they were later demonstrated to be positively charged particles of varying sizes. The smallest of these particles had a charge which was equal in size to that of the electron, while its weight was about 1840 times as great. The positively charged particle was called a proton, and its discovery reinforced the belief that matter was constructed of oppositely charged particles.

By this time experiments with Crooke's tubes had become as popular as campaign slogans in an election year. Wilhelm Roentgen dis-

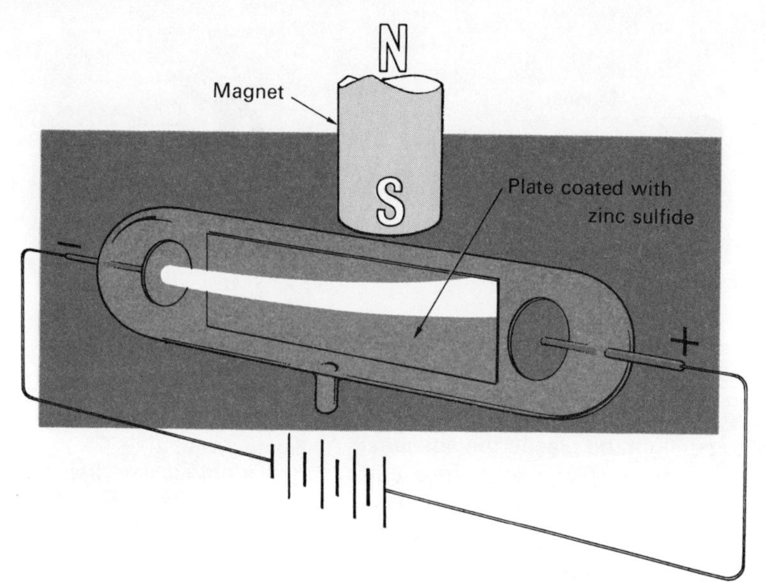

Figure 3.5
Behavior of cathode rays in a magnetic field. The rays are attracted toward the south (+) pole, showing them to have a negative charge.

covered that increasing the voltage produced invisible radiations that not only affected photographic film, but were capable of penetrating the opaque film wrapper as well. Unlike the cathode and canal rays, these radiations had neither mass nor charge, so they were called *x-rays*. (Their characteristics will be discussed later in this chapter, and their use will be discussed in Chapter 16). Like cathode rays, x-rays were capable of stimulating zinc sulfide to give off visible light; but in the case of x-rays, the mineral continued to emit light even after the radiation was turned off. This phenomenon, which involves the delayed conversion of invisible radiation to visible light, is called *phosphorescence* (Figure 3.6).

In 1896 Henri Becquerel was investigating the possibility that phosphorescence might occur in reverse, that is, that visible light might cause certain minerals to emit x-rays. His procedure called for exposure of the mineral to intense sunlight and then placing it in contact with film wrapped in black paper. Subsequent blackening of the film would be proof of reverse phosphorescence. Once, when cloudy skies

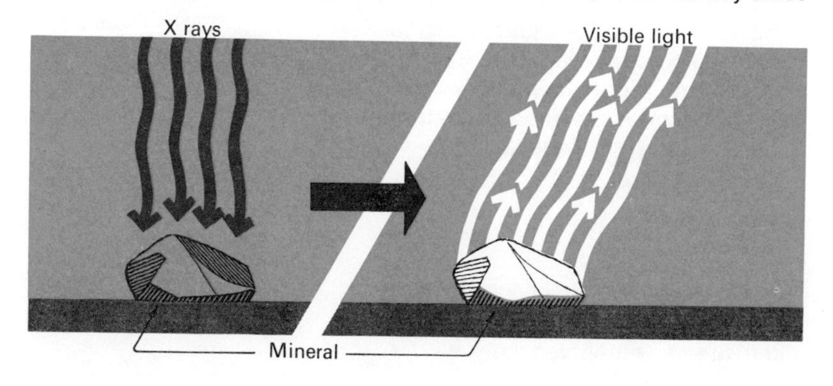

Figure 3.6
Phosphorescence. The mineral absorbs invisible radiation and later emits visible light.

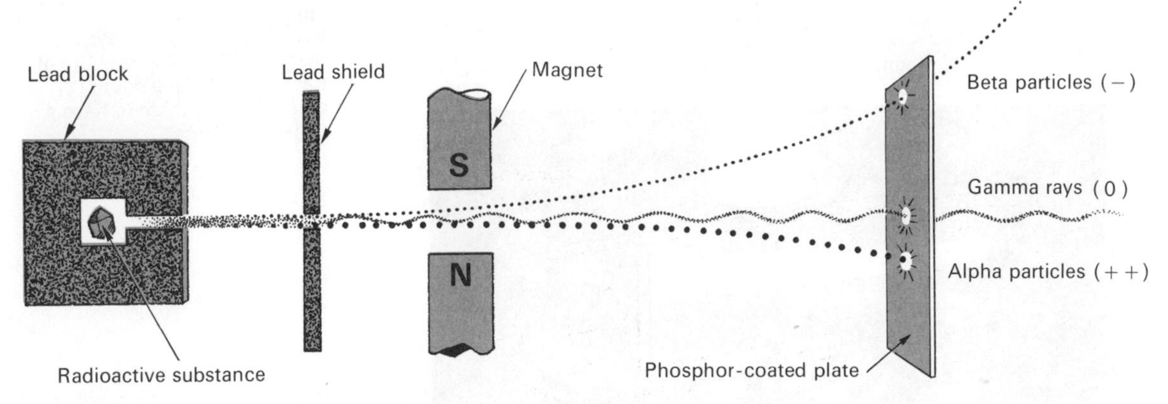

Figure 3.7
Rutherford's demonstration of the charged nature of alpha and beta particles from radioactive elements.

prevented exposure of the mineral to sunlight, he placed the specimen (which was called pitchblende) on top of the film packet in a cupboard. He later developed the film and discovered on it an exact outline of the rock. From further investigation he learned that the emission of penetrating radiation was an inherent property of a rather small number of minerals and that prior exposure to light or x-rays was not necessary. The spontaneous emission of penetrating radiation was called *radioactivity*.

The radiation from pitchblende was able to penetrate wood, glass, and most other materials, but was almost completely absorbed by lead. In 1899 Lord Rutherford placed some of the mineral in a hollowed-out block of lead which was then sealed except for a small hole through which the radiation could escape. When the stream of radiation struck a zinc-sulfide-coated screen, a glowing spot appeared, just as with cathode and canal rays. However, in a magnetic field the stream was split, and three separate spots appeared on the screen (Figure 3.7). One part of the stream was attracted toward the north (negative) pole, which suggested that it was made of particles similar in charge to the proton. A second part of the stream was bent toward the south pole, indicating particles with a negative charge, while the third stream was undeflected. The radiations were called alpha (α), beta (β), and gamma (γ), respectively, the first three letters of the Greek alphabet. Additional work confirmed that the beta particles were indistinguishable from electrons, and that an alpha particle had a mass four times that of a proton but only twice the charge. Gamma radiations, having neither mass nor charge, were similar to x-rays, only more energetic. The characteristics of these radiations, together with those of the electron and proton, are shown in Table 3.1.

3.5 THOMSON'S "PLUM-PUDDING" MODEL OF THE ATOM

With the discovery of the subatomic particles it became clear that Dalton's atomic theory was in jeopardy, since it asserted that atoms were indivisible. However, since the theory accounted so well for

TABLE 3.1 Symbol, Charge, and Relative Mass of Some Subatomic Particles

Name	Symbol	Relative Charge	Relative Mass[1]
Electron	e^-	−1	0.0005
Proton	p^+ or ${}^1_1H$	+1	1.0
Alpha particle	α	+2	4.0
Beta particle	β	−1	0.0005
Gamma ray	γ	0	0.0
Neutron	n	0	1.0

[1] Relative to the mass of the proton, which is assigned an arbitrary mass of one unit.

quantitative aspects of chemical combinations, it was hoped that at least part of it could be salvaged. It was in 1900 that J. J. Thomson suggested a way out of the dilemma—a modification that would retain the essence of Dalton's theory and yet include the observations concerning cathode rays, canal rays, and static electricity production.

Thomson suggested that the atom is a sphere in which protons are embedded, and that electrons float on its surface. Since matter is neutral except under special conditions, he further suggested that the number of protons in an atom equaled the number of electrons. Static electricity and cathode rays were explained as the removal or ejection of electrons from the surfaces of atoms, leaving the atoms with a positive charge. Canal rays were streams of atoms following the removal of one or more electrons. And the amber acquired a negative charge because stroking removed electrons from the fur and they accumulated on the amber.

Despite the fact that the model did not account fully for the observations concerning radioactivity, it was a major step in the evolution of the modern atomic theory.

3.6 RUTHERFORD'S NUCLEAR ATOM

By 1911 Madame Curie had isolated from Becquerel's radioactive ore an intensely radioactive element, called radium, which emitted alpha, beta, and gamma radiations. Rutherford and his associates used this element to direct a stream of alpha particles at a piece of gold foil only a few microns in thickness. A coated screen behind the foil registered the impact of the particles as tiny flashes of light. On the assumption that Thomson's model was correct, Rutherford had expected the energetic alpha particles to pass easily through the foil, driving ahead of them the smaller protons and electrons with which they collided. He was amazed, therefore, to see flashes of light on the screen when it was moved to other positions off to the side of the alpha stream (Figure 3.8). Rutherford explained this by assuming that protons were concentrated in the center of a gold atom and that the electrons were some relatively large distance away; thus the atom was mostly empty space and the majority of alphas passed through without hindrance. However, an alpha particle passing close to the center of positive charge would be repelled, since according to Coulomb's Law, the

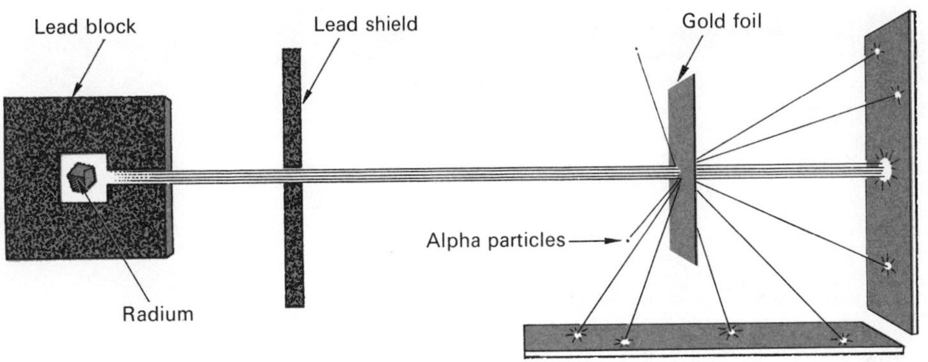

Figure 3.8
Rutherford's alpha-scattering experiment. The scattering of the charged particles by the gold indicated the presence of a concentration of positive charges in each atom. Rutherford called this region of the atom the nucleus.

repulsion increases in direct proportion to the charge on either body. Rutherford called the dense positively charged atomic center the *nucleus* and calculated its diameter to be roughly 1×10^{-13} cm or about 1×10^{-5} Å. For the entire atom he calculated a diameter of 1×10^{-8} cm or about 1 Å. On a relative scale, if the nucleus were as big as a dime, the electrons would be about one mile away!

The discovery of the nucleus led to the oversimplified idea that an atom was like a solar system in miniature, in which fast-moving electrons orbited a nuclear body. The nucleus attracted the electrons because of their opposite charge, and this attraction was exactly balanced by the tendency of the electron to fly off into space.

Rutherford's planetary model soon drew the fire of critics who argued that a system composed of a charged particle orbiting an oppositely charged one was unstable. The moving electron would radiate energy and gradually be drawn into a spiral path ending in collision with the nucleus. The solution to this puzzle lay in developments having to do with the nature of light.

3.7 LIGHT AND ELECTROMAGNETIC RADIATION

When white light is passed through a triangular glass solid called a prism, it is broken up into bands of color which have no definite boundaries. A similar effect is produced by sunlight shining through raindrops, and the array of colors is called a rainbow. Although the colors blend into each other, certain major colored areas are distinguishable, as shown in Figure 3.9. Visible light is a type of radiant energy which takes the form of waves and travels at a velocity of 3×10^{10} cm/sec (about 186,000 miles/sec) in a vacuum. The length of the wave is different for each color, and varies from about 7,000 Å for red light to slightly less than 4,000 Å for violet. Figure 3.10 shows a comparison of two waves, one of which is exactly twice as long as the other.

Since the wavelength of the violet light is only half that of the red, it takes two wavelengths of violet to cover the same distance as one wave of red. Since their velocities are the same, the number of violet

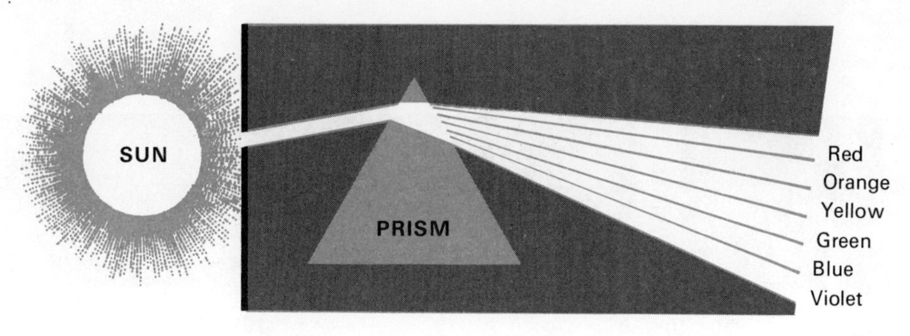

Figure 3.9
The visible spectrum. White light is broken up as it passes through the prism, producing a rainbow.

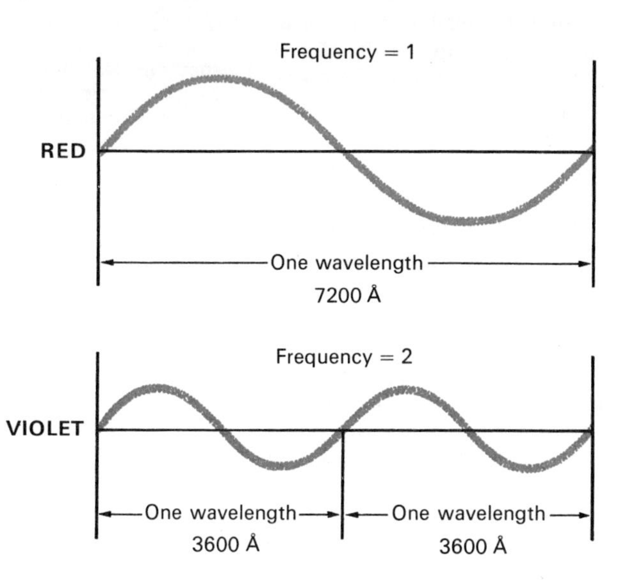

Figure 3.10
A comparison of the waves that make up red and violet light. Since the wavelength of violet light is only half that of red light, its frequency is twice as great.

waves per second (known as the frequency) must be twice that of red. Thus the frequency of the wave increases gradually as the waves become shorter. Another feature of light waves is that the waves become more energetic as the frequency increases—thus violet waves have more energy than red.

Sunlight also contains radiations whose wavelengths are shorter or longer than visible light, but which travel at the same velocity. Radiations which have wave character and travel at the velocity of light are collectively referred to as *electromagnetic radiation*. The electromagnetic radiation spectrum is shown in Figure 3.11.

Notice that the length of an x-ray is about 0.1 Å. This is about 1/10,000 the length of the average visible light wave, but considerably more energetic. The great energy and small size of an x-ray makes it a very penetrating kind of radiation. In contrast, typical radio waves are about as long as a football field and have very low energies.

An interesting consequence of the difference in energy of various light rays concerns the working of the human eye. The eye responds

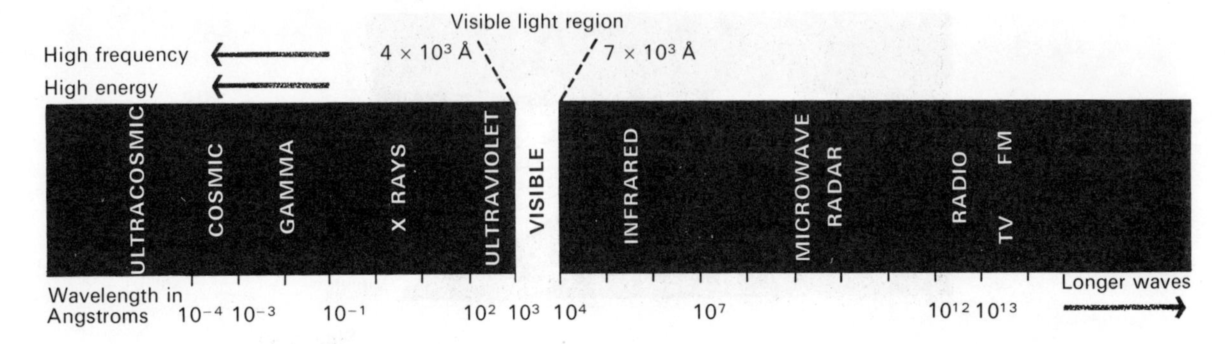

Figure 3.11
The electromagnetic radiation spectrum. Note that as the waves become longer, they become less energetic and have lower frequencies.

to visible light because the waves which comprise visible light are of the correct energy for stimulating a chemical change in the visual pigment. Larger waves (infrared) or shorter (ultraviolet) cannot be seen because their energies are either too low or too high.

The color of a substance is due to its absorption of some wavelengths and its transmission or reflection of others. For example, a chlorophyll solution appears greenish-yellow because it absorbs light in the red-orange and blue-violet regions and transmits the remainder. When an object absorbs most of the wavelengths in the visible spectrum, it is said to be black.

The color of flame changes when different chemicals are placed in it. The different colors of fireworks and the sometimes striking appearance of the flame of a laboratory burner are due to specific elements. A rapid analytical procedure called a flame test is often used in the laboratory to identify individual elements by their characteristic flame colors. For example, flames are turned red by calcium, strontium, and lithium, while barium and copper produce a green flame and potassium a violet flame. When ordinary glass is heated intensely, the flame turns yellow-orange because the glass contains compounds of the element sodium.

Some elements in the gaseous state also emit visible light when stimulated by electrical energy: the use of neon in advertising signs is a well-known example. However, when the light from an excited element is passed through a prism, it does not emerge as a single color, or even as a continuous spectrum, but as a few discrete bands. This discontinuous, or line, spectrum is called an *emission spectrum*. The pattern of the emission spectrum is unique for individual elements, and an elemental analysis of blood, urine, or other chemicals can be obtained by vaporizing the sample in an intense flame and then recording and analyzing the emission spectra. The instrument that is used for this analytical procedure is called a flame photometer. Such analyses are extremely rapid and accurate, and may often be carried out on samples weighing only a few micrograms. In many cases the quantity of an element can also be determined from its emission spectrum. Sodium, potassium, and lithium in blood or urine are routinely determined in this manner.

The emission spectrum produced by the element hydrogen is composed of four lines in the visible region—two blue, one green, and one

Figure 3.12
The emission spectrum of hydrogen in the visible region—the Balmer series.

orange-red (Figure 3.12). This set of four wavelengths is called the Balmer series after its discoverer. The element sodium, by contrast, shows seven lines in the visible region. Excited hydrogen atoms also produce another set of lines, called the Lyman series, which are in the ultraviolet region, as well as two sets of lines in the infrared.

The discovery of hydrogen's emission spectra raised several questions about the atom. What was the source of the light? Why were only a few lines seen? And what was the significance of the spectra in terms of atomic structure? The answers to these questions were provided by Niels Bohr in 1913.

3.8 THE BOHR ATOM

The model proposed by Bohr retained the concept of a small, dense, positively charged nucleus with satellite electrons, but restricted the electrons to definite orbits having specific amounts of energy. He suggested that these orbits corresponded to stable conditions for the electrons, and that electrons could move from one orbit to another but could not occupy a position between orbits. The orbit closest to the nucleus was lowest in energy, with successively higher energies for more distant orbits.

Bohr assumed that when an electron absorbed energy (from flame, electricity, etc.), it was temporarily promoted to an orbit with a higher energy level, representing an excited state. The return to a lower energy level was accompanied by the release of energy in the form of electromagnetic radiation. Each of the lines in the emission spectrum of hydrogen corresponded to a movement from one orbit to another. The very small number of lines indicated that only a few stable orbits existed.

If the atom is visualized in three dimensions, the orbits become concentric shells which were originally designated by the letters K, L, M, etc. In more modern treatment the energy level of an orbit is indicated by an integer called the *principal quantum number* (symbol n), where $n = 1$ corresponds to the K shell, $n = 2$ to the L shell, and so on. The first six energy levels of the hydrogen atom are shown in Figure 3.13, which also indicates the source of the lines in the Balmer series.

3.9 DIFFICULTIES WITH THE BOHR ATOM

Despite its ability to account for the spectral lines of hydrogen, the Bohr model was far from satisfactory. It could not, for example, account for the increased number of spectral lines which were observed

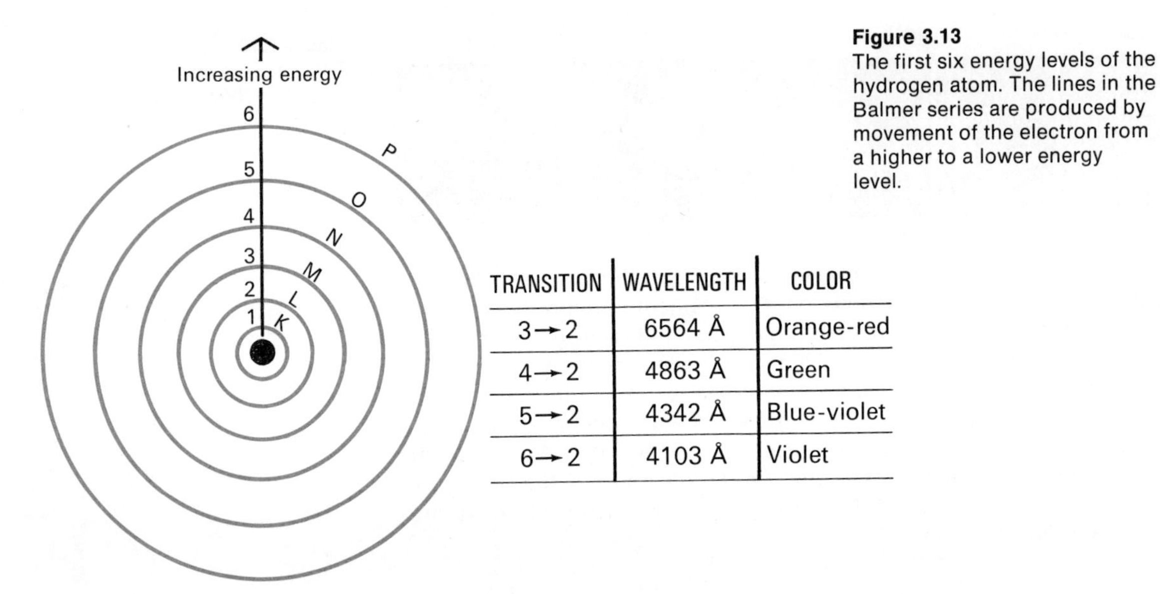

TRANSITION	WAVELENGTH	COLOR
3→2	6564 Å	Orange-red
4→2	4863 Å	Green
5→2	4342 Å	Blue-violet
6→2	4103 Å	Violet

Figure 3.13
The first six energy levels of the hydrogen atom. The lines in the Balmer series are produced by movement of the electron from a higher to a lower energy level.

when hydrogen was examined in a magnetic field. But its most serious deficiency was the fact that it worked *only* for hydrogen: spectral lines observed for atoms with more than one electron could not be explained. Obviously a model that was valid for only one element left something to be desired. One possible way out of the dilemma was to measure the position and energy of each electron in an atom and use this information in constructing a new model. As will be seen, this simple problem did not have a solution.

3.10 THE UNCERTAINTY PRINCIPLE

The experimental determination of an electron's position is made exceedingly difficult by its small size. It is too small to be seen because the waves reflected by it are not in the visible region of the spectrum. An object reflects radiant energy only if the size of the object and the length of the wave are approximately the same. For example, an airliner may be tracked through a cloud layer by radar because the radar waves are roughly the same size as the plane, but too large to be reflected by the water in the cloud. Thus a particle the size of an electron reflects waves that are so small that they lie in the x-ray region of the spectrum. Such waves have extremely high frequencies; therefore they also have very high energies. When an electron is struck by an energetic wave, it may be deflected from its initial path and may also absorb some of the energy of the wave. The problem was expressed by Werner Heisenberg in his *Uncertainty Principle,* which states that it is not possible to know with certainty both the position and energy level of an electron. Thus if we know the position, we cannot be certain of the energy level, and vice versa.

An answer to this problem was worked out in 1926 by Erwin Schrödinger. His solution incorporated certain ideas of Einstein, De Broglie, and Max Planck which indicated that the electron was not only a particle, but had a wave nature as well. Since the characteristics of waves were well known, the assumption of a wave nature meant that it was possible to calculate the energy and position of an electron—but only if one were willing to accept an answer that was given in terms of probability rather than stated exactly. His mathematical proposition, called the Schrödinger wave equation, ushered in the quantum mechanical model of the atom.

3.11 THE QUANTUM ATOM

By the use of the Schrödinger wave equation it is possible to specify the energy level of an electron and to obtain an extremely large number of probable locations for an electron with that amount of energy. For the hydrogen atom, a plot of these probable locations resembles a diffuse cloud with the nucleus at its center. The probability of finding the electron is quite low in the vicinity of the nucleus, increasing to a maximum at a radius of about 1 Å, and then decreasing again at greater distances. The region where there is a very high (usually specified as 90%) probability of finding the electron is called an *orbital*. A comparison between a Bohr orbit and a quantum orbital for the hydrogen atom is shown in Figure 3.14. The principal difference between them is that the quantum model makes no attempt to describe the path taken by the electron. A second major difference is that the successive orbits of the Bohr atom are concentric circles,

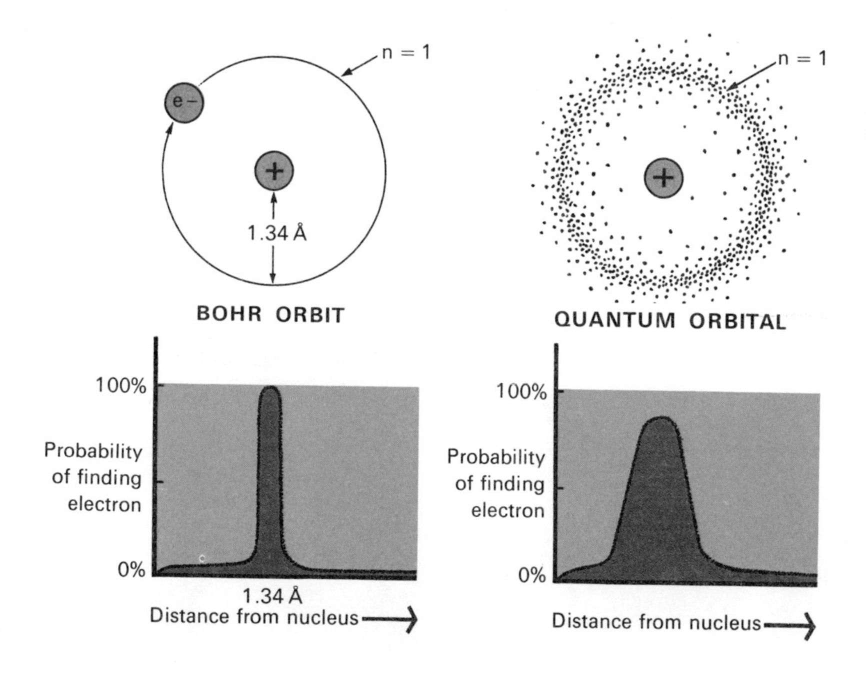

Figure 3.14
Comparison of the hydrogen atom according to the Bohr and quantum models.

while the quantum orbitals of multiple-electron atoms have several different shapes.

The shapes of the orbitals are indicated by the *azimuthal* or *orbital quantum number* (l). The various shapes are designated by coded values of l. For example, when $l = 0$, the orbital resembles a diffuse sphere with the nucleus at its center, as in the hydrogen atom. This orbital shape is also referred to as an *s* orbital. A different orbital shape results when $l = 1$: it is formed from two lobes which are on opposite sides of the nucleus and is called a *p* orbital. The shapes of the *s* and *p* orbitals are shown in Figure 3.15. The other two orbital shapes are indicated by l values of 2 and 3 and are called *d* and *f* orbitals, respectively. They are not shown because their shapes are rather complex for representation by two-dimensional drawings and because an understanding of the exact shape of such orbitals is not necessary for our discussion. The orbital quantum numbers and values are shown in Table 3.2.

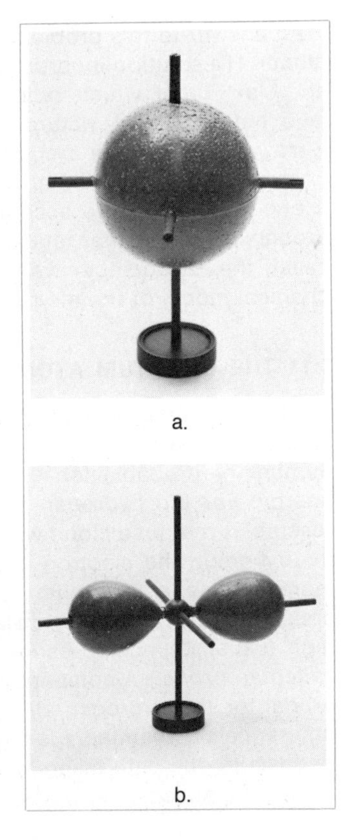

Styrofoam and wood representations of (a) an *s* orbital, (b) a *p* orbital. (Photographs by J. Asdrubal Rivera)

TABLE 3.2 Relationship between l Values and Orbital Type

Value of Orbital Quantum Number	Type Orbital	Shape of Orbital
0	*s*	Spherical
1	*p*	Two-lobed
2	*d*	Complex (generally 4-lobed)
3	*f*	Very complex

In addition to the principal quantum number which corresponds to the energy level, and the orbital quantum number which describes the shape of the orbital, each electron has a *magnetic quantum number* (m) and a *spin quantum number* (s). The magnetic quantum number is necessary because the orientation of the *p, d,* and *f* orbitals is affected by a magnetic field. For example, there are three possible positions for a *p* orbital—one on each of three mutually perpendicular axes passing through the nucleus (Figure 3.16). One way to illustrate the orientation of these axes is to place the point of your pencil at the place where they intersect. If your pencil comes straight out of the paper toward

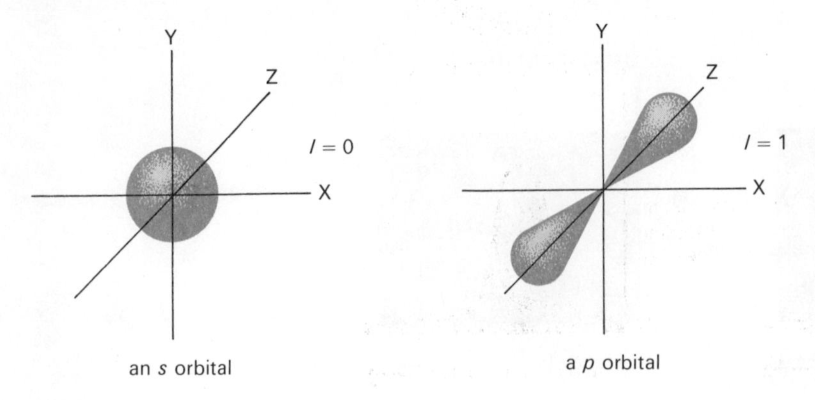

Figure 3.15
The *s* and *p* orbital shapes. The nucleus should be pictured at the junction of the three reference axes.

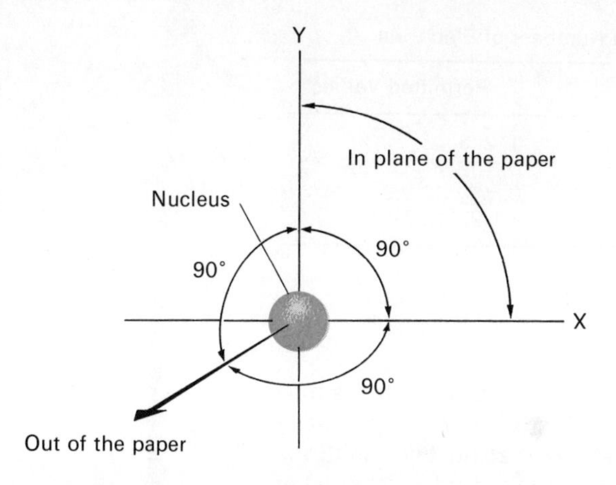

Figure 3.16
Imaginary axes passing through the nucleus.

you, it represents the third axis. The axes are designated *x, y,* and *z,* and the orbitals on the respective axes are p_x, p_y, and p_z, as shown in Figure 3.17. These three orbitals are identical in energy because they have the same principal quantum number.

The spin quantum number is necessary because electrons have been found to have a rotational or spinning motion as well as a translational, or place-to-place, motion. There are only two possible directions of spin and these are arbitrarily designated $+1/2$ and $-1/2$. If it is less confusing to think of $+1/2$ as being clockwise and $-1/2$ as counter-clockwise, please do so. The spinning motion of a particle creates a magnetic field around it: thus a particle with a $+1/2$ spin might generate a north magnetic field, and a $-1/2$ spin a south magnetic field.

The value of the principal quantum number determines the value of the orbital quantum number which, in turn, determines the value of the magnetic quantum number, as shown in Table 3.3. For example, when $n = 1$, the electron being described is in the first, or lowest, energy level.

The orbital quantum number, which gives the orbital shape, may have any value up to and including $(n - 1)$. Since $n = 1$, the value of $(n - 1)$ is zero, which indicates an *s* orbital (see Table 3.2). This means that an electron in the first energy level must also be in an *s* orbital.

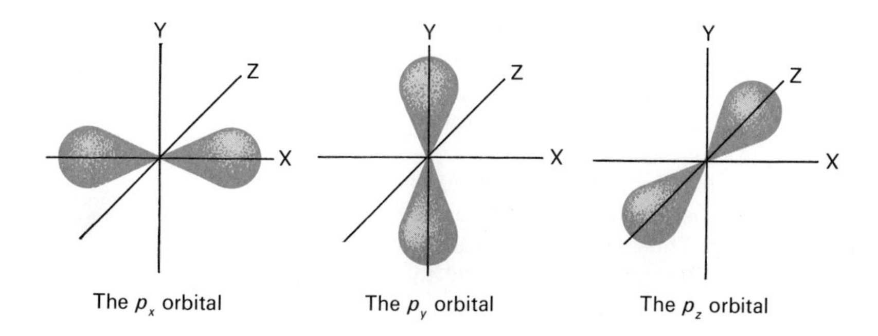

Figure 3.17
The effect of a magnetic field on the *p* orbital. The result is to produce three possible orientations of the orbital.

TABLE 3.3 Permitted Values for the Quantum Numbers of Electrons

Symbol	Designation	Permitted Values
n	Principal	$1, 2, 3, 4 \ldots \infty$
l	Orbital	$0, 1, 2, 3 \ldots (n-1)$
m	Magnetic	$0, \pm 1, \pm 2, \pm 3 \ldots (\pm l)$
s	Spin	$+1/2, -1/2$

Pauli Exclusion Principle

It is logical at this point to ask the question, "How many electrons may there be in the first energy level?" This question is answered indirectly by the *Pauli Exclusion Principle,* which states that no two electrons within the same atom may have the same set of quantum numbers. Let us apply this rule to the case under discussion.

We have seen that, for any electron in the first energy level, the orbital quantum number is zero and the magnetic quantum number is limited to ± 0, or simply zero. The scoreboard shows at this point that all of the electrons in the first energy level appear to have an identical set of quantum numbers ($n = 1$, $l = 0$, and $m = 0$) which would be a violation of the exclusion principle. However, the complete quantum description includes the spin quantum number, which may have either of two values. This means that there can be two electrons in the first energy level: one which is described by the quantum set $n = 1$, $l = 0$, $m = 0$, $s = +1/2$; and the other by $n = 1$, $l = 0$, $m = 0$, and $s = -1/2$. These quantum descriptions tell us that both electrons would be in the same orbital and would be indistinguishable but for their opposite spins. Since the element helium has an atomic number of two—that is, two protons and two electrons—the description just given is that of a helium atom.

3.12 THE AUFBAU PRINCIPLE

We have seen that the first energy level can hold only a pair of electrons, and that the spins must be paired—that is, opposite in sign. In addition to the fact that this situation conforms to the requirements of the Exclusion Principle, it is a logical arrangement from another standpoint. Because all electrons have a negative charge, two electrons in the same region would tend to repel each other. However, if they are spinning in opposite directions, they will generate two different magnetic fields, which will result in an attractive force to offset the repulsion. It should be apparent from this that no orbital can contain more than two electrons because a third electron must have the same spin as one of the first two and would be repelled, not only by its charge, but also by a similar magnetic field. Thus an atom with three electrons (lithium) would have two of its electrons in the first energy level and the third electron in a region of higher energy.

The idea that the placement of electrons begins with the level of lowest energy and that additional electrons fill successively higher energy levels is known as the *Aufbau,* or "building-up" principle. The addition of electrons is regulated by *Hund's Rule,* which states that electrons shall not be paired in any orbital so long as there is a vacant orbital of the same energy. This situation is somewhat analogous to the operation of a motel: just as the manager would not ask two strangers to share the same room if there were similar rooms vacant, so we do not put two electrons in the same space so long as there are orbitals having the same energy which are unoccupied. The operation of the Aufbau principle will be made clear when we consider the second energy level.

For electrons in the second energy level the value of the principal quantum number is 2. This results in two possible values for the orbital quantum number: $l = 0$ and $l = 1$. Remembering that $l = 0$ specifies an *s* orbital and $l = 1$ denotes a *p* orbital, we can see that the second energy level contains two different sublevels or orbital shapes. And when $l = 1$, the magnetic quantum number may have values of 0, +1, and −1. In other words, we can predict that the second level will consist of a single *s* orbital together with three identical *p* orbitals—one on each of the three axes. Each of the four orbitals can hold two electrons with opposite spins, making a population of eight electrons for the second energy level when it is filled.

The arrangement of electrons which is predicted for an atom by the quantum theory is referred to as the electron configuration and may be represented in a number of ways. The first is an energy-level diagram with circles for orbitals and arrows whose point indicates the direction of spin. The energy-level diagram for the hydrogen atom is

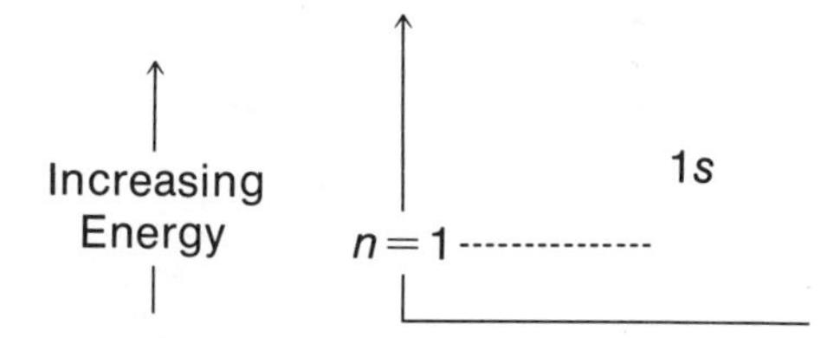

The second convention is known as orbital notation and is a combination of letters and numbers which take the form aB^c, where *a* is the principal quantum number or energy level, *B* is the type of orbital (*s, p, d,* etc.) and *c* is the number of electrons in the orbital at that time. The orbital notation of the hydrogen atom would be

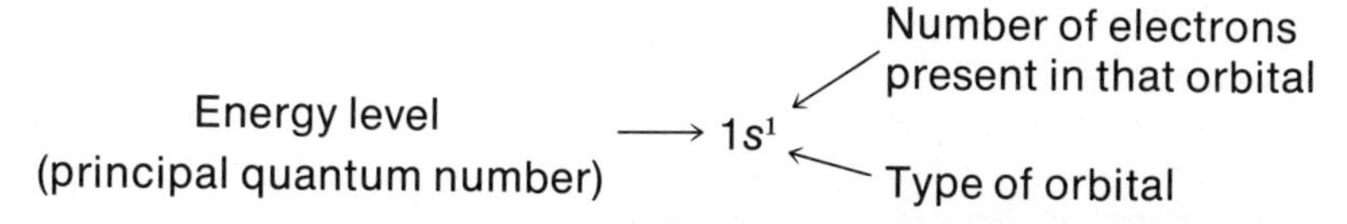

The electron configurations for the remainder of the first ten elements are given by both methods in Figure 3.18.

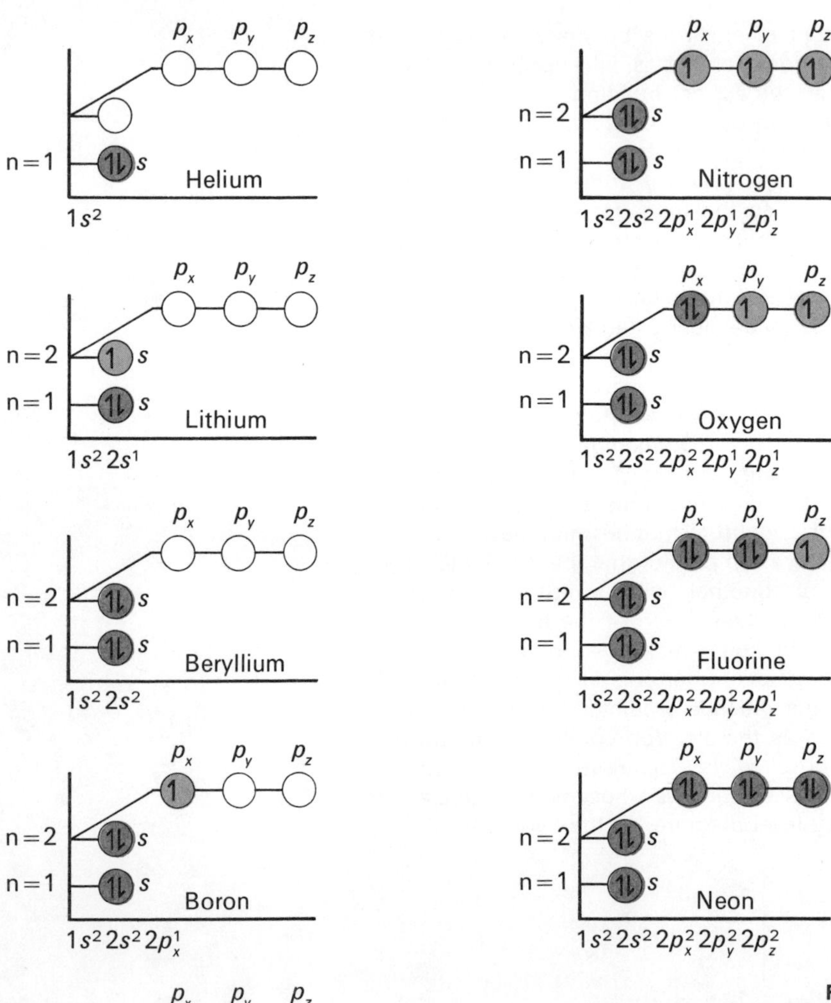

Figure 3.18
Energy-level diagrams and orbital notations of elements with atomic numbers 2 through 10.

Note that, although there are unoccupied *p* orbitals in the second energy level, the third and fourth electrons of beryllium are paired in the 2*s* orbital. This is because the vacant 2*p* orbitals require higher energy than the 2*s*. Note also the operation of Hund's Rule beginning with carbon. The fifth electron of boron has gone into one of the three equal-energy *p* orbitals, leaving two vacancies. According to Hund's Rule, the next electron will enter one of the vacant orbitals rather than that *p* orbital which already contains an electron. The same is true for nitrogen: the seventh electron also prefers a vacant orbital. The eighth, ninth, and tenth electrons are then paired with those already in the *p* orbitals, and the second energy level is completed.

The Schrödinger equation predicts that the number of sublevels in a shell will be the same as the primary quantum number: thus the third level should contain *s, p,* and *d* orbitals. The predicted composition of the first four levels is shown in Table 3.4.

TABLE 3.4 Predicted Composition of the First Four Energy Levels

Energy Level	Number of Sublevels	Types of Orbitals and Their Electron Complements	Total Number of Electrons
First	1	*s*(2)	2
Second	2	*s*(2)*p*(6)	8
Third	3	*s*(2)*p*(6)*d*(10)	18
Fourth	4	*s*(2)*p*(6)*d*(10)*f*(14)	32

The filling of the third level matches that of the second level up to a point. The interactions between electrons result in a shifting of orbitals so that the 4*s* orbital has a slightly lower energy than the 3*d* and so is filled first—thus potassium has a single electron in the 4*s* orbital rather than in the 3*d*.

As the number of electrons in an atom increases, so does the irregularity. Authors of chemistry texts have used a variety of diagrams to show this irregularity. One such diagram is shown in Figure 3.19. An equally satisfactory device for determining the order in which the orbitals are filled is called the "$n + l$" *Rule*. It states that those orbitals fill first whose sum of n and l is lowest, and that in case of a tie the orbital with the lowest n value fills first. The use of the rule in deciding between the 3*d* and 4*s* orbitals is shown below.

orbital	3*d*	4*s*
n value	3	4
l value (from Table 3.3)	2	0
sum of n and l	5	4

Since the sum of n and l is lowest for the 4*s* orbital, it will fill first.

3.13 THE OCTET RULE

When the electron configurations were correlated with chemical behavior, it appeared that those elements whose outer shells were either filled, or contained filled *s* and *p* orbitals, were chemically inert. This group is composed of helium, neon, argon, krypton, radon, and xenon. Since all of these elements are gases under ordinary conditions, they were for years referred to as the inert gases. However, within the last ten years several compounds of xenon have been prepared, and so the name "inert" gases is no longer appropriate.

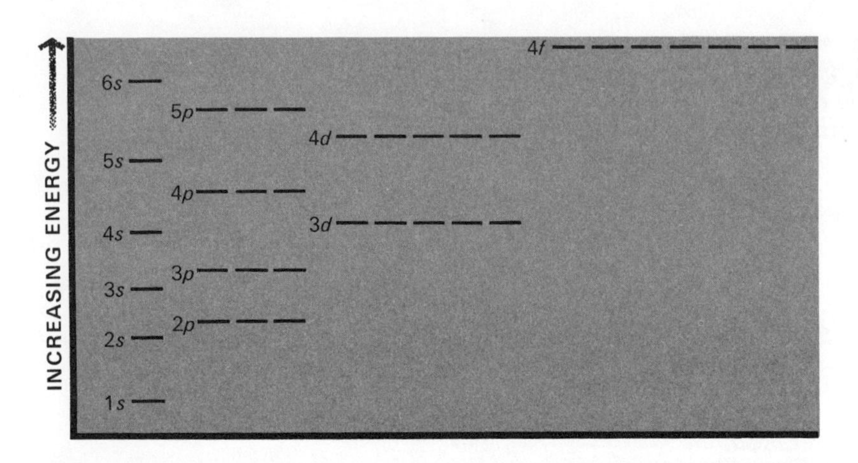

Figure 3.19
Relative energies of the orbitals in a multi-electron atom.

These gases are known as "noble" gases. The greatest majority of chemical reactions, however, involve only those electrons in the outer energy level, and there is an unusual stability associated with having eight electrons in the outer level. This generalization concerning stability is known as the *Octet Rule* and will be used in connection with chemical bonding in Chapter 4.

3.14 LEWIS (ELECTRON DOT) DIAGRAMS

Electrons that occupy the outer energy level are called valence electrons. One convenient way of representing the valence electrons is by the use of an electron dot, or Lewis, diagram. Since the electrons in the outer energy level are generally the only ones involved in chemical reactions, the underlying levels are ignored. The rules for Lewis diagrams are:

1. The symbol for the element represents the nucleus and all electrons except those in the outer shell.
2. A dot (or "x") is used to represent each electron in the outer energy level.
3. The dot may be placed on any of the four sides of the symbol—no significance is attached to any of the four positions.
4. If two electrons are in different orbitals, their dots must be separated, and if in the same orbital, the dots are shown as a pair.
5. An element which has either a filled outer shell or filled *s* and *p* orbitals in the outer shell (the Octet Rule) is represented by the symbol alone.

The Lewis diagrams of the first 11 elements are shown in Figure 3.20.

3.15 NEUTRONS AND ISOTOPES

The discovery of the *neutron* by Chadwick in 1932 necessitated another minor change in Dalton's theory and led to the revision of ideas

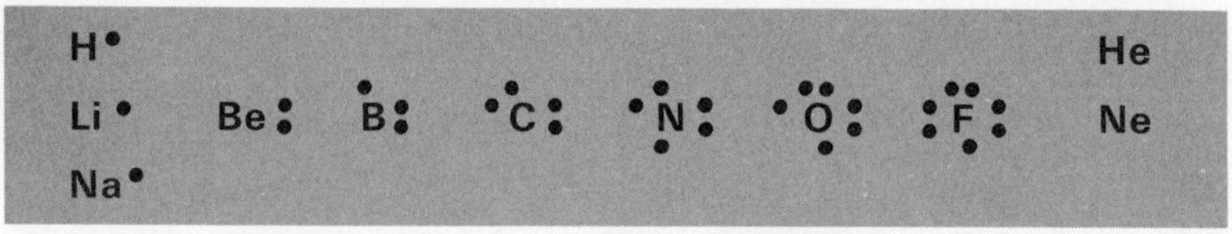

H•
Li•   Be: B: •C: •N: •O: :F: He   Ne
Na•

Figure 3.20
Lewis diagrams of elements with atomic numbers 1 through 11.

concerning the composition of the nucleus. The existence of the neutron was indicated when it was discovered that, contrary to Dalton's theory, all atoms of a given element did not have the same mass. Since all of the atoms of a given element were chemically indistinguishable and it was known that the chemical identity of an element was determined by the number of protons and electrons, it was concluded that the difference in mass was due to a variable number of neutral particles in the nucleus. This conclusion proved to be correct when an uncharged particle having almost the same mass as the proton was found.

Neutrons are found in all atoms except some of those of the element hydrogen. It has been suggested that neutrons act as a sort of insulation to reduce the strong repulsive forces which arise from crowding protons together in the nucleus and, since hydrogen atoms contain only one proton, neutrons are not needed. Be that as it may, there are neutrons in some hydrogen atoms; and there are none in others.

If hydrogen atoms could be weighed and sorted, about 9,998 of every 10,000 would weigh 1 atomic mass unit (amu)—that is, they are composed of 1 proton and 1 electron. However, about 2 of every 10,000 would contain a neutron and would have a relative mass of 2 amu. And approximately 1 atom of every billion contains 2 neutrons and has a relative mass of 3 amu. These atoms of the same element which have different masses are called *isotopes*. Isotopes may also be defined as atoms having the same atomic number (number of protons) but different mass numbers (number of protons plus neutrons).

A convention has been adopted which makes it possible to distinguish between the different isotopes of an element: the atomic number is written to the left and slightly below the symbol, and the mass number to the left and above.

${}^{A}_{Z}X$

A = mass number (number of protons plus number of neutrons)
Z = atomic number (number of protons or electrons)
X = symbol of element

The isotopes of hydrogen have been so much studied that the different isotopes have individual names: ordinary hydrogen (1 amu) is called protium, while that with a mass of 2 amu is deuterium and the heaviest isotope is called tritium. The three isotopic forms of hydrogen are shown diagrammatically in Figure 3.21. Protium and deuterium are stable isotopes; that is, they do not spontaneously emit radiation (see Section 3.4). Tritium, however, is radioactive. The stable isotopes of some other elements are shown in Table 3.5.

TABLE 3.5 Stable Isotopes of Representative Elements

${}^{12}_{6}C$	${}^{16}_{8}O$	${}^{35}_{17}Cl$	${}^{50}_{24}Cr$	${}^{28}_{14}Si$
${}^{13}_{6}C$	${}^{17}_{8}O$	${}^{37}_{17}Cl$	${}^{52}_{24}Cr$	${}^{29}_{14}Si$
	${}^{18}_{8}O$		${}^{53}_{24}Cr$	${}^{30}_{14}Si$
			${}^{54}_{24}Cr$	

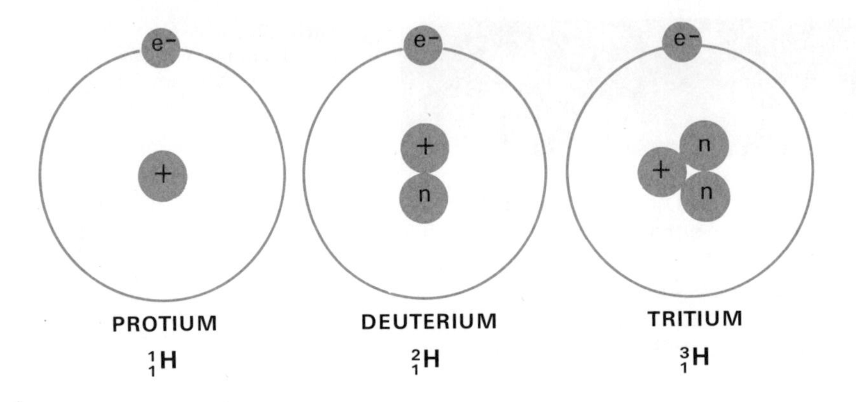

Figure 3.21
The three isotopes of hydrogen.

APPLICATION OF PRINCIPLES

1. Which has the lower energy, yellow or blue light?
2. Write the orbital notations for magnesium, chlorine, sulfur, and argon.
3. Explain why the orbital notation $1s^2 2s^2 2p_x{}^2 2p_y{}^2$ is incorrect.
4. Using Hund's Rule, determine the order in which the 3*d*, 4*p*, and 5*s* orbitals will receive electrons.
5. Draw Lewis diagrams for silicon, phosphorus, aluminum, and calcium.
6. Propose an explanation for the following observation: the colors produced when lithium, sodium, and potassium are burned in a flame are red, yellow, and violet, respectively.
7. What are the permissible values for *l* and *m* when the value of the principal quantum number is 3?
8. Indicate how the electron configuration of each of the following elements might be changed in order to make it conform to the Octet Rule: carbon, potassium, oxygen, nitrogen.
9. How many neutrons are there in an atom of $^{234}_{92}U$?
10. Explain why **:** N̤ **:** is an incorrect Lewis diagram.
11. Figure 3.14 represents the region around the nucleus where the probability of finding an electron is 90%. Describe the appearance of a similar figure showing a 95% probability.
12. A solution of chlorophyll a is blue-green. What regions of the visible spectrum of electromagnetic radiation does it absorb?
13. The relative charge of an alpha particle is +2 and its mass is 4 amu. Propose a model of the alpha particle that would account for these observations.
14. Explain how Dalton's atomic theory was modified by the discovery of the neutron.
15. What is the difference between gamma radiation and a neutron?
16. Propose a model to explain why you sometimes get an electrical shock when you touch the inside door handle after sliding across a plastic seat cover in an automobile.

An ice crystal. The characteristic hexagonal shape reflects the pattern of the water molecules in the crystal. (Courtesy Buffalo Museum of Science)

4

ATOMIC INTERACTIONS

LEARNING OBJECTIVES

1. Describe the interactions of the charged particles within an atom.
2. Describe the variation in ionization energy that is observed as the atomic number increases.
3. Explain the concept of periodicity.
4. Describe the construction of the modern form of the Periodic Table.
5. Describe the model of the covalent bond.
6. Explain the concept of valence.
7. Describe the hybridization of the atomic orbitals of a carbon atom.
8. Summarize the main points of the Electron Pair Repulsion Theory, and use examples to show how the theory accounts for observed molecular shapes.
9. Explain the concept of electronegativity.
10. Describe the process of ion formation using appropriate examples.
11. Describe the model of the ionic bond.
12. Explain how the type of bonding determines the physical properties of a substance.
13. Compare hydrogen bonding and van der Waals forces.

KEY TERMS AND CONCEPTS

- anion
- bond length
- cation
- chemical family (group)
- coordinate covalent bond
- covalent bond
- dipole
- electrolyte
- electronegativity
- Electron Pair Repulsion theory
- hybridization
- hydrogen bonding
- ion
- ionic bond
- ionization energy
- lone pair of electrons
- metal
- metalloid
- mole
- monoatomic ion
- multiple covalent bond
- nonmetal
- nonpolar
- period
- Periodic Law
- polarity
- polyatomic ion
- radical
- sigma bond
- valence
- valence electrons
- van der Waals forces

Dalton proposed in his atomic theory that all atoms of a given element are alike, but that they are different from atoms of all other elements. This proposal has been modified to accommodate the discovery of isotopes, as noted in Chapter 3, but the basic premise remains unchanged. Each element *is* chemically unique, and research spanning more than a century has shown that this individuality stems from two factors: the number of protons or electrons (the atomic number) in an atom, and the way in which the electrons are arranged in the atom. The arrangement of the protons does not appear to have any effect on chemical properties but does have some bearing in the stability of the nucleus, and lack of nuclear stability is associated with radioactivity (Chapter 16).

The electron configuration of a particular element depends upon the total number of electrons, but it is also influenced by energy considerations. For example, two unpaired electrons in equivalent orbitals exist in a lower energy condition than if they were paired in the same orbital (Hund's Rule). The lower energy condition is not only more stable, but the availability of half-filled orbitals plays an important role in atomic interactions, as we will see presently.

4.1 FORCES WITHIN THE ATOM

Protons and electrons, being charged particles, interact with each other in a manner described by Coulomb's Law. These interactions are another piece of the puzzle of chemical reactivity. Three types of charged-particle interactions may be described for a given atom: electron-electron and proton-proton repulsions and proton-electron attractions.

Electron-Electron Repulsions

Orbital electrons repel each other because they have the same charge. When two electrons are in different orbitals, the repulsion is reduced by distance; when they are in the same orbital, it is minimized by spin pairing.

Proton-Proton Repulsions

The mutual repulsion of the protons in the nucleus is minimized by spin pairing (similar to the pairing of electron spins) and by the insulating effect of neutrons. The exact way in which the presence of neutrons brings about a reduction in proton-proton repulsion is still under investigation.

Proton-Electron Attractions

The electrons in the outer energy level of an atom are usually referred to as *valence electrons.* Since chemical reactions generally involve only the valence electrons, the attractions between these electrons and the protons within the nucleus are of special importance.

Three factors affect the strength of proton-electron attraction within an individual atom:

1. The size of the nuclear charge (i.e., the atomic number).
2. The distance between the electron and the nucleus.
3. The screening effect of intervening electrons.

The volume of the nucleus is so small that for all practical purposes we can consider the combined charge of the protons to be concentrated at one point. Thus each orbital electron experiences a pull that is directly related to the atomic number.

According to Coulomb's Law, the attraction between the nucleus and an electron declines sharply as their separation increases. Assuming the same nuclear charge, an electron in the 4*s* orbital, for example, would be less strongly attracted than one in the 3*s* orbital.

The force of attraction between two charged bodies is also diminished by the presence of other charged bodies in the space between them. Consider the attraction between the nucleus and the 4*s* electron in potassium. All of the electrons in the first three levels act to screen the 4*s* electron so that it experiences less attraction for the nucleus. Electrons in the same orbital or in equivalent orbitals of the same energy level do not screen each other.

The net result of these factors is that the attraction between the nucleus and the valence electrons is slightly different for each element, matching the difference in chemical properties.

4.2 IONIZATION ENERGY

A measure of the net attractive force is the ionization energy of the element. The *ionization energy* is the number of kilocalories of energy required for the removal of one electron from every atom of a standard number of atoms. This standard number is called a *mole* and is equal to 6.023×10^{23} atoms. The ionization energies of the elements with atomic numbers 1 through 20 are shown along with the electron configurations of their outer energy levels in Table 4.1. Through an examination of these data we may gain some understanding of the forces at work within the atom.

TABLE 4.1 Ionization Energies of the First Twenty Elements

Element	Ionization energy (kcal/mole)	Valence Electrons	Element	Ionization energy (kcal/mole)	Valence Electrons
H	313	$1s^1$	Na	119	$3s^1$
He	567	$1s^2$	Mg	176	$3s^2$
Li	124	$2s^1$	Al	138	$3s^23p^1$
Be	215	$2s^2$	Si	188	$3s^23p^2$
B	191	$2s^22p^1$	P	254	$3s^23p^3$
C	260	$2s^22p^2$	S	239	$3s^23p^4$
N	336	$2s^22p^3$	Cl	300	$3s^23p^5$
O	314	$2s^22p^4$	Ar	363	$3s^23p^6$
F	402	$2s^22p^5$	K	100	$4s^1$
Ne	497	$2s^22p^6$	Ca	141	$4s^2$

Except for minor fluctuations, the ionization energy follows a cyclic pattern: it is low for elements having a single *s* electron and increases until that energy level is filled. For example, the first energy level contains only two electrons, and the single *s* electron of hydrogen is more easily removed than either of the two equivalent *s* electrons of helium. Lithium also has a single *s* electron in its outer energy level and a correspondingly low ionization energy. There is a gradual increase from lithium to neon as the *s* and *p* orbitals are filled, and then another low value for sodium which, like lithium, has only an *s* electron in its outer shell. The pattern of the third level resembles that of the second, and a fourth cycle begins with potassium.

The increase in ionization energy within a given energy level can be attributed primarily to the increase in nuclear charge. For example, the removal of an electron from an atom of helium requires more energy than for an atom of hydrogen because the nuclear charge of helium is twice as large as that of hydrogen. Although the number of electrons in helium is also twice that of hydrogen, the additional electron takes up a position in the same orbital as the first, and so the distance between the nucleus and the valence electrons is not appreciably different for the two elements.

Although the nuclear charge of lithium is greater than that of helium, the distance is also greater because the valence electron of lithium is in the second shell. According to Coulomb's Law, we would expect the increased distance to have a greater effect than the increased charge. The 2*s* electron of lithium is also screened by the electrons in the first level. The result is that the valence electron of lithium is less strongly held than those of helium, and so the ionization energy of lithium is lower.

The effect of increased distance is clearly shown by the series Li, Na, K. Each element has a single valence electron, but in a different energy level. The valence electron of potassium is least strongly held despite the fact that potassium has the highest nuclear charge. The lower ionization energy of potassium is due mainly to the greater distance between its nucleus and valence electron, but is also due in part to the screening effect of electrons in the three underlying levels. The ionization energy of sodium is less than that of lithium for the same reasons.

The outer electrons in atoms of the alkali metals (lithium, sodium, potassium, rubidium, and cesium) are loosely held and can be dislodged easily. The energy of visible light falling on a piece of cesium displaces electrons and produces an electric current. Photoelectric cells containing cesium can be used to automatically set the lens opening of a camera to match the available light, and to turn equipment on and off.

4.3 RECURRING PROPERTIES: THE CONCEPT OF PERIODICITY

A cyclic or periodic behavior is observed not only for ionization energy but for other properties such as melting point, boiling point, and density. One way of illustrating the cyclic variation in melting points is to use a graph (Figure 4.1). Recurring patterns in the properties of the elements were recognized by Lothar Meyer and Dmitri Mendeleyev around 1870 as they attempted to draw some order out of a mass of accumulated data. Their efforts led to various arrangements of these data in forms called periodic tables. Their work also helped in the formulation of the *Periodic Law,* which states that *the properties of elements are a periodic function of their atomic numbers.* The periodic nature of the ionization energy and the melting point are two illustrations of the Periodic Law.

The most common form of the Periodic Table is the tabular one shown in Figure 4.2. The horizontal rows, called *periods,* correspond to energy levels; and elements in vertical columns, called *groups* or *families,* have similar electron arrangements. When the elements are arranged in this order, the trends or patterns are more easily observed. Ionization energy increases within a period and decreases within a group as the atomic number increases. The trend is similar for melting points.

Other versions of the Periodic Table are possible (Figure 4.3). Each different version emphasizes some particular aspect of the Periodic Law. The form we commonly use is simply a useful way of emphasizing the relationships of certain properties.

The tabular form of the Periodic Table may also be used to display specific information. Electron configurations are shown in Figure 4.4, and in Figure 4.5 the elements are divided into categories based on their chemical properties.

The metals constitute the largest group of elements. Although most metals have a shiny appearance and are good conductors of elec-

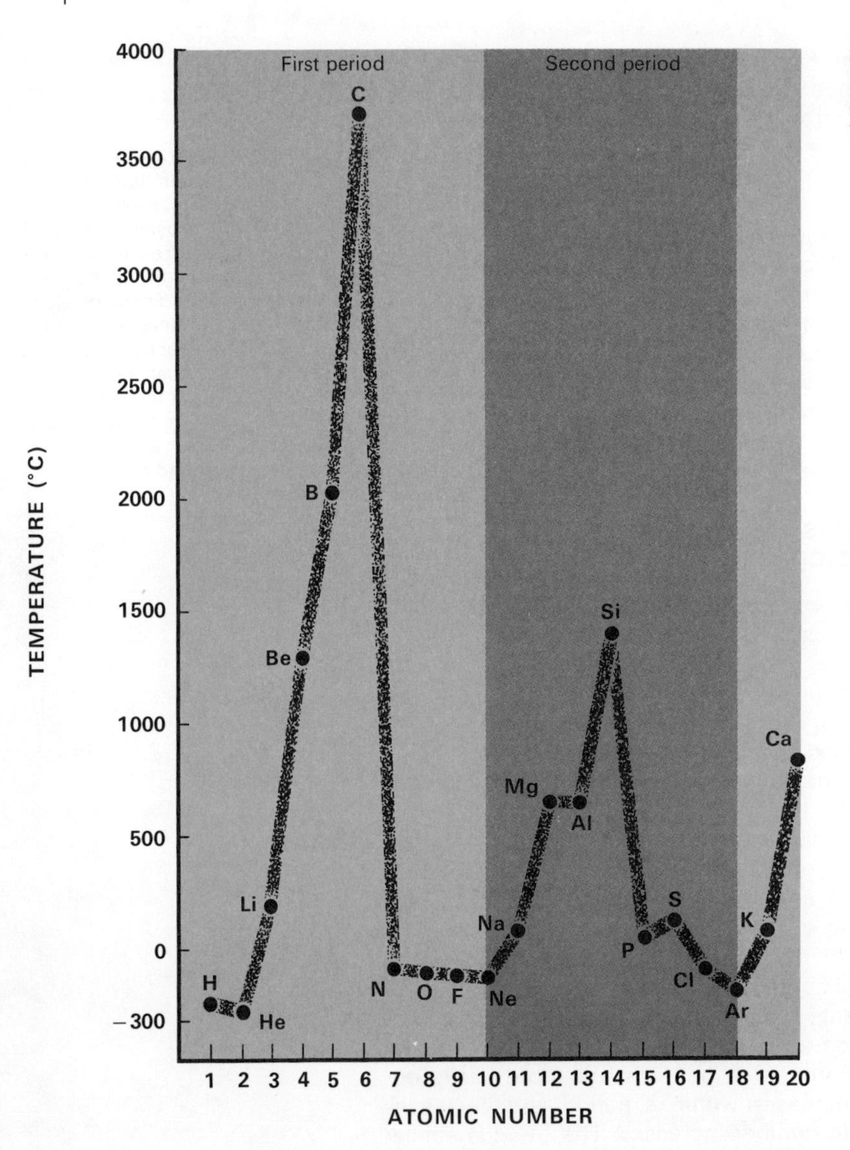

Figure 4.1
Melting points of the first 20 elements. Note the cyclic variation.

tricity, a better chemical criterion is the ease with which they lose electrons. A *metal* is an element that tends to lose electrons, while a *nonmetal* tends to gain them. Recall that the ionization energy is a measure of this tendency, and that the ionization energy increases within a period and decreases within a group. This means that the elements also become less metallic within a period and more metallic within a group.

The border between metallic and nonmetallic elements is not sharply defined; thus the elements on either side of the dividing lines are called *metalloids*.

I A	II A	III B	IV B	V B	VI B	VII B	VIII	VIII	VIII	I B	II B	III A	IV A	V A	VI A	VII A	VIII A	Period
1 H																	2 He	1
3 Li	4 Be											5 B	6 C	7 N	8 O	9 F	10 Ne	2
11 Na	12 Mg											13 Al	14 Si	15 P	16 S	17 Cl	18 Ar	3
19 K	20 Ca	21 Sc	22 Ti	23 V	24 Cr	25 Mn	26 Fe	27 Co	28 Ni	29 Cu	30 Zn	31 Ga	32 Ge	33 As	34 Se	35 Br	36 Kr	4
37 Rb	38 Sr	39 Y	40 Zr	41 Nb	42 Mc	43 Tc	44 Ru	45 Rh	46 Pd	47 Ag	48 Cd	49 In	50 Sn	51 Sb	52 Te	53 I	54 Xe	5
55 Cs	56 Ba	57 La	72 Hf	73 Ta	74 W	75 Re	76 Os	77 Ir	78 Pt	79 Au	80 Hg	81 Tl	82 Pb	83 Bi	84 Po	85 At	86 Rn	6
87 Fr	88 Ra	89 Ac	104 ?	105 Ha														7

TRANSITION ELEMENTS

														Period
58 Ce	59 Pr	60 Nd	61 Pm	62 Sm	63 Eu	64 Gd	65 Tb	66 Dy	67 Ho	68 Er	69 Tm	70 Yb	71 Lu	6
90 Th	91 Pa	92 U	93 Np	94 Pu	95 Am	96 Cm	97 Bk	98 Cf	99 Es	100 Fm	101 Md	102 No	103 Lr	7

Figure 4.2
The Periodic Table.

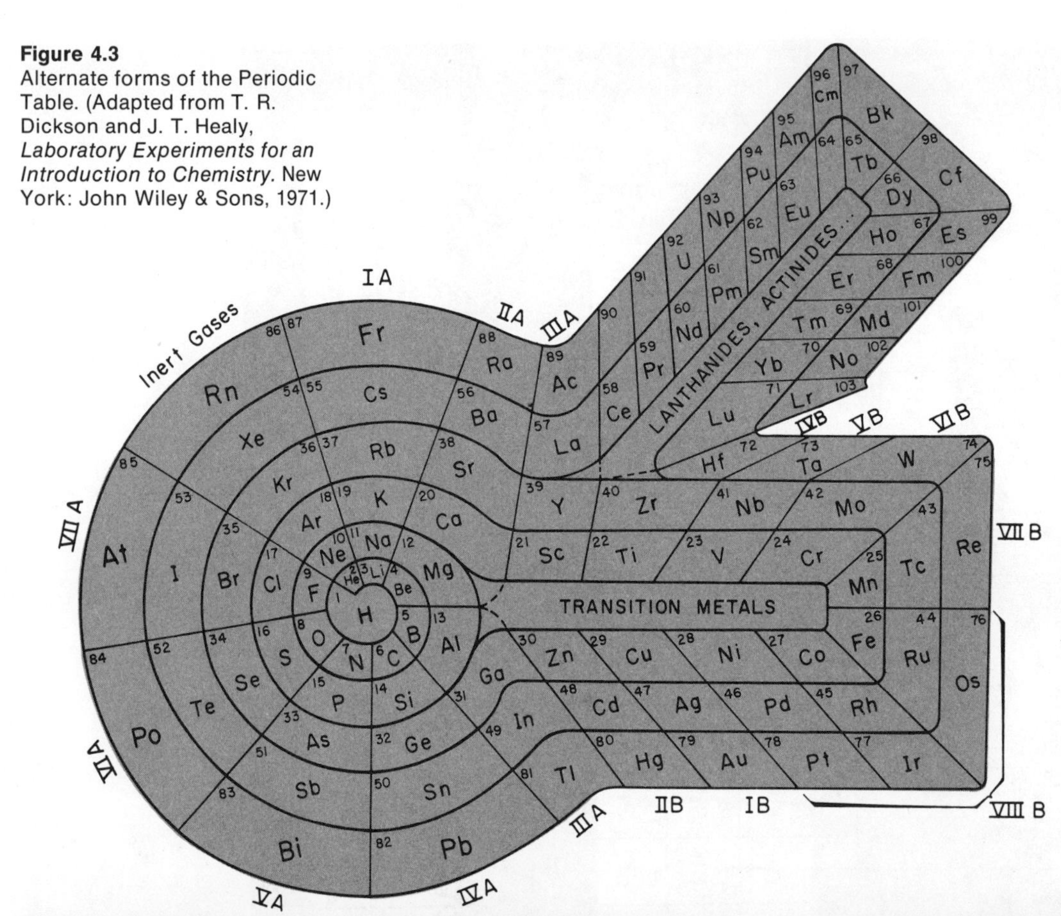

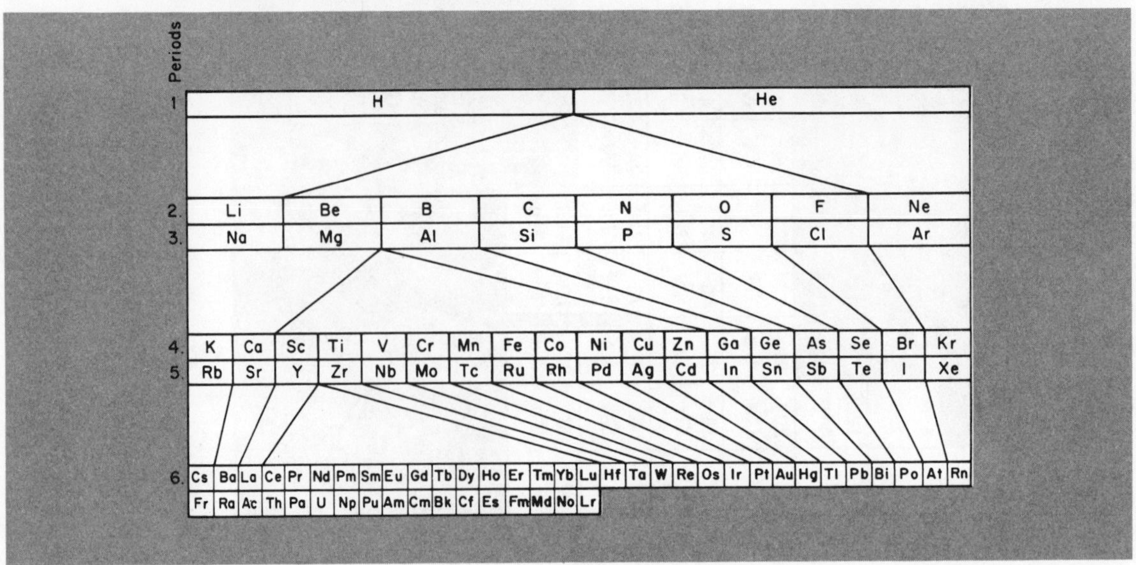

Figure 4.3
Alternate forms of the Periodic Table. (Adapted from T. R. Dickson and J. T. Healy, *Laboratory Experiments for an Introduction to Chemistry.* New York: John Wiley & Sons, 1971.)

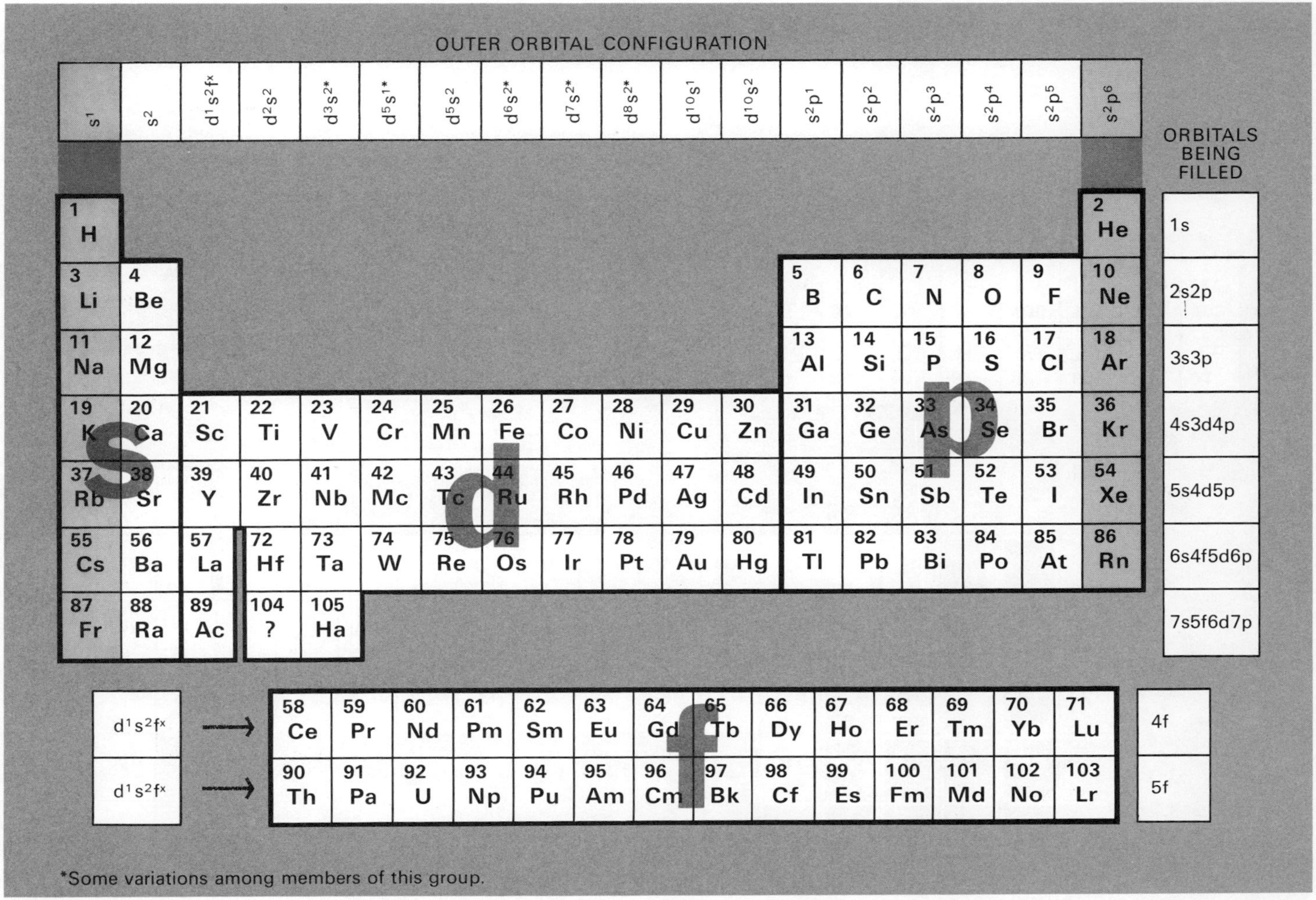

OUTER ORBITAL CONFIGURATION

s^1	s^2	$d^1s^2f^x$	d^2s^2	d^3s^{2*}	d^5s^{1*}	d^5s^2	d^6s^{2*}	d^7s^{2*}	d^8s^{2*}	$d^{10}s^1$	$d^{10}s^2$	s^2p^1	s^2p^2	s^2p^3	s^2p^4	s^2p^5	s^2p^6	ORBITALS BEING FILLED
1 H																	2 He	1s
3 Li	4 Be											5 B	6 C	7 N	8 O	9 F	10 Ne	2s2p
11 Na	12 Mg											13 Al	14 Si	15 P	16 S	17 Cl	18 Ar	3s3p
19 K	20 Ca	21 Sc	22 Ti	23 V	24 Cr	25 Mn	26 Fe	27 Co	28 Ni	29 Cu	30 Zn	31 Ga	32 Ge	33 As	34 Se	35 Br	36 Kr	4s3d4p
37 Rb	38 Sr	39 Y	40 Zr	41 Nb	42 Mc	43 Tc	44 Ru	45 Rh	46 Pd	47 Ag	48 Cd	49 In	50 Sn	51 Sb	52 Te	53 I	54 Xe	5s4d5p
55 Cs	56 Ba	57 La	72 Hf	73 Ta	74 W	75 Re	76 Os	77 Ir	78 Pt	79 Au	80 Hg	81 Tl	82 Pb	83 Bi	84 Po	85 At	86 Rn	6s4f5d6p
87 Fr	88 Ra	89 Ac	104 ?	105 Ha														7s5f6d7p

$d^1s^2f^x$	→	58 Ce	59 Pr	60 Nd	61 Pm	62 Sm	63 Eu	64 Gd	65 Tb	66 Dy	67 Ho	68 Er	69 Tm	70 Yb	71 Lu	4f
$d^1s^2f^x$	→	90 Th	91 Pa	92 U	93 Np	94 Pu	95 Am	96 Cm	97 Bk	98 Cf	99 Es	100 Fm	101 Md	102 No	103 Lr	5f

*Some variations among members of this group.

Figure 4.4
The Periodic Table showing outer orbital configurations of each group and the orbitals being filled in each energy level.

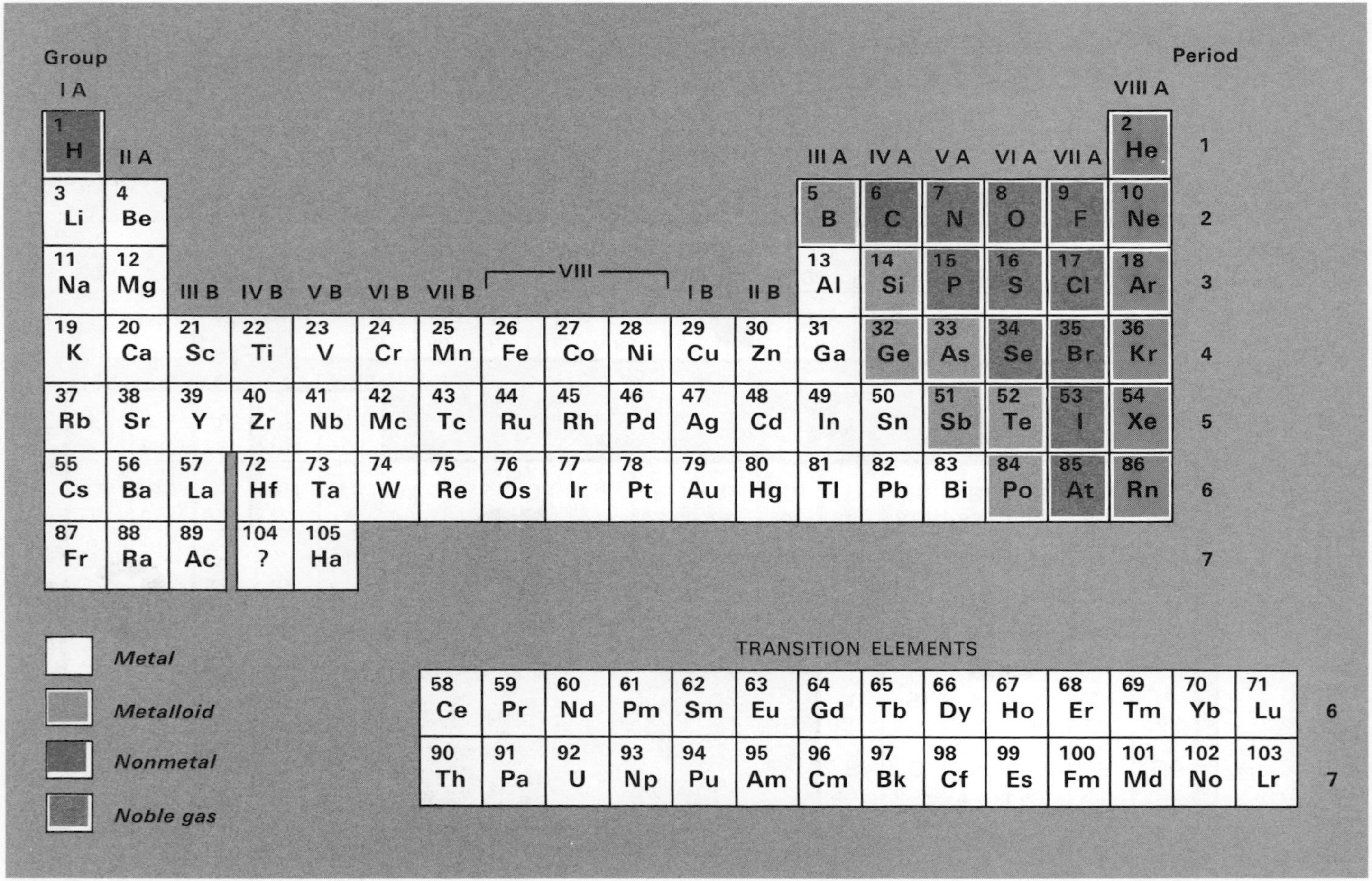

Group I A	II A	III B	IV B	V B	VI B	VII B	VIII	VIII	VIII	I B	II B	III A	IV A	V A	VI A	VII A	VIII A	Period
1 H																	2 He	1
3 Li	4 Be											5 B	6 C	7 N	8 O	9 F	10 Ne	2
11 Na	12 Mg											13 Al	14 Si	15 P	16 S	17 Cl	18 Ar	3
19 K	20 Ca	21 Sc	22 Ti	23 V	24 Cr	25 Mn	26 Fe	27 Co	28 Ni	29 Cu	30 Zn	31 Ga	32 Ge	33 As	34 Se	35 Br	36 Kr	4
37 Rb	38 Sr	39 Y	40 Zr	41 Nb	42 Mc	43 Tc	44 Ru	45 Rh	46 Pd	47 Ag	48 Cd	49 In	50 Sn	51 Sb	52 Te	53 I	54 Xe	5
55 Cs	56 Ba	57 La	72 Hf	73 Ta	74 W	75 Re	76 Os	77 Ir	78 Pt	79 Au	80 Hg	81 Tl	82 Pb	83 Bi	84 Po	85 At	86 Rn	6
87 Fr	88 Ra	89 Ac	104 ?	105 Ha														7

TRANSITION ELEMENTS

														Period
58 Ce	59 Pr	60 Nd	61 Pm	62 Sm	63 Eu	64 Gd	65 Tb	66 Dy	67 Ho	68 Er	69 Tm	70 Yb	71 Lu	6
90 Th	91 Pa	92 U	93 Np	94 Pu	95 Am	96 Cm	97 Bk	98 Cf	99 Es	100 Fm	101 Md	102 No	103 Lr	7

Metal

Metalloid

Nonmetal

Noble gas

Figure 4.5
The Periodic Table showing a classification of the elements based on chemical characteristics.

4.4 FORCES BETWEEN ATOMS: THE COVALENT BOND

Atoms that strongly attract and hold their own electrons can be expected to strongly attract the electrons of a neighboring atom as well. However, the close proximity of two atoms generates forces of repulsion as well as forces of attraction, as shown in Figure 4.6. The hydrogen nuclei and the two electron clouds are mutually repulsive, but each nucleus attracts the electron of the other atom. When the atoms are relatively far apart, the attractive force is weak. However, should the atoms move closer together, the attractive force will increase more than the repulsive force until the atoms reach a position in which the forces are exactly balanced. When this condition is reached, the atoms are said to be *covalently bonded.* If the nuclei attempt to move closer together, there will be an increase in the force of repulsion; and if they move apart, the attractive force will fall off. The distance between the nuclei of bonded atoms is called the *bond length.* Covalent bond lengths range from about 0.7 Å to 3.0 Å.

4.5 OVERLAPPING ORBITALS

The explanation just proposed for the bonding of two hydrogen atoms is an example of the use of models. This model is simply an explanation of bonding in terms of interatomic forces. Another model of bond formation incorporates the concept of overlapping orbitals.

Each hydrogen atom has an *s* orbital containing a single electron. Orbital notation for the isolated hydrogen atoms is

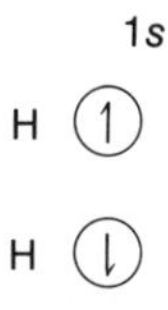

If the two atoms approach each other to such a position that the *s* orbitals overlap, each atom acquires for part of the time an electron

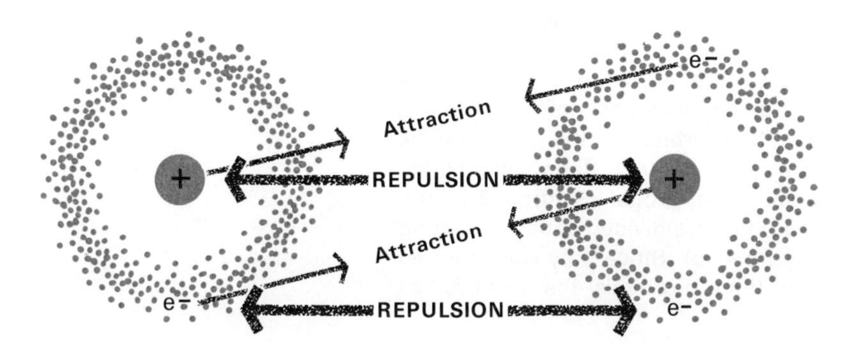

Figure 4.6
Forces of attraction and repulsion between two hydrogen atoms. At this distance the repulsive forces are greater than the attractive forces. As the atoms move closer together the attractive force will be dominant.

which fills its *s* orbital, thus giving it the very stable electron configuration of the element helium.

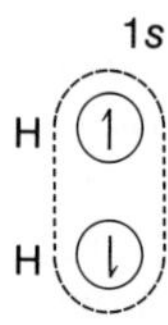

The same idea can be represented by Lewis (electron dot) diagrams:

$$H\cdot + \cdot H \longrightarrow H:H$$

and by electron cloud drawings, as shown in Figure 4.7. When the orbitals of two atoms overlap, they form a new orbital that encompasses the entire molecule. The bond produced by the overlapping of two half-filled atomic orbitals is called a *sigma* bond.

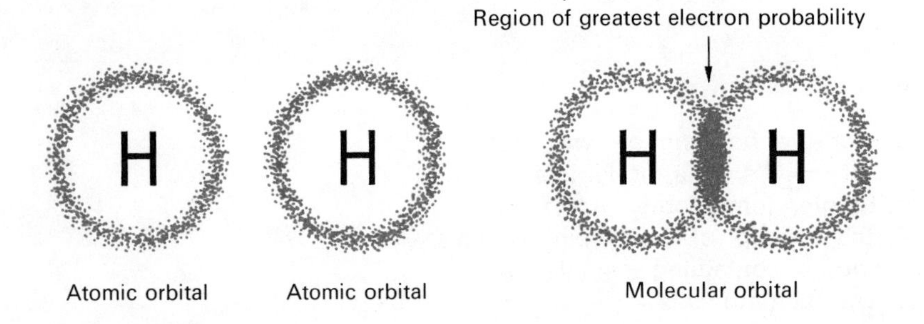

Figure 4.7
Formation of molecular orbital by overlap of two atomic orbitals.

4.6 ATOMIC ORBITAL HYBRIDIZATION

It would appear from its energy-level diagram that the carbon atom is capable of forming two equivalent bonds (Figure 4.8). The diagram shows two half-filled *p* orbitals, each capable of overlapping with an orbital of a different atom to form a sigma bond. All experimental evidence indicates, however, that in most of its compounds carbon forms *four* equivalent bonds. Since the majority of compounds in the body contain carbon, this phenomenon deserves further study.

A model can be constructed that accounts for the existence of four equivalent bonds. The model suggests the formation of new orbitals by a process called *hybridization.* In hybridization new orbitals are formed by combining the energies of the orbitals in the ground state. Our model proposes that there are two distinct steps: one of the 2*s* electrons is promoted to the vacant p_z orbital (Figure 4.9a), and then the total energies of the new orbitals are divided equally, creating four equivalent hybrid orbitals (Figure 4.9b). Since they were formed by pooling the energy of one *s* and three *p* orbitals, the new orbitals are designated sp^3 hybrids.

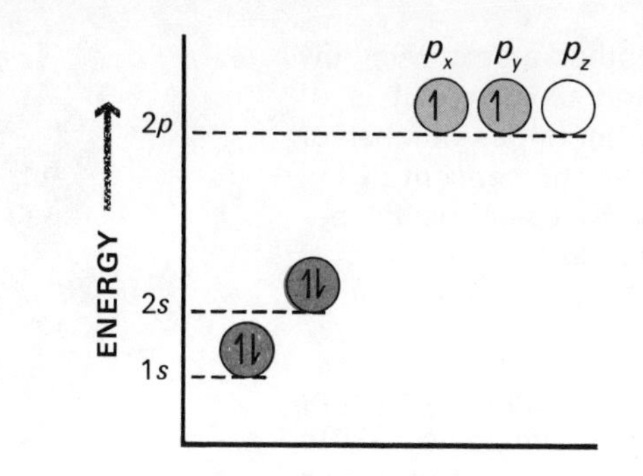

Figure 4.8
Energy-level diagram of carbon in the ground state.

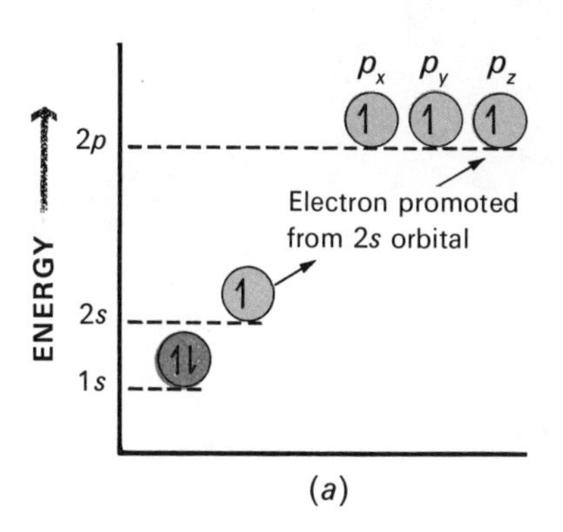

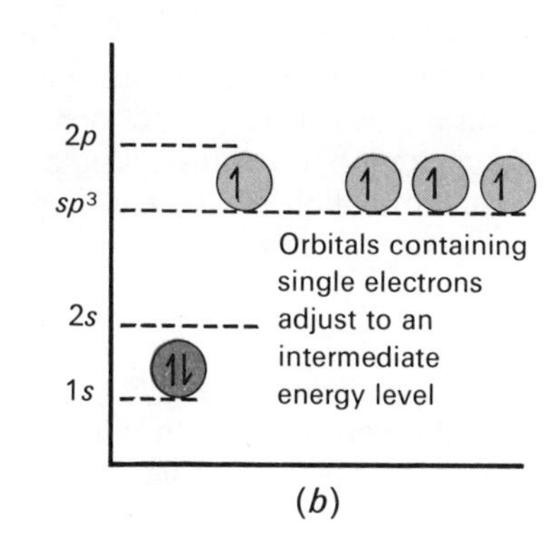

Figure 4.9
Model of hybridization process in carbon. (a) One of the 2*s* electrons in the carbon atom is promoted to a *p* orbital. (b) The total energy of the orbitals in the second energy level is divided equally, producing four equivalent orbitals.

The electron cloud of an individual sp^3 hybrid orbital may be pictured as having the asymmetrical shape shown in Figure 4.10. A complete set of four hybrid orbitals is arranged so that the large lobes point toward the corners of a regular tetrahedron (Figure 4.11). This arrangement provides the maximum separation between orbitals so that the repulsions between electrons are minimized. The angle between any two orbitals is 109.5°.

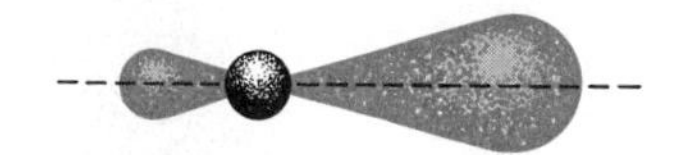

Figure 4.10
An sp^3 hybrid orbital.

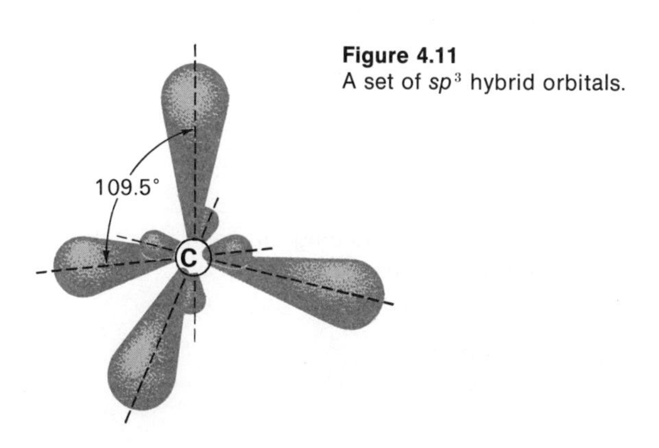

Figure 4.11
A set of sp^3 hybrid orbitals.

Hybridization of atomic orbitals occurs in other atoms with low atomic numbers such as beryllium, boron, and oxygen, but is of greatest importance in carbon because of the enormous number of compounds containing hybridized carbon atoms. The concept of hybridization, while only a theoretical model, is extremely useful because it explains a puzzling experimental observation.

A styrofoam model representing a set of sp^3 hybrid orbitals. (Photograph by J. Asdrubal Rivera)

4.7 MULTIPLE COVALENT BONDS

A covalent bond may form between two atoms of the same element or between two unlike atoms, or an atom may bond simultaneously to several different atoms. Sharing of more than one pair of electrons between atoms (i.e., overlap of more than one pair of orbitals) is also possible. A bond in which several pairs of electrons are shared is called a *multiple covalent bond.* Sharing two pairs of electrons constitutes a double bond, and three shared pairs form a triple bond. Sharing of four pairs is not possible. Formation of single, double, and triple bonds is illustrated by the following examples:

Single Bonds

Recall that an octet of electrons—specifically the s^2p^6 configuration—is especially stable. Note that the carbon atom in methane, the chlorine atom in hydrogen chloride, and both of the chlorine atoms in the chlorine molecule have acquired an octet of electrons through sharing. Hydrogen is also stabilized because it has the configuration of the inactive element helium.

H° ×C× °H (with H° above and H° below C) ⟶ H:C:H (with H above and H below C)

Isolated atoms Methane molecule

H° ×Cl: ⟶ H:Cl:

Isolated atoms Hydrogen chloride molecule

:Cl° ×Cl: ⟶ :Cl:Cl:

Isolated atoms Chlorine molecule

Double Bonds

To show the formation of multiple bonds, we use Lewis diagrams and move the electrons around so that more than one pair are shared by some atoms, and all atoms acquire the most stable configuration.

H° ×C× °C° ×H (with H below each C) ⟶ H:C::C:H (with H below each C)

Isolated atoms Ethene molecule

We can explain the double bond in ethene by assuming that one of the 2*s* electrons is promoted to the vacant $2p_z$ orbital and that the remaining 2*s* electron mixes with the other 2*p* electrons to form a new hybrid orbital (designated sp^2) that has three lobes. In ethene two of the sp^2 lobes bond with hydrogen atoms while the third lobe, together with the unhybridized $2p_z$ electron, form a double bond between the two carbon atoms. Carbon dioxide also contains a double bond.

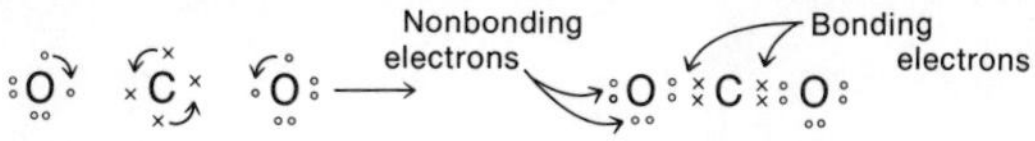

Nonbonding electrons — Bonding electrons

:O: ×C× :O: ⟶ :O::C::O:

Isolated atoms — Carbon dioxide molecule

Note that two pairs of electrons belonging to each oxygen atom are not used in bonding with the carbon. These nonbonding electrons influence the shape of the molecules, as will be shown later.

Triple Bonds

The triple bond in ethyne involves hybridization of one *s* and one *p* orbital in each carbon atom (*sp* hybridization). One lobe of each *sp* orbital bonds to hydrogen. The triple bond is formed from the other lobe and the two unhybridized *p* orbitals.

H° ×C× :C° ×H ⟶ H:C:::C:H

Isolated atoms — Ethyne molecule (acetylene)

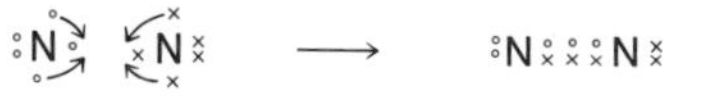

:N: :N: ⟶ :N:::N:

Isolated atoms — Nitrogen molecule

Coordinate Covalent Bonds

A bond in which one of the bonding atoms furnishes both of the pair of electrons is called a *coordinate covalent* bond. The ability of oxygen and nitrogen to form coordinate covalent bonds is of importance and will be discussed in connection with acids and bases, and the reactions of water.

4.8 VALENCE

Atoms are joined by covalent bonds in definite ratios. For example, the combining ratio between hydrogen atoms and chlorine atoms is 1:1, and the ratio between hydrogen and carbon is 4:1. Dalton noted the existence of these fixed ratios between combining atoms, and modern atomic theory has supplied an explanation for them in terms of electron configurations.

The number of bonds an atom is capable of forming is a measure of its combining capacity and is called the *valence* of that element. The valence of an element is related to its electron configuration and is constant for most metals. The valence of a nonmetal tends to be influenced by the nature of its bonding partner. From the examples just cited we can determine that the valence of both hydrogen and chlorine is 1, while that of hybridized carbon is 4. The general rule is that the valence of an element is equal to the difference between its atomic number and that of the closest noble gas. For example, the atomic number of sodium is 11 and that of neon is 10: the predicted valence is 1. The difference between the atomic numbers of sulfur and argon is 2, which agrees with the usual valence of sulfur.

4.9 ENERGY RELATIONSHIPS IN BOND FORMATION

Every orbital in an atom corresponds to a specific amount of energy. When an orbital of one atom overlaps that of another atom, a covalent bond is formed. Formation of a bond is favored when the combined orbital contains less energy than the sum of the two separate orbitals. In other words, if energy must be added in order to bring the electrons of the two atoms up to their original energy levels, they will tend to remain in the overlapping orbital.

Another generalization concerning bond formation is that multiple bonds are stronger than single bonds. This means that more energy is needed to separate atoms joined by multiple bonds. For example, the energy required to rupture the double bond between the two carbon atoms in ethene is 1.76 times as much as for the single bond between carbons in ethane.

In addition to being stronger, multiple bonds are also shorter than single bonds. The bond length of ethene is only 1.34 Å compared with 1.54 Å for ethane.

4.10 MOLECULAR SHAPES

Covalently bonded molecules have definite shapes which are determined by the kinds and locations of the orbitals used in bonding. For example, the *s* orbital is spherically symmetrical and so cannot have a preferred direction or orientation for bonding with other atoms. An individual *p* orbital also forms a bond that has no specific orientation. However, when there are two bonding *p* orbitals in an atom, a bond angle of 90° is expected. Since oxygen fits this description, we can predict that the two hydrogens in a water molecule will be separated by a right angle. This agrees reasonably well with a measured angle of 104.5°. The reason for the discrepancy will be explained shortly.

When an atom is bonded through all three of its *p* orbitals, bond angles of 90° are again expected. Nitrogen has three bonding *p* orbitals, and its reaction with hydrogen can be predicted from either the Lewis diagram or an energy-level diagram:

Lewis Diagram

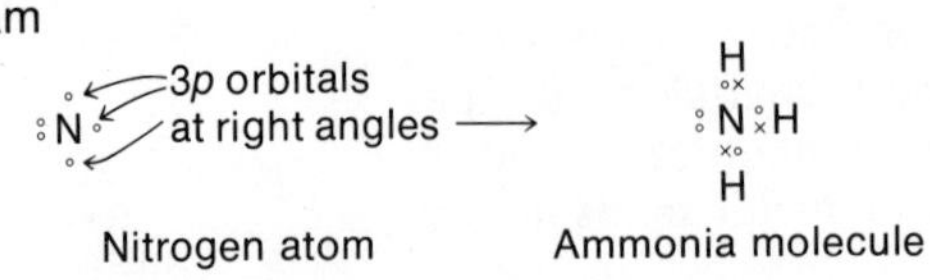

:N (3*p* orbitals at right angles) ⟶ H:N:H with H above and below

Nitrogen atom — Ammonia molecule

Energy-Level Diagram

N 2s (↑↓) 2p_x (↑) 2p_y (↑) 2p_z (↑) ⟶ N 2s (↑↓) 2p_x (↑) 2p_y (↑) 2p_z (↑); H 1s (↓) H (↓) H (↓)

Nitrogen atom — Ammonia molecule

We would expect the ammonia molecule to resemble a pyramid with the nitrogen atom at the apex and a base of three hydrogen atoms. While this is an accurate description of the overall shape, the measured angle is 107° rather than 90°.

The variation between the predicted and observed bond angles in molecules containing only single bonds is accounted for by the *Electron Pair Repulsion* (EPR) theory. According to this theory, pairs of electrons attached to a central atom are repelled by each other and try to move as far apart as possible. The greatest possible separation occurs when the angle between any two pairs of electrons is 109.5°, as in methane. Nonbonding pairs of electrons (also called "lone" pairs) are also included. The general rule is that any molecule whose Lewis diagram has an octet of electrons, either as shared pairs, or in a combination of lone and shared pairs around a central atom, will approach the tetrahedral shape (Figure 4.12).

The orbitals of the lone pairs, however, will expand to a greater volume of space than the orbitals of bonded electrons because the lone pairs are not pulled in by two different nuclei. The expanded orbitals of the lone pairs squeeze the orbitals of the bonded electrons closer together, causing the bond angle to be slightly less than the predicted value. Since the oxygen atom in a water molecule has two lone pairs of electrons, while the nitrogen atom in ammonia has only one pair, EPR theory predicts that the bonding pairs of electrons will be squeezed more tightly together than those in ammonia. This prediction is borne out by observed angles of 104.5° for water and 107° for ammonia. (Figure 4.13).

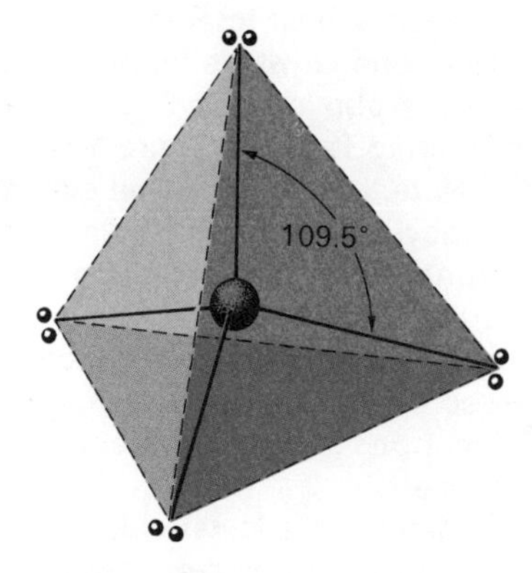

Figure 4.12
Electron Pair Repulsion theory and molecular shape. EPR theory predicts that pairs of electrons will assume a tetrahedral configuration around a central atom.

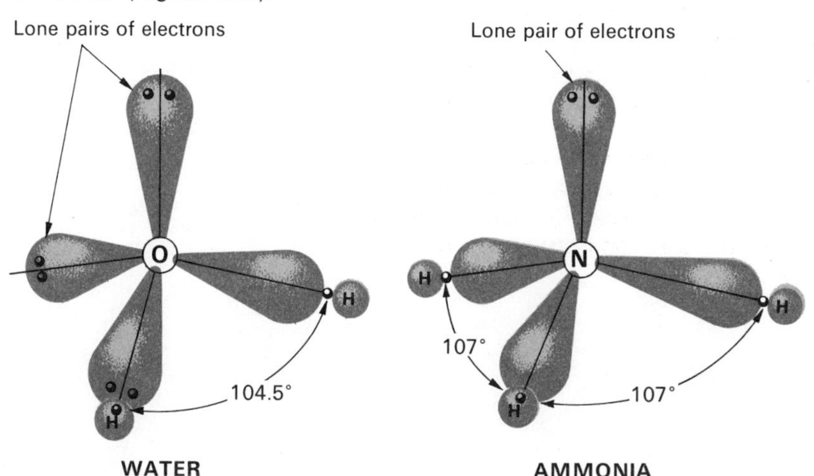

Figure 4.13
The effect of nonbonding electron pairs on molecular shape. The bond angle in water is slightly less than the bond angle in ammonia because the two lone pairs of electrons repel the bonding electrons more strongly.

4.11 ELECTRONEGATIVITY

The *electronegativity* of an element is an indication of its ability to attract electrons in a chemical bond. The concept of electronegativity was first proposed by Linus Pauling. He suggested that each element could be assigned a rating on a relative scale commensurate with its power to attract electrons. The electronegativity of an element is a calculated value, not an experimental value. It is derived, however, from measurements of ionization energy and other experimental data. The noble gases are assigned electronegativity values of zero because of their almost total lack of reactivity. The element having the highest electronegativity is fluorine, with a rating of 4.0. The electronegativity values obtained by Pauling are shown against a background of the Periodic Table in Figure 4.14.

Note that the trends in electronegativity almost parallel the periodic behavior described earlier for ionization energy. In general, electronegativity increases from left to right within a period, and decreases from top to bottom within a family. The increase within a period is primarily a result of the increase in nuclear charge, while the decrease within a family reflects the screening effect of additional electrons and the greater separation of the valence electrons and the nucleus.

Metals have lower electronegativities than nonmetals and thus tend to lose electrons more readily. An element having an electronegativity of less than 2.0 is usually considered to be a metal.

Electronegativity is a valuable concept because it simplifies a number of complex relationships within the atom. The following section illustrates an important use of electronegativity ratings.

Polarity and Symmetry

When a covalent bond is formed between two atoms of the same element, the electrons involved in the bond are attracted equally by the two nuclei. In such molecules the center of positive charge coincides with the center of negative charge, and the molecule is said to be nonpolar. Examples of nonpolar molecules are O_2, H_2, and Cl_2. Each of these molecules is perfectly symmetrical and the centers of positive and negative charge are located at the midpoint of a line connecting the two nuclei.

A nonpolar molecule may also be formed from elements having quite different electronegativities if the molecule is symmetrical. As an example, let us consider a molecule composed of carbon (electronegativity 2.5) and chlorine (3.0). The electronic structure predicted for this compound is

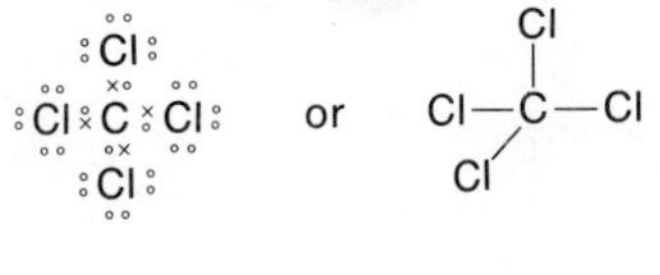

```
     ∘∘                    Cl
    ∘Cl∘                   |
  ∘∘ ×∘ ∘∘           Cl—C—Cl
 ∘Cl×C×Cl∘    or         /
  ∘∘ ∘× ∘∘             Cl
    ∘Cl∘
     ∘∘
```

H 2.1																	He 0
Li 1.0	Be 1.5											B 2.0	C 2.5	N 3.0	O 3.5	F 4.0	Ne 0
Na 0.9	Mg 1.2											Al 1.5	Si 1.8	P 2.1	S 2.5	Cl 3.0	Ar 0
K 0.8	Ca 1.0	Sc 1.3	Ti 1.5	V 1.6	Cr 1.6	Mn 1.5	Fe 1.8	Co 1.8	Ni 1.8	Cu 1.9	Zn 1.6	Ga 1.6	Ge 1.8	As 2.0	Se 2.4	Br 2.8	Kr 0
Rb 0.8	Sr 1.0	Y 1.2	Zr 1.4	Nb 1.6	Mo 1.8	Tc 1.9	Ru 2.2	Rh 2.2	Pd 2.2	Ag 1.9	Cd 1.7	In 1.7	Sn 1.8	Sb 1.9	Te 2.1	I 2.5	Xe 0
Cs 0.7	Ba 0.9	La 1.1	Hf 1.3	Ta 1.5	W 1.7	Re 1.9	Os 2.2	Ir 2.2	Pt 2.2	Au 2.4	Hg 1.9	Tl 1.8	Pb 1.8	Bi 1.9	Po 2.0	At 2.2	Rn 0
Fr 0.7	Ra 0.9	Ac 1.1	?	Ha													

Ce 1.1	Pr 1.1	Nd 1.1	Pm 1.1	Sm 1.1	Eu 1.1	Gd 1.1	Tb 1.2	Dy 1.2	Ho 1.2	Er 1.2	Tm 1.2	Yb 1.2	Lu 1.2
Th 1.3	Pa 1.5	U 1.1	Np 1.3	Pu 1.3	Am 1.3	Cm 1.3	Bk 1.3	Cf 1.3	Es 1.3	Fm 1.3	Md 1.3	No 1.3	Lw 1.3

Figure 4.14
Electronegativity values. Electronegativity increases from left to right within a period and from bottom to top within a family.

Since chlorine has the higher electronegativity, we would expect the bonding electrons to be more strongly attracted to the chlorine atoms. In other words, the electrons around the carbon atom move away from it toward the chlorine atoms. Each chlorine atom thus acquires a greater share of the negative charge. This additional charge is represented by the symbol $\delta-$ (called delta minus), as shown in Figure 4.15. Delta is often used in mathematics to indicate a small increment. At the same time the displacement of electrons away from the carbon atom leaves it with a slight excess of positive charge, as indicated by the $\delta+$. When the positive and negative charge centers of a pair of bonded atoms do not coincide, we describe the situation by saying that a dipole exists.

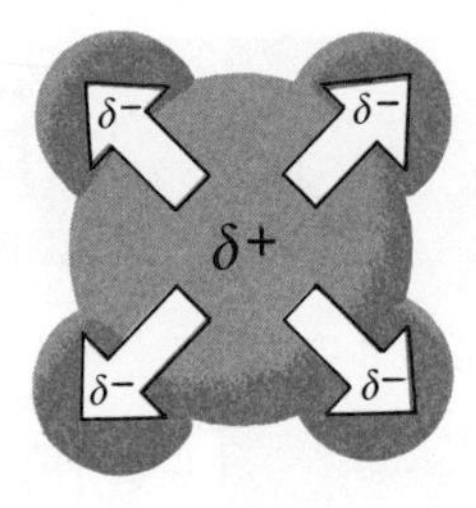

Figure 4.15
A molecule of carbon tetrachloride. Note the unequal distribution of charge due to the higher electronegativity of chlorine.

We must remember, however, that the four bonds are equal and that the bonding orbitals are separated by identical angles of 109.5°, and so a dipole created in any direction will be matched and counterbalanced by a dipole in the opposite direction. The result is that the dipoles cancel each other, and the center of negative charge coincides with the center of the molecule—which is also the geometric center of the nuclear charges.

Another example of a nonpolar molecule is carbon dioxide (Figure 4.16). Although the electronegativity of oxygen is considerably higher than that of carbon, the dipole created by the displacement of electrons toward the oxygen atom on one side of the molecule is counteracted by the dipole on the opposite side, and the charge centers coincide.

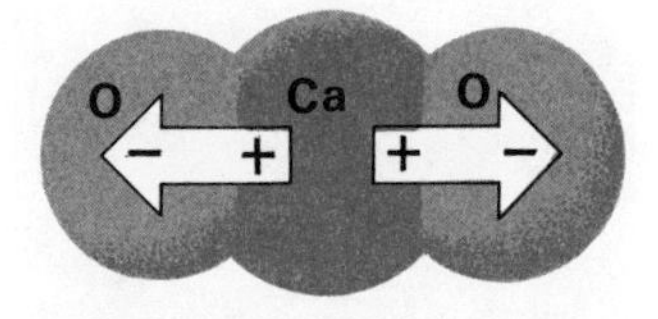

Figure 4.16
Carbon dioxide—a nonpolar molecule. Despite the bonding between atoms of different electronegativities, the molecule is nonpolar because the dipoles cancel each other.

A molecule containing an unbalanced dipole is called a polar molecule. Water is a polar molecule, and many of its properties may be traced to this fact.

Recall that the angle between the two hydrogen atoms in a water molecule is 104.5°. Because oxygen has a higher electronegativity than hydrogen, the electrons shared between the two atoms are displaced toward the oxygen atom, as shown in Figure 4.17. Movement of the electrons away from the hydrogen atoms leaves each one with a slight positive charge and creates two dipoles (Figure 4.18). Because the angle between the dipoles is 104.5°, the dipoles do not cancel each other. We can, however, resolve them into a single dipole, as shown in Figure 4.19, and this resolution of dipoles clearly shows that water is a polar molecule. Other examples of polar covalent molecules are hydrogen fluoride (Figure 4.20) and formaldehyde (Figure 4.21).

The knowledge that a molecule is either polar or nonpolar is important because it enables us to predict some of the properties of that compound, or to explain some aspects of its behavior.

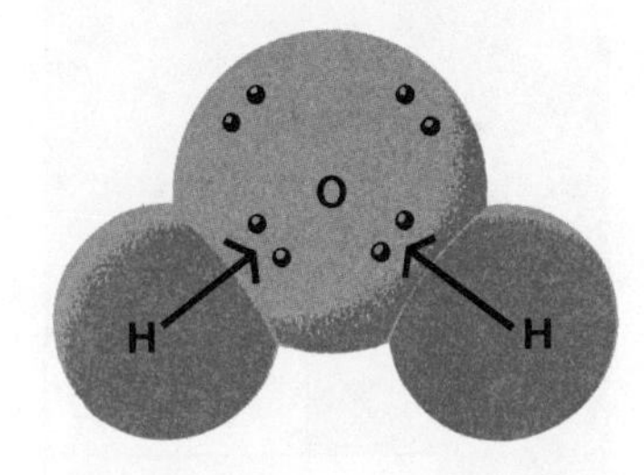

Figure 4.17
The source of the dipoles in a water molecule. The shared pairs of electrons are displaced toward the oxygen atom because of its higher electronegativity.

4.12 PROPERTIES OF COVALENT MOLECULES

A definite relationship exists between the properties of a covalent molecule and the arrangement of its charge centers. For this reason the properties of polar molecules are different from those of nonpolar molecules. In general, a nonpolar substance has low melting and

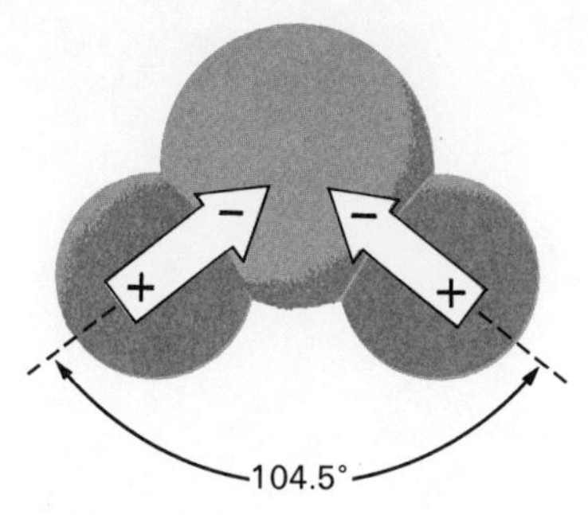

Figure 4.18
The effect of molecular shape on the dipoles in a water molecule. Because the hydrogen atoms do not lie opposite each other, the dipoles do not cancel each other.

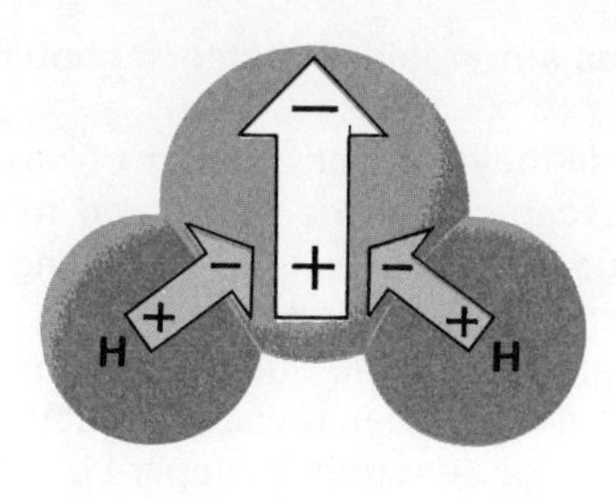

Figure 4.19
Resolution of the dipoles in a water molecule. The effect of the two dipoles is the same as that of a single dipole represented by the larger arrow.

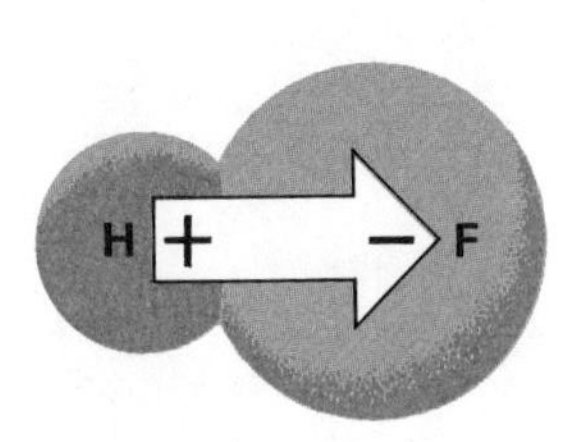

Figure 4.20
Hydrogen fluoride—a polar molecule.

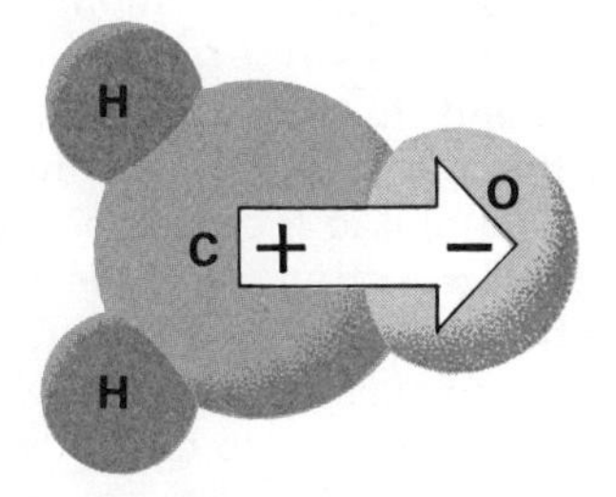

Figure 4.21
Formaldehyde—a polar molecule.

boiling points, a low density, and an extremely low solubility in water. Polar molecules, on the other hand, have higher melting and boiling points, higher densities, and slight-to-moderate solubility in water.

A comparison of the models of polar and nonpolar molecular structure reveals the reason for these differences. A nonpolar molecule is characterized by a specific number of rather strong bonds. Each bond has a fixed length, and the bonding electrons are highly localized; thus the molecule has a definite size. The molecule tends to assume a configuration such that repulsions between valence electrons are minimized, which means that it has a definite shape as well. Since they do not contain uncompensated dipoles, the molecules of a nonpolar substance exist as discrete units having very little attraction for each other. With such low attractions the substance should exist in the solid state as loosely packed individual units—that is, it should have a low density. This being the case, a small amount of external energy should be able to counterbalance the attractive forces and allow the individual molecules to move around freely—in other words, to melt or become a liquid. Thus a nonpolar substance should have a low melting point.

Following the addition of a slightly greater amount of energy, the molecules should be able to escape from the liquid and to exist as separated units in the vapor state. This implies that the boiling point of a nonpolar substance should also be rather low.

The extremely low solubility of nonpolar substances in water is predicted by our model because the dipole in a water molecule cannot attract a nonpolar molecule strongly enough to separate it from its neighbors.

The very low solubility of gaseous nonpolar molecules such as O_2 and CO_2 has important implications for most members of the animal kingdom. Animals require oxygen in order to obtain energy from their food, and the release of this energy is accompanied by the production of carbon dioxide. Except in the case of very small organisms composed of only a few tissues, these respiratory gases are transported within the body by fluids which are mostly water. Since only small amounts of either gas will dissolve in water under normal conditions, the fluids of the body cannot carry enough gas in solution to meet its needs. In man, and in most higher animals, the respiratory gases become attached by weak chemical bonds to specific molecules in the body fluid and are transported in this manner. Certain other chemicals, such as carbon monoxide, are harmful to the body because they compete for the carrier molecules and form stronger bonds with them.

The limited solubility of oxygen has additional significance for animals that live in water. Most aquatic animals have specialized membranes through which the dissolved gas passes into the body, and the rate of this transfer depends upon the concentration of oxygen. Warming the water decreases the solubility of oxygen, and reduces the amount of oxygen available to the organism. The use of water for cooling in industrial processes such as steel-making, and in thermonuclear power plants for the generation of electricity, may raise the temperature of a body of water to the point that it contains very little dissolved oxygen, and the organisms living in it are adversely affected. This is what is meant by thermal pollution.

The model of polar molecular structure differs only slightly from the one just described. The concept of a definite molecular size and shape is common to both models because both types of substances result from a sharing of electrons. Polar molecules, however, contain uncompensated dipoles and thus attract each other like small magnets. The more pronounced the dipole, the stronger the attraction. Because of these attractions, the spaces between molecules of a polar substance are smaller and the density is higher. A substantial amount of energy must be supplied before these attractions can be overcome, and an even greater amount before the molecules can exist in the gaseous state, and these high energy requirements are reflected in higher melting and boiling points.

Our model also predicts increased solubility for polar substances because of attractions between the dipole of a water molecule and that of other polar molecules.

Dead shad on the Anacostia River. The oxygen content of the water was reduced by a combination of pollution and hot weather. (Courtesy Bureau of Sport Fisheries and Wildlife, Washington, D.C.)

A comparison of some of the physical properties of two polar and two nonpolar molecules of approximately the same size is shown in Table 4.2.

TABLE 4.2 Physical Properties of Some Covalent Molecules

Substance	Classification	Melting Point	Boiling Point	Solubility in Water (g per 100 cc water)	
O_2	Nonpolar	−218.4°C	−183.0°C	0.0007	Roughly equal size
HCl	Polar	−114.8°C	−84.9°C	82.3	
CH_4	Nonpolar	−182.5°C	−161.5°C	Very slight	Roughly equal size
NH_3	Polar	−77.7°C	−33.4°C	89.9	

4.13 ION FORMATION

Covalent compounds result from the sharing of electrons between atoms. There is a smaller class of compounds in which bonding takes the form of electrostatic attractions between oppositely charged particles called ions. An *ion* is a chemical unit composed of one or more atoms and bearing a net positive or negative charge. A monoatomic ion is derived from a single atom, while a polyatomic ion is an aggregate of covalently bonded atoms and is often referred to as a *radical.*

Positively charged ions are called *cations* because they are attracted to a cathode—which is the negative electrode of a battery. Negatively charged ions are called *anions* because the electrode that attracts them is the anode. Table 4.3 lists the important common ions.

TABLE 4.3 Common Ions of Importance

Cations		Anions	
Name	Symbol	Name	Symbol
Aluminum	Al^{3+}	Acetate	$C_2H_3O_2^-$
Ammonium	NH_4^+	Bromide	Br^-
Barium	Ba^{2+}	Carbonate	CO_3^{2-}
Calcium	Ca^{2+}	Chlorate	ClO_3^-
Copper (II) or cupric	Cu^{2+}	Chloride	Cl^-
Copper (I) or cuprous	Cu^+	Chlorite	ClO_2^-
Iron (III) or ferric	Fe^{3+}	Chromate	CrO_4^{2-}
Iron (II) or ferrous	Fe^{2+}	Cyanide	CN^-
Hydrogen	H^+	Dichromate	$Cr_2O_7^{2-}$
Hydronium	H_3O^+	Hydrogen carbonate (bicarbonate)	HCO_3^-
Lead (IV) or plumbic	Pb^{4+}	Hydrogen sulfate (bisulfate)	HSO_4^-
Lead (II) or plumbous	Pb^{2+}	Hydroxide	OH^-
Magnesium	Mg^{2+}	Iodide	I^-
Manganese (II) or manganous	Mn^{2+}	Nitrate	NO_3^-
Mercury (II) or mercuric	Hg^{2+}	Nitride	N^{3-}
Mercury (I) or mercurous	Hg_2^{2+}	Nitrite	NO_2^-
Potassium	K^+	Oxalate	$C_2O_4^{2-}$
Silver	Ag^+	Permanganate	MnO_4^-
Sodium	Na^+	Phosphate	PO_4^{3-}
Zinc	Zn^{2+}	Sulfate	SO_4^{2-}
		Sulfide	S^{2-}
		Sulfite	SO_3^{2-}

The process of ion formation involves the gain or loss of one or more electrons. When an individual atom becomes a monoatomic ion, its electron configuration is usually changed to that of the nearest noble gas. Hydrogen is an exception to this generalization and will be described later. The formation of a polyatomic ion is similar in that most of the atoms acquire a stable configuration by sharing electrons, and the loss or gain of additional electrons results in a noble gas configuration for all of them. The formation of monoatomic and polyatomic ions will be discussed separately.

Monoatomic Cations

Monoatomic cations are formed from metallic elements by the loss of electrons, as illustrated by the example of sodium. The electron configuration of sodium is $1s^22s^22p^63s^1$; thus a sodium atom has a single valence electron in its third energy level. The loss of this electron leaves sodium with an octet of electrons in the second level—which is the same as the stable configuration of neon. Recall that sodium has a low ionization energy, which suggests that the electron can be removed rather easily. The nuclear structure is unaffected by the loss of the electron, and so the nuclear charge remains 11+ as compared with 10− for the electrons. This gives the sodium atom an overall charge of 1+ and converts it to an ion of sodium. Because of its greater stability and the change in its electron configuration, a sodium ion is chemically quite different from an atom of sodium.

Note that lithium and potassium are in the same family as sodium and that each one forms an ion with a 1+ charge, which shows the periodic nature of the ionic charge.

In general, a cation is formed by the loss of all of the electrons from the valence shell of an atom. The aluminum atom, for example, loses three electrons to form the Al^{3+} ion, which also has the same electron configuration as neon.

Although hydrogen behaves like a metal in that it loses an electron from its valence shell to form a positive ion, it is an exception because ion formation does not give it the configuration of a noble gas, but leaves it without any electrons.

Monoatomic Anions

Nonmetallic elements (i.e., elements with high electronegativities) tend to gain electrons and form anions. Chlorine, for example, has an electronegativity of 3.0 and the electron configuration $1s^22s^22p^63s^23p^5$. There are seven electrons in the valence shell. The addition of one electron gives chlorine the same configuration as argon; but the nucleus still has a charge of 17+, and so the chlorine ion has a charge of 1−. Fluorine, bromine, and iodine are in the same family and also form 1− ions.

Why doesn't chlorine lose seven electrons and become like neon in its electron configuration? The answer is that chlorine has a high ionization energy and the removal of seven electrons would require extraordinary amounts of energy. Because of the energy involved, the gain or loss of more than three electrons is extremely rare.

Polyatomic Ions (Radicals)

Although there are a few exceptions, the common polyatomic ions are anions. The general procedure for deriving the electronic structure of a polyatomic ion is to begin with the Lewis diagrams of the elements and to assign the available electrons with a goal of giving each atom a stable configuration. The number of additional electrons needed after the available electrons have been assigned is the ionic charge. The formation of the cyanide ion from carbon and nitrogen is representative.

Step 1

The Lewis diagram is shown for each element.

C   N

Step 2

The available electrons are assigned.

C N

Vacancy

Step 3

One electron is added to fill the vacancy.

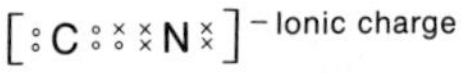

[C N]⁻ Ionic charge

Another common anion is the hydroxyl ion, which consists of a single atom each of hydrogen and oxygen.

Step 1

O   H

Step 2

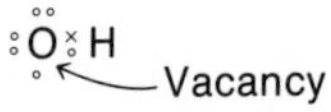

O H

Vacancy

Step 3

[O H]⁻

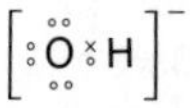

One of the few polyatomic cations is the ammonium ion, which also contains a coordinate covalent bond. The ammonium ion can be constructed from an ammonia molecule and a hydrogen ion. Its ionic charge is 1+.

Step 1

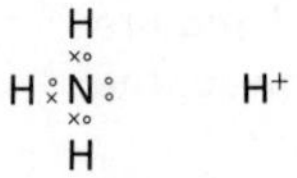

H
H N H^+
H

Step 2

H
H N H
H
Coordinate covalent bond

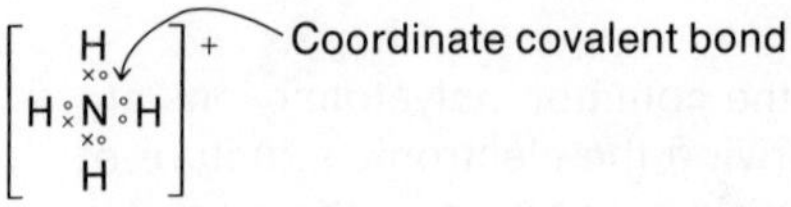

4.14 THE IONIC BOND

An *ionic compound* is composed of oppositely charged ions held together in a lattice arrangement by electrostatic attractions. The formation of an ionic compound may be visualized as taking place in three steps although the distinct steps are not detectable. In other words, a model of ionic bonding will be presented. An example will show the formation of a compound containing two monoatomic ions, K^+ and Cl^-.

Step 1

The metallic element loses an electron and forms a positive ion.

$$K \longrightarrow \text{Electron} + \underset{\text{Potassium ion}}{K^+}$$

Step 2

The electron is acquired by an atom of the nonmetal, turning it into an ion.

$$Cl + \text{Electron} \longrightarrow \underset{\text{Chloride ion}}{Cl^-}$$

Step 3

Positive and negative ions attract each other.

$$K^+ + Cl^- \longrightarrow K^+Cl^-$$

In actuality, an ionic compound is not composed of single ions but is an aggregate of ions called a crystal lattice. The ions that make up a crystal lattice have a definite pattern and combine in a fixed ratio. The spacing of the ions is determined by their shapes and sizes, and the ratio by their charges. The potassium and chloride ions are roughly the same size and shape and each carries a charge of one electrostatic unit. Equal numbers of potassium and chloride ions combine to form an ionic crystal in which each ion is equidistant from six other ions of opposite charge, as shown in Figure 4.22. If one of the ions in a crystal is polyatomic, or carries a greater charge than the other, the lattice structure will be different from that just described. In general, ions tend to pack together in such a way that the spaces between them are minimized.

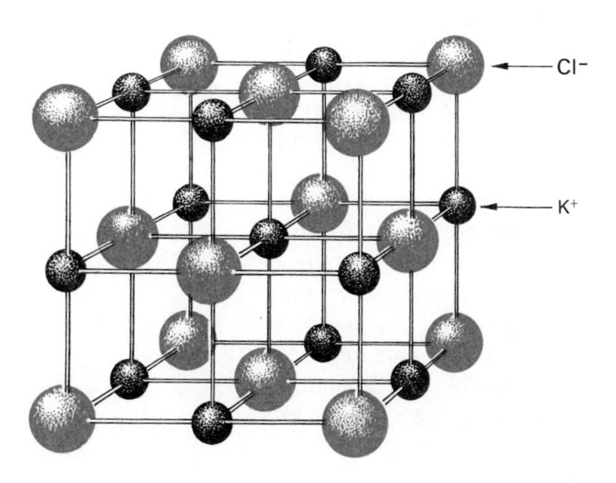

Figure 4.22
The crystal lattice structure of potassium chloride. The lattice is held together by attractions between oppositely charged ions.

In summary, our model of ionic bonding proposes that an ionic substance in the solid state consists of oppositely charged ions held in a definite lattice pattern by electrostatic forces of attraction. Unlike covalent substances, ionic compounds are not composed of discrete units that can be called molecules. Since each individual ion in a lattice is bonded to all of its oppositely charged neighbors, the situation somewhat resembles polygamy. It is just as arbitrary to select two neighboring ions from the lattice and to refer to this pair as "the molecule" as it would be to pair one of the wives with the husband and call them "the couple." For this reason some chemists insist that the term molecule should not be applied to ionic substances.

4.15 CHARACTERISTICS OF IONIC COMPOUNDS

Each individual ion carries a charge of at least one electrostatic unit, which is greater than the partial charge on even the most polar of molecules. Because of this, the attraction between two adjacent ions is considerably greater than the attraction between two molecules. This results in a greater density for most ionic substances. Since a large amount of energy is needed to overcome the attractions between ions, an ionic compound tends to have very high melting and boiling points. Most ionic compounds melt at a temperature above 300°C.

Ionic substances are generally quite soluble in water and other polar solvents. An ionic compound will dissolve in a polar solvent if the solvent molecules attract an ion more strongly than two ions attract each other. The process of solvation (dissolving) will be discussed in detail in Chapter 6.

4.16 ELECTRONEGATIVITY DIFFERENCES AND BOND CHARACTER

It would appear from what has been said up to this point that every chemical bond is the result of either electron sharing or electron transfer. However, the distinction between covalent and ionic bonding is not that sharp, and in almost every covalent bond there is some ionic character. In other words, only when the two bonded atoms are identical will the properties of the substance be unaffected by ionic influence. The extent to which this ionic influence is felt depends upon the difference in electronegativity of the two elements: the greater the difference, the more ionic the bond.

An example may help to illustrate the point. Chlorine (electronegativity 3.0) combines in a 1:1 atomic ratio with both sodium (0.9) and silver (1.9). The elements in the sodium compound have an electronegativity difference of 2.1, which is considerably more than the difference between silver and chlorine, and suggests that the bond in sodium chloride should have a greater ionic character. This prediction is confirmed by the observation that sodium chloride is much more soluble in water and melts at a higher temperature than silver chloride.

TABLE 4.4 The Relationship between the Electronegativity Difference of Bonded Elements and the Character of the Bond between Them

Electronegativity Difference	% Ionic Character	Electronegativity Difference	% Ionic Character
0.2	1	2.0	63
0.4	4	2.2	70
0.6	9	2.4	76
0.8	15	2.6	82
1.0	22	2.8	86
1.2	30	3.0	89
1.4	39	3.2	92
1.6	47	3.4	96
1.8	55		

0.0–0.4 bond is essentially nonpolar
0.6–2.4 bond is polar covalent
2.6–3.4 bond is essentially ionic

The relationship between the electronegativity difference and the ionic character of a bond is shown in Table 4.4. Note that when two elements differ in electronegativity by 0.5 units or less, the bond between them is essentially nonpolar, while differences greater than 2.5 units indicate ionic compounds.

4.17 VAN DER WAALS FORCES

Molecules of a nonpolar substance are attracted to each other by very weak forces which exist—often fleetingly—between the nucleus of an atom in one molecule and electrons in a different molecule. Since the atoms in these molecules already have stable electron configurations, the attractions are not the result of electron sharing, but rather of induced dipoles. The induction of an instantaneous dipole in a hydrogen molecule is shown in Figure 4.23. The attractions between nonpolar molecules are called van der Waals forces and, although they are weak, their cumulative effect is noticeable.

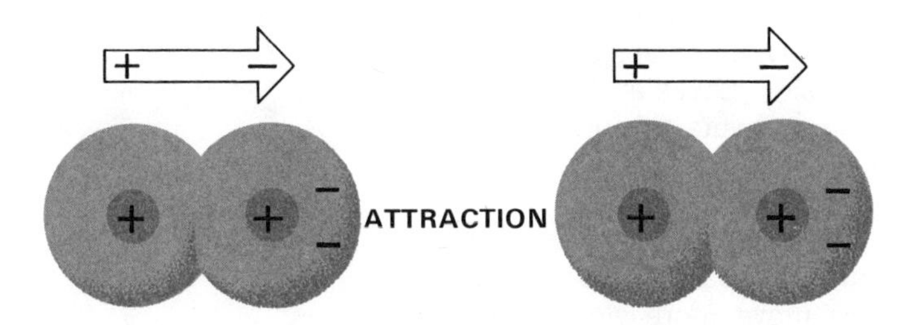

Figure 4.23
An example of van der Waals forces. An instantaneous dipole is created when both electrons in the hydrogen molecule are on the same side.

4.18 HYDROGEN BONDING

Molecules of polar substances attract one another because of dipole-dipole interactions. The strongest intermolecular attractions of this

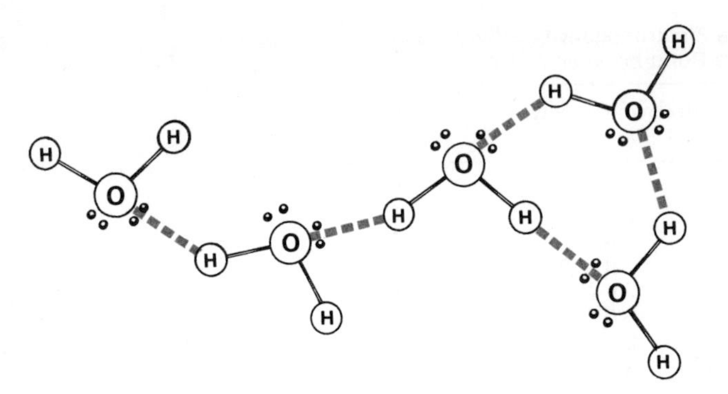

Figure 4.24
Hydrogen bonding between water molecules. Hydrogen bonds are represented by dotted lines.

type are observed when the compound contains hydrogen that is covalently bonded to a small atom of extremely high electronegativity such as nitrogen, oxygen, or fluorine. Such molecules are strongly polar because the electrons spend most of their time in the vicinity of the more electronegative element, leaving the hydrogen with a partial positive charge. The negative portion of one molecule attracts the hydrogen part of another across fairly long distances to form intermolecular bonds (Figure 4.24). Bonds of this type are not covalent because electrons are not being shared, and yet they are roughly ten times as strong as van der Waals forces. Their rather unique nature is recognized by naming them hydrogen bonds. A *hydrogen bond* may be considered as the sharing of a hydrogen nucleus between two atoms in different molecules. Hydrogen bonding occurs in HF, H_2O, NH_3, alcohols, and proteins.

One effect of hydrogen bonding is to increase the energy needed for separation of the molecules so that melting and boiling points are higher than expected. Hydrogen bonding is also responsible for water's unusually high heat of vaporization and for the fact that ice floats on its surface. It is interesting to speculate on what the consequences would be if water had a lower heat of vaporization or if ice had a higher density. A lower heat of vaporization would mean that water in the temperate regions of the earth would probably exist as vapor and there would be no bodies of liquid water. And if water behaved like other substances, the density of its solid state would exceed that of the liquid and ice would sink. Lakes and rivers would freeze from the bottom up, and life within the waters would cease. The way in which hydrogen bonding affects the density of ice is worthy of discussion.

The formation of hydrogen bonds in a sample of water is offset by random molecular motions which tend to break the bonds. As the temperature decreases, the molecules move more slowly and the same amount of water occupies a progressively smaller volume. As molecular motion decreases, hydrogen bonding increases. The distribution of electrons in the water molecule is such that each molecule can form hydrogen bonds with four others. These bonds are highly directional, as shown in Figure 4.25, and the molecules must separate slightly before all of them can form. The result is that the space

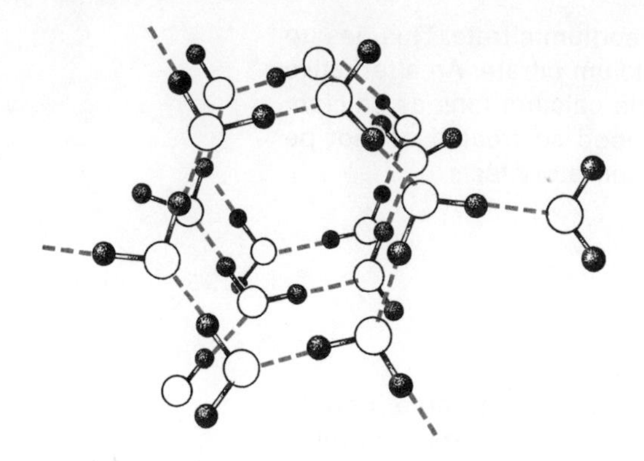

Figure 4.25
The crystal structure of water (ice). The hydrogen bonds hold the water molecules together in a specific pattern.

between water molecules in the crystalline state is greater than in liquid water, which means that ice has a lower density.

4.19 IONS OF PHYSIOLOGICAL IMPORTANCE

The ionic substances found in living organisms are often referred to as electrolytes, although this is a rather specific use of the term. In its broader sense an *electrolyte* is any substance whose water solution conducts electricity. For purposes of this discussion the distinction is not important, and it is made simply for the sake of clarification. What is important is that electrolytes are involved in many physiological processes. Calcium ions, for example, play a role in blood clotting, muscle contraction, and digestion; potassium and sodium ions in transmission of nerve impulses; chloride ion in the digestion of starch; phosphate and bicarbonate ions in maintaining acid-base balance; and various other ions are involved in osmotic pressure regulation. Blood clotting is discussed in some detail in this section, while osmotic pressure regulation is covered in Chapter 6 and acid-base balance in Chapter 8.

The fluid portion of the blood is called plasma. Among the constituents of plasma are calcium ions, and large protein molecules called prothrombin and fibrinogen. Small, rather fragile, blood cells called platelets contain the protein thromboplastin. When a blood vessel is damaged, the platelets rupture, releasing thromboplastin into the plasma. In the presence of calcium ions the thromboplastin causes the conversion of prothrombin to thrombin.

$$\text{Prothrombin} \xrightarrow[\text{Ca}^{2+}]{\text{thromboplastin}} \text{Thrombin}$$

Thrombin, in turn, changes fibrinogen into fibrin. If additional calcium ions are available, the fibrin molecules form tiny threads which are the framework of the clot.

$$\text{Fibrinogen} \xrightarrow{\text{thrombin}} \text{Fibrin} \xrightarrow{\text{Ca}^{2+}} \text{Fibrin threads}$$

Blood can be kept from clotting by removing the calcium ions or making them unavailable. If the blood is to be used for transfusion, clotting can be retarded by the addition of sodium citrate. This ties up the calcium ions by forming un-ionized calcium citrate. An alternative is to add oxalate ion, which precipitates the calcium ions as calcium oxalate. Since oxalate ion is poisonous, blood so treated cannot be used for transfusion, but can be used for laboratory tests.

APPLICATION OF PRINCIPLES

1. The ionization energy for sodium atoms is 119 kcal/mole. Should the ionization energy for sodium ions be more or less than this? Why?
2. A compound is formed by the combination of 1 atom of element *A* and 1 atom of element *B*. The electronegativity of element *A* is 2.5 and that of element *B* is 3.5. Predict the type of bond between the two elements and the physical properties of the compound.
3. The electronegativity values for a family of nonmetals are 3.5, 2.5, 2.4, 2.1, and 2.0. Which of these elements has the lowest atomic number? How can you tell?
4. Draw possible Lewis diagrams for compounds composed of the following elements: (a) hydrogen and phosphorus, (b) carbon and sulfur, (c) phosphorus and chlorine.
5. Use orbital notation to predict the structure and charge of the ion that could be formed from an atom of each of the following elements: (a) nitrogen, (b) magnesium, (c) sulfur, and (d) oxygen.
6. Compound *X* is a colorless gas having a rather pungent odor and a high water solubility. From this information what can you predict about (a) the shape of the molecule, and (b) the type of bonding?
7. Arrange the following elements in order, starting with the most metallic: carbon, oxygen, nitrogen.
8. From the following list of elements select the one(s) whose atoms contain a single lone pair of electrons: aluminum, oxygen, fluorine, and phosphorus.
9. An element from Group IIA on the Periodic Table combines with an element from Group VIIA. Will the compound be covalently bonded or ionically bonded? What will be the probable ratio between the two elements in the compound?
10. The elements S, Se, and Te are members of the same chemical family. Which of the elements is the most metallic?
11. The elements carbon, nitrogen, and oxygen are members of the same period. Select the one whose atoms have the largest diameter and justify your choice.
12. Give the most probable valence for each of the following elements: oxygen, nitrogen, calcium, beryllium.
13. Assume that a compound is formed from each of the following pairs of elements: (a) hydrogen and sulfur, (b) hydrogen and oxygen, and (c) hydrogen and nitrogen. In which of the compounds

would you expect to find hydrogen bonding? In which compound would hydrogen bonding be strongest? Why?

14. What does an oxygen ion have in common with an atom of neon?
15. Would you expect van der Waals forces to be strongest in solids, liquids, or gases? Why?

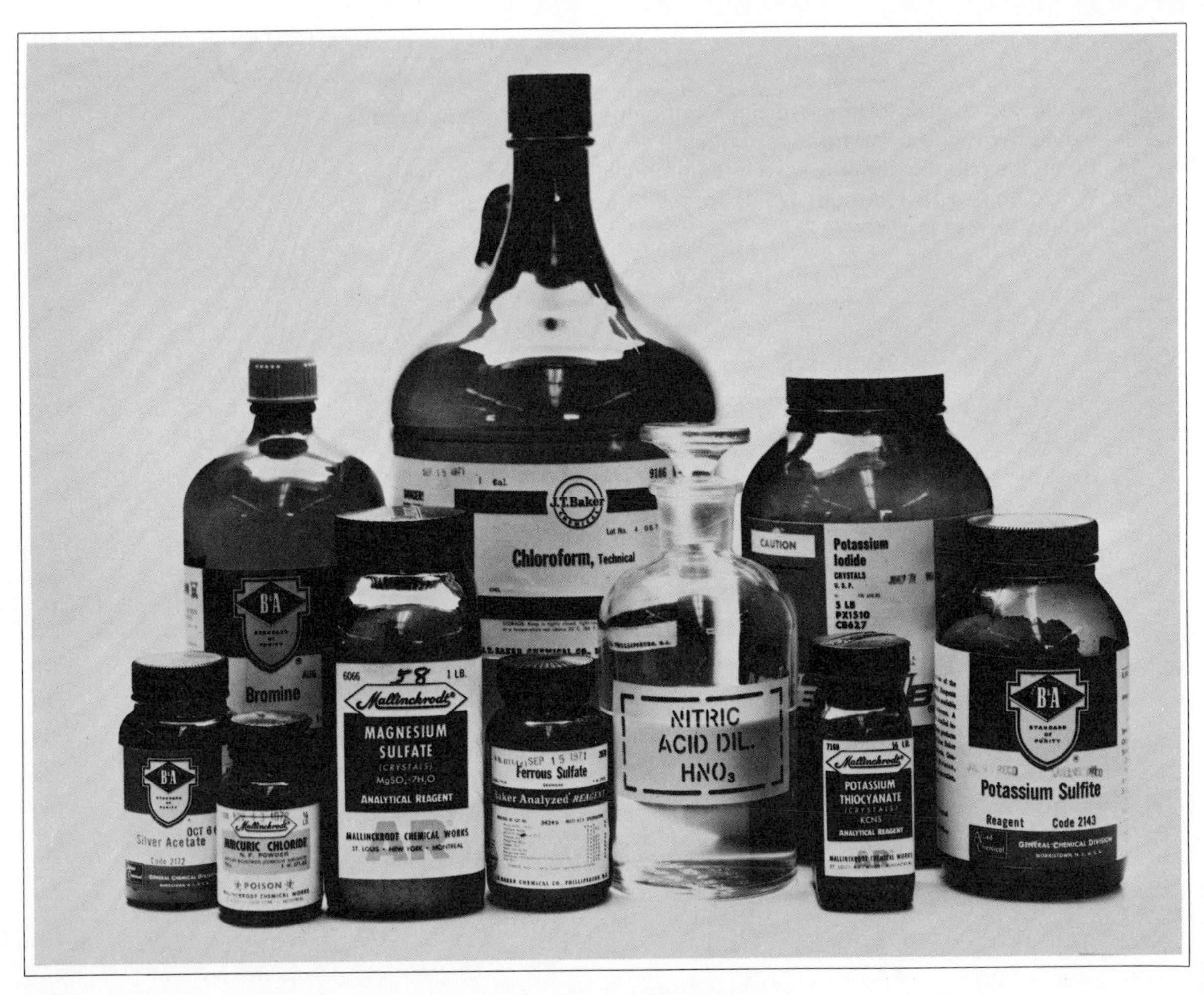

Containers of some common chemicals. Note that several of the labels show both the chemical name and the formula. (Photograph by J. Asdrubal Rivera)

5

CHEMICAL SHORTHAND AND SOME ADDITIONAL CONCEPTS OF MATTER

LEARNING OBJECTIVES

1. Interpret the various kinds of chemical formulas.
2. Derive the names of simple compounds from their formulas and vice versa.
3. Describe the derivation and use of atomic weights.
4. Explain the mole concept.
5. Derive the molecular weights of compounds from their formulas.
6. Interpret word equations and symbolic equations.
7. Recognize the various types of chemical reactions from their equations.
8. Balance symbolic equations, given the formulas of their reactants and products.
9. Determine the quantitative relationship (either in moles or mass) between any two participants in a chemical reaction.

KEY TERMS AND CONCEPTS

alkane series
atomic weight
Avogadro's number
binary compound
catalyst
decomposition
electronic formula
empirical formula
equation
equilibrium
formula weight
gram atomic weight
hydrocarbon
inorganic compound
Law of Conservation of Matter
metathesis
mole
molecular formula
molecular weight
net ionic equation
organic compound
oxidation number
product
reactant
reversible reaction
spectator ion
Stock system of nomenclature
structural formula
substitution reaction
synthesis
word equation

There is no doubt that communication between two people is hampered if they do not speak the same language. Pictures can be helpful for the transmission of simple ideas, but they become less effective as the concepts become more abstract. Chemistry encompasses a number of abstract ideas, and so chemists have developed special devices for representing them. Each compound is assigned an appropriate name and a combination of symbols that describes its composition. A special format is used to summarize the changes that take place during a chemical reaction. Familiarity with these devices is essential to an understanding of chemistry.

5.1 CHEMICAL FORMULAS

The formation of ionic and molecular compounds was described in the previous chapter. Various aspects of bond formation were illustrated by the use of symbolic devices such as Lewis diagrams and orbital notation. Despite their simplicity, they manage to convey complex ideas quite clearly. This section describes another useful and fundamental form of chemical shorthand called the formula.

A *formula* is a combination of symbols that tells something about either the composition or structure of a compound. Although they were not identified as such, formulas were used in the previous chapter to show the combining ratios of atoms or ions in a compound, and to show clearly which atoms were connected by a bond. Altogether there are four common types of formulas: electronic, structural, molecular, and empirical.

Electronic Formulas

An *electronic formula* shows the distribution of valence electrons within a molecule. It is produced by combining the Lewis diagrams of bonded atoms in such a way that each atom acquires its most stable electron configuration. Electronic formulas ignore the three-dimensional nature of molecules. Some examples are

```
       °°
      °Cl°                 ×O×
   °°  ×°  °°              ×°
  °Cl×C×Cl°               H×C×H          ×N×××N°
   °°  ×°  °°
      °Cl°
       °°
```

Carbon tetrachloride   Formaldehyde   Nitrogen

Structural Formulas

Structural formulas are electronic formulas in which a dash is used instead of a pair of dots to represent each bonding pair of electrons, and the nonbonding electrons are not shown. The structural formulas for the examples used above are

```
      Cl              O
      |               ‖
  Cl—C—Cl          H—C—H          N≡N
      |
      Cl
```

Carbon tetrachloride   Formaldehyde   Nitrogen

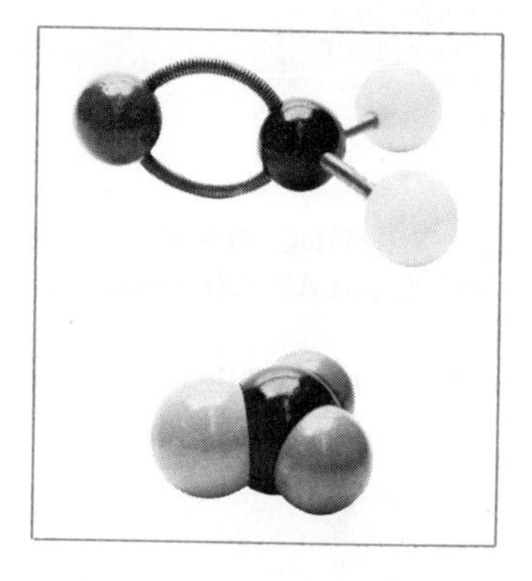

Models representing the formaldehyde molecule: (a) a ball and stick model; (b) a scale model. (Photographs by J. Asdrubal Rivera)

Molecular Formulas

A *molecular formula* is a combination of elemental symbols and numbers that shows the exact composition of a compound. Each element contained in the compound is represented in the formula by its symbol, and a subscript number to the right of the symbol tells how many atoms of that element are in one molecule. A subscript of 1 is not shown. The symbol for the least electronegative (most metallic) element usually appears first in the formula although there are exceptions to this rule.

The molecular formula for carbon tetrachloride is CCl_4. The elements contained in this compound are carbon and chlorine. The symbol for carbon appears first because it is the more metallic of the two elements. The subscript 4 indicates that there are four atoms of chlorine in each molecule. Other examples of molecular formulas are

N_2	CH_2O	$C_6H_{12}O_6$	H_2O
Nitrogen	Formaldehyde	Glucose	Water

Empirical Formulas

The word *empirical* means based solely on experiment and observation. An *empirical formula* shows the kinds of atoms or ions in a compound and their simplest ratio as determined by experimental methods. Since ionic compounds exist as aggregates of ions rather than as individual molecules, their composition is best described by the simplest ratio between their ions. For instance, the ionic compound composed of sodium and chlorine has the empirical formula NaCl because its crystal lattice contains equal numbers of sodium and chloride ions.

Empirical formulas are also used for molecular substances, but their use sometimes results in a loss of information, as shown by the following examples:

Compound	Molecular Formula	Empirical Formula
Glucose	$C_6H_{12}O_6$	CH_2O
Nitrogen	N_2	N
Benzene	C_6H_6	CH

5.2 WRITING FORMULAS FOR IONIC AND MOLECULAR COMPOUNDS

Ionic Compounds

As mentioned earlier, the formula for an ionic compound is the simplest ratio of its ions. Since every compound is neutral, its net charge must be zero, which means that the total charge on the positive ions must equal the charge on the negative ions. The following guidelines should simplify the procedure:

1. Write the symbols of the two ions in sequence with the positive ion first.
2. Determine the least common multiple of the ionic charges (i.e., the smallest number into which both will divide without a remainder).
3. Divide the least common multiple by the charge on the positive ion and use the result as a subscript for that ion.
4. Repeat step 3 for the negative ion.
5. If the subscript for an ion is 1, the subscript is omitted.

Sample problem 5.1 Write the formula for the ionic compound of chlorine and magnesium.

Solution Step 1. The symbols for the two ions are Cl^- and Mg^{2+}. They are written in the order $Mg^{2+}Cl^-$.

Step 2. The least common multiple of the charges is 2.

Step 3. Dividing the least common multiple by the charge on the positive ion (2 divided by 2) gives a result of 1. This is the required number of magnesium ions and would also be the subscript of the magnesium symbol except that a subscript of 1 is not used.

Step 4. Repeating step 3 for the negative ion (2 divided by 1) gives a result of 2. Two chloride ions are needed, and this is indicated by a subscript of 2. The formula is $MgCl_2$, showing that the crystal lattice contains two chloride ions for every magnesium ion.

Sample problem 5.2 Write the formula for the ionic compound of aluminum and sulfate ions.

Solution Step 1. Write the symbols in the order $Al^{3+}SO_4^{2-}$.
Step 2. The least common multiple of 3 and 2 is 6.
Step 3. Dividing the least common multiple by the charge on the aluminum ion gives a result of 2.
Step 4. Dividing the least common multiple by the charge on the sulfate ion gives a result of 3. The formula of the compound is $Al_2SO_{4_3}$. However, the subscript of the sulfate ion may be confused with the number 4 that is part of the ion, and so the symbol is enclosed within parentheses, as $Al_2(SO_4)_3$.

Parentheses are used *only* with polyatomic ions, and only when the subscript is 2 or more. For example, parentheses would not be required for the single sulfate ion in the formula $CaSO_4$.

Molecular Compounds

The procedure for writing the formula for a simple molecular compound is identical with that used for ionic compounds with one modification: since molecular compounds are not composed of ions, oxidation numbers are used in place of ionic charges. The *oxidation number* (also called the oxidation state) of an element is the charge its atoms *appear* to have when combined with other atoms. The following rules govern the assignment of oxidation numbers:

1. Each atom of an element in the free or uncombined state is neutral and has an oxidation number of zero.
2. Compounds are also neutral, and so the oxidation numbers of all the atoms in a compound must have an algebraic sum of zero.
3. Oxygen in its compounds usually has an oxidation number of 2−, and hydrogen is normally 1+.
4. The oxidation number of an atom in a compound is usually the same as the charge on an ion of that element.
5. The algebraic sum of the oxidation numbers for all atoms in a polyatomic ion is equal to the charge on the ion.

Here are some examples of the application of these rules.

1. The oxidation number of oxygen in O_2 is zero because it is in its free or elemental state.
2. The charge on the sulfide ion is 2−. Its oxidation number is also 2−.
3. The hydroxyl ion has a charge of 1−. This represents the algebraic sum of the oxidation numbers of the oxygen and hydrogen. If the oxygen is assigned an oxidation number of 2− and the hydrogen a 1+, the algebraic sum is 1−.

4. The molecule N_2O_3 is neutral. If each of the three oxygen atoms is given an oxidation number of 2−, the negative oxidation numbers total 6−. This must be equal algebraically to the combined oxidation numbers of the two nitrogen atoms; thus each nitrogen atom has an oxidation number of 3+.
5. In the compound CCl_4 each chlorine atom is assigned an oxidation number of 1− (the same as the normal charge on an ion of chlorine). Since the algebraic sum of oxidation numbers is zero, the oxidation number of carbon must be 4+.

Sample problem 5.3 Write the formula for the molecular compound of hydrogen and sulfur.

Solution The oxidation number of hydrogen in its compounds is assumed to be 1+, while that of sulfur is the same as the charge on the sulfide ion, 2−.
Step 1. The symbols are written in order H^{+} S^{2-}.
Step 2. The least common multiple is 2.
Step 3. Dividing the least common multiple by the oxidation number of hydrogen gives a result and a subscript of 2.
Step 4. Repeating the procedure for sulfur gives a result of 1. The correct formula for the compound is thus H_2S.

The writing of formulas for molecular compounds is complicated by the fact that the sharing of electrons between the same elements may take place in a variety of ways, which means that some elements have two or more oxidation numbers. The two different compounds of carbon and oxygen, for example, have formulas of CO and CO_2. You will not be expected to write molecular formulas for all of the different possible combinations.

Oxidation numbers are also used in assigning appropriate names to formulas in a later section.

5.3 NAMING COMPOUNDS

When atoms of two or more elements are joined chemically, they form a compound. Each individual compound results from a specific combination of elements and has both a unique set of properties and a distinctive name. The name is assigned according to rules agreed upon by an international committee of chemists, and so is referred to as a systematic name. Despite an intensive campaign to bring all chemical names into conformity with the rules, some compounds are still widely known by nonsystematic or common names. Table 5.1 contains a short list of compounds whose common names may be familiar.

Note that some of the compounds have several common names. Washing soda, for example, is also called soda ash and sal soda in addition to its systematic name.

Chemical compounds may be divided into two major categories for purposes of naming. The larger of these categories is composed en-

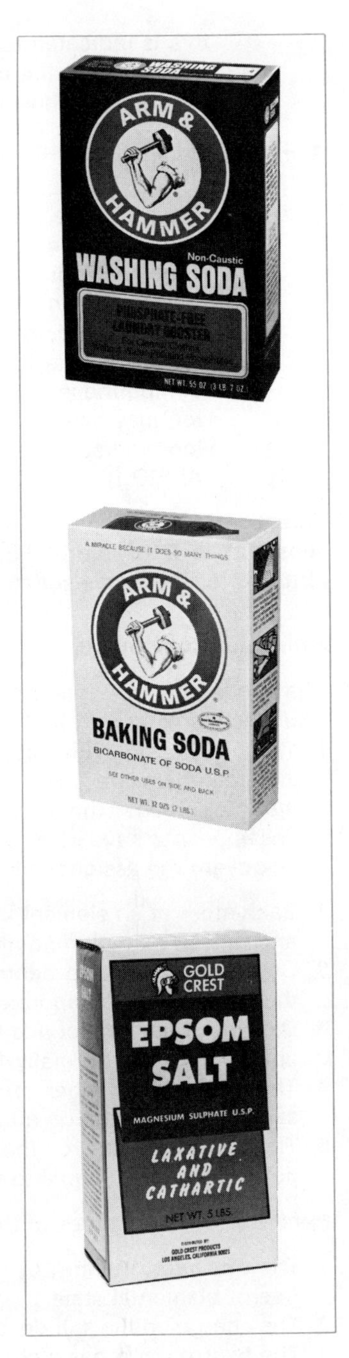

Chemicals frequently used in the home. (Courtesy Church & Dwight Co., New York; photograph by J. Asdrubal Rivera)

TABLE 5.1 Common and Systematic Names of Familiar Compounds

Common Name	Systematic Name	Formula	Uses
Baking soda	Sodium hydrogen carbonate	$NaHCO_3$	Baking powders, antacids
Cream of tartar	Potassium hydrogen tartrate	$KHC_4H_4O_6$	Baking powder
Epsom salt	Magnesium sulfate	$MgSO_4$*	Purgative
Ether	Diethyl ether	C_2H_5O	General anaesthetic
Grain alcohol	Ethanol	C_2H_5OH	Disinfectant, solvent
Laughing gas (nitrous oxide)	Dinitrogen monoxide	N_2O	General anaesthetic
Milk of magnesia	Magnesium hydroxide	$Mg(OH)_2$	Antacid, laxative
Plaster of paris	Calcium sulfate	$CaSO_4$*	Casts, molds
Washing soda (Sal soda, soda ash)	Sodium carbonate	Na_2CO_3*	Water softener

* These compounds also contain water molecules as part of their structure and are called hydrates. They will be discussed in Chapter 6.

tirely of covalent molecules containing hybridized carbon atoms. The members of this group number in the millions and are collectively referred to as *organic* compounds. All other compounds may be placed in the smaller group which is called *inorganic.* Because of the chain-forming ability of hybridized carbon atoms, fairly complex molecules are possible, and so the rules for naming organic molecules are also a bit more involved. A few of the basic rules will be demonstrated in the following sections so that the organic substances discussed in the next few chapters may be more readily referred to, and a more complete treatment will accompany the detailed discussion of organic compounds in Chapters 10 and 11.

5.4 NAMING INORGANIC COMPOUNDS

Compounds Composed of Two Elements

The simplest inorganic compounds are those composed of two elements, and are called *binary* compounds. The procedure for naming binary compounds varies slightly with the chemical classification of its constituent elements.

Recall from Chapter 4 that elements are classified as metals or nonmetals according to their tendency to lose or gain electrons. On the Periodic Table the metals are shown to the left of the diagonal line starting between boron and aluminum. To the right of this line are the nonmetals. The possible binary combinations are metal + metal, metal + nonmetal, and nonmetal + nonmetal. Combinations of two metals are so rarely encountered that they will be disregarded in the discussion that follows. However, since a number of the metals may also have variable oxidation states, three categories will be considered:

1. Metal + nonmetal combinations in which the metal has a fixed oxidation number.
2. Metal + nonmetal combinations in which the metal has a variable oxidation number.
3. Combinations of two nonmetals.

Procedure for Naming Binary Compounds Containing a Metal with a Fixed Oxidation Number

Step 1. Write the name of the metallic element followed by the name of the nonmetal.
Step 2. Change the ending of the nonmetal to -ide.

Sample problem 5.4 Write the name of the binary compound of potassium and bromine.

Solution Inspection of the Periodic Table shows that potassium is the metal and bromine the nonmetal.
Step 1. Write the names in order: potassium bromine.
Step 2. Change the ending of the nonmetal to -ide. The name of the compound is thus potassium bromide.

Sample problem 5.5 Write the name of the binary compound whose molecular formula is H_2S.

Solution Although hydrogen has few of the properties normally associated with metals, it is shown on the left side of the Periodic Table, and it behaves in many cases like a metal—especially in combinations with a nonmetal.
Step 1. Write the names in order: hydrogen sulfur.
Step 2. Change the ending of the nonmetal to -ide. The correct name is hydrogen sulfide.

Naming Binary Compounds Containing a Metal with a Variable Oxidation Number

There are two different systems for naming these compounds. Each involves the addition of a third step to the procedure outlined above. The process is easily demonstrated by use of an example.

Sample problem 5.6 Write the name of a binary compound composed of copper and oxygen.

Solution Copper may have oxidation numbers 1+ and 2+. Recall that the oxidation state of oxygen in its compounds is 2−. The two formulas predicted by these oxidation numbers are

Cu_2O (Copper in 1+ oxidation state) and CuO (Copper in 2+ oxidation state)

The older, and less preferred, system involves the use of the suffixes "-ous" and "-ic". Step 3 in this system is: the suffix -ous is added to the Latin name of the metal in its lower oxidation state, and the suffix -ic is used for the higher oxidation state.

The Latin name for the element copper is cuprum. The compound represented by the formula Cu_2O is thus called cuprous oxide, and CuO is cupric oxide. Some other examples of names derived by this procedure are:

Metal	Oxidation State	Formula	Name
Iron	2+	FeI_2	Ferrous iodide
	3+	FeI_3	Ferric iodide
Lead	2+	PbO	Plumbous oxide
	4+	PbO_2	Plumbic oxide
Mercury	1+	Hg_2Cl_2*	Mercurous chloride
	2+	$HgCl_2$	Mercuric chloride

* Mercury in the 1+ oxidation state does not occur as the species Hg^+, but as the ion Hg_2^{2+} in which two Hg^+ ions are covalently bonded. It is an exception among metals.

The newer, preferred method for naming compounds of the types shown above is called the *Stock system.* Step 3 in the stock system is: the oxidation state of the metal is indicated by a Roman numeral following the name of the metal. When the previous examples are named according to the Stock system, they become:

Metal	Oxidation State	Formula	Name
Copper	1+	Cu_2O	Copper (I) oxide
	2+	CuO	Copper (II) oxide
Iron	2+	FeI_2	Iron (II) iodide
	3+	FeI_3	Iron (III) iodide
Lead	2+	PbO	Lead (II) oxide
	4+	PbO_2	Lead (IV) oxide
Mercury	1+	Hg_2Cl_2	Mercury (I) chloride
	2+	$HgCl_2$	Mercury (II) chloride

Other metals which exhibit variable oxidation states are shown in Table 4.3 along with their Latin names.

Naming Binary Compounds Composed of Two Nonmetals

The same basic procedure applies to the naming of binary compounds of two nonmetals, but is additionally complicated by the fact that many nonmetals have more than two oxidation states. Thus the same two nonmetals may combine in different ratios to produce a number of different compounds. The way out of this dilemma is through the use of numerical prefixes to specify the number of atoms of each element. The numerical prefixes for the numbers 1 through 10 are shown in Table 5.2.

TABLE 5.2 Numerical Prefixes Used in Chemical Nomenclature

Prefix	Meaning	Prefix	Meaning
Mono-	1	Hexa-	6
Di-	2	Hepta-	7
Tri-	3	Octa-	8
Tetra-	4	Nona-	9
Penta-	5	Deca-	10

One point remains to be settled: that of the order of the two elements. In binary combinations of a metal and a nonmetal, the name of the metal (i.e., the less electronegative element) was written first. The same order is followed for two nonmetals. If the two elements are located within the same period (e.g., carbon and oxygen) the element farthest to the left(carbon) is the least electronegative. If the two elements are in the same group (e.g., sulfur and oxygen), the lower one is less electronegative. If the elements do not belong to either the same group or the same period, the one farther to the left is usually less electronegative.

Having settled these points, we may now consider the entire procedure. The first two steps are the same as described previously. Step 3 is: indicate by numerical prefixes the number of atoms of each element.

Sample problem 5.7 Name the compound whose formula is N_2O_3.

Solution Step 1. The symbol appearing first in the formula is that of the least electronegative element; thus nitrogen appears first in the name: nitrogen oxygen.

Step 2. The ending of oxygen is changed to -ide: nitrogen oxide.

Step 3. The prefix di- is used to indicate the two atoms of nitrogen, and tri- for the three atoms of oxygen. The correct name is dinitrogen trioxide.

Sample problem 5.8 Name the compound whose formula is CO.

Solution Step 1. Carbon is first in the formula and also first in the name: carbon oxygen.

Step 2. The ending of oxygen is changed to -ide.

Step 3. Since there is a single atom of each element, the prefix mono- would be appropriate for both carbon and oxygen. However, in actual practice the prefix mono- is not used for the first element. The correct name is carbon monoxide.

Examples of other binary compounds named according to these rules are:

SF_4	silicon tetrafluoride	CCl_4	carbon tetrachloride
CS_2	carbon disulfide	SO_3	sulfur trioxide
CO_2	carbon dioxide	P_2O_3	diphosphorus trioxide

There is one notable exception to the rules for writing the formulas of binary compounds which is also an exception to the rules for nomenclature: NH_3 is called ammonia and is written backwards.

Compounds of More than Two Elements

Compounds of more than two elements usually include a polyatomic ion as part of the structure. The name of the ion becomes part of the name of the compound. The rule is quite simple: write the name of the cation followed by the name of the anion.

Sample problem 5.9 Write the name of the compound whose formula is $MgCO_3$.

Solution One atom of carbon and three of oxygen combine to form the carbonate anion. Since this is preceded in the formula by the symbol of the magnesium cation, the name of the compound is magnesium carbonate.

Sample problem 5.10 Write the name of the compound whose formula is $CuSO_4$.

Solution The anion made of one sulfur and four oxygen atoms is called the sulfate radical, and has an oxidation number of 2−. Since this is balanced by a single ion of copper, the oxidation number of the copper must be 2+. Remember that copper in the 2+ oxidation state may be called either cupric or copper(II). Thus there are two possible names for the compound: cupric sulfate and copper(II) sulfate.

Examples of other compounds named according to this procedure are:

$Ca(OH)_2$, calcium hydroxide (commonly known as slaked lime)
$AgNO_3$, silver nitrate (also called lunar caustic)
$Pb(C_2H_3O_2)_2$, plumbous acetate or lead(II) acetate

Special Classes of Compounds: Inorganic Acids

Among the hundreds of inorganic compounds there are several small groups whose members are characterized by marked chemical similarities. Each of these groups will be discussed later in some detail, but members of the group called acids are among the most common chemicals in the laboratory and thus deserve special consideration.

The pure compounds which are classed as acids are named according to the rules previously set forth. However, the special characteristics of acids become apparent when the compound is dissolved in water, and so special names are given to their aqueous solutions. The rules that follow are for naming these aqueous solutions.

Most acids contain the element oxygen as part of the anion and are referred to as "oxy" acids. A small number contain no oxygen. These will be considered first.

Procedure for Naming Acids that Do Not Contain Oxygen

Step 1. The first part of the name is the prefix "hydro-."
Step 2. The prefix is followed by the name of the anion.
Step 3. The ending of the anion is changed to -ic and followed by the word "acid." Examples:

HCl hydrochloric acid
H_2S hydrosulfuric acid
HCN hydrocyanic acid

Naming the Oxyacids

Although all oxyacids contain oxygen, some of them contain an amount of oxygen which is arbitrarily considered to be either more or less than the "normal" amount. The names of acids that contain the normal amount of oxygen are given in Table 5.3. Other oxyacids derive their names from these.

1. The suffix of an oxyacid having one less oxygen than normal is changed to -ous. Examples:

   Normal oxygen is H_2SO_4 = Sulfuric acid
   One less oxygen is H_2SO_3 = Sulfurous acid
   Normal oxygen is HNO_3 = Nitric acid
   One less oxygen than normal is HNO_2 = Nitrous acid

2. Oxyacids having two oxygen atoms less than normal are given the prefix "hypo-" and the suffix "-ous." Example:

   Normal oxygen is $HClO_3$ = Chloric acid
   Two atoms less than normal is $HClO$ = Hypochlorous acid

3. Oxyacids having one oxygen atom more than normal are given the prefix "per-" and the suffix "-ic." Example:

   Normal oxygen is $HClO_3$ = Chloric acid
   One atom more than normal is $HClO_4$ = Perchloric acid

TABLE 5.3 Oxyacids with Normal Oxygen Content

Formula	Name	Formula	Name
H_3BO_3	Boric	HNO_3	Nitric
H_2CO_3	Carbonic	H_3PO_4	Phosphoric
$HClO_3$	Chloric	H_2SO_4	Sulfuric
HIO_3	Iodic		

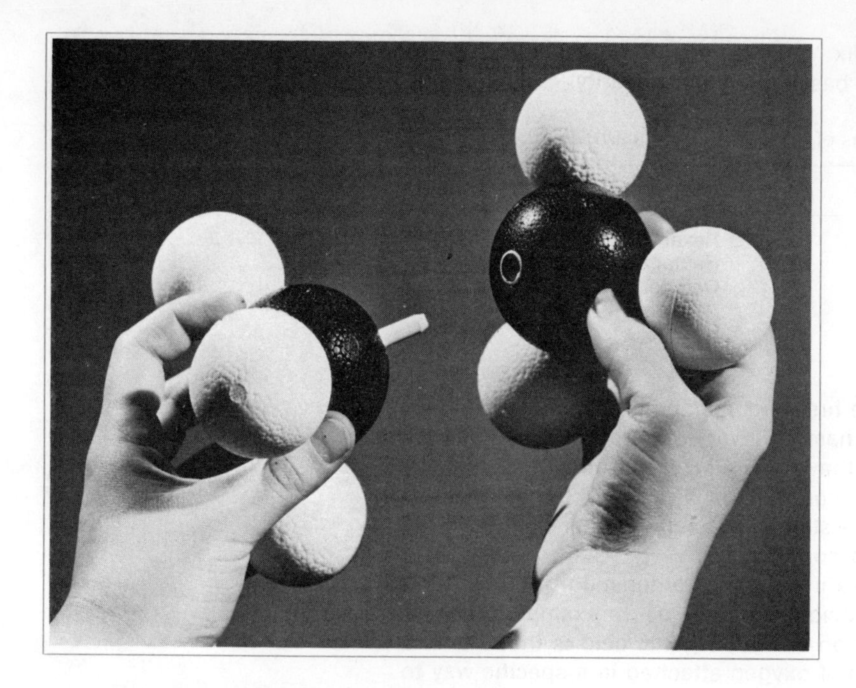

Using models to demonstrate the hypothetical formation of ethane from two methane residues. (Courtesy Ealing Corporation, Cambridge, Massachusetts)

5.5 NAMING ORGANIC COMPOUNDS

From the standpoint of composition and structure, the least complex organic molecules are *hydrocarbons*—that is, compounds consisting entirely of hydrogen and carbon. The simplest hydrocarbon is methane (CH_4). The C-H bonds in methane are all equivalent. If one hydrogen atom is removed from each of two methane molecules and the residues are joined together, the result is the molecule called ethane (Figure 5.1). The molecular formula for ethane is C_2H_6. The difference between methane and ethane is $C_2H_6 - CH_4 = CH_2$. Increasing the size of the ethane molecule by this amount produces a different hydrocarbon called propane (C_3H_8), and successive increases by this constant amount result in a series of hydrocarbons called the *alkane series.* The names and formulas of the first ten alkanes are shown in

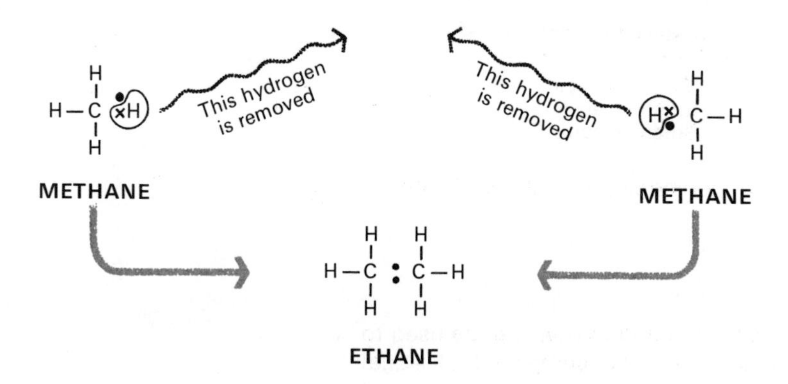

Figure 5.1
Hypothetical formation of ethane from two methane residues.

Table 5.4. The common suffix "-ane" indicates membership in the alkane series and suggests a basic chemical similarity.

TABLE 5.4 Names and Formulas of the Alkane Hydrocarbons

Name	Formula	Name	Formula
Methane	CH_4	Hexane	C_6H_{14}
Ethane	C_2H_6	Heptane	C_7H_{16}
Propane	C_3H_8	Octane	C_8H_{18}
Butane	C_4H_{10}	Nonane	C_9H_{20}
Pentane	C_5H_{12}	Decane	$C_{10}H_{22}$

Although the names of the first four members of the alkane series have no clear meanings, the name of every other alkane specifies the number of carbon atoms in the chain. Recall the numerical prefixes used in naming certain binary compounds. The prefix "penta-," which means five, is used here as the stem of the name pentane to show that the molecule contains five carbon atoms.

The names of more complex molecules are derived from the names of the alkanes. The organic acids will serve as an example. The difference between a hydrocarbon and an organic acid is that the acid molecule contains two atoms of oxygen attached in a specific way to one of the terminal carbon atoms in the chain. The structures of pentane and the organic acid containing five carbons are shown for comparison:

$$\begin{array}{ccccccccccc} & & H & & H & & H & & H & & H & & \\ & & | & & | & & | & & | & & | & & \\ H & - & C & - & C & - & C & - & C & - & C & - & H \\ & & | & & | & & | & & | & & | & & \\ & & H & & H & & H & & H & & H & & \end{array} \qquad \begin{array}{ccccccccccc} & & H & & H & & H & & H & & O & & \\ & & | & & | & & | & & | & & \| & & \\ H & - & C & - & C & - & C & - & C & - & C & - & OH \\ & & | & & | & & | & & | & & & & \\ & & H & & H & & H & & H & & & & \end{array}$$

Pentane   Pentanoic acid

An acid is named by changing the ending of the root word to "-oic" and adding the word "acid." The names of the first ten members of the series of organic acids are shown in Table 5.5. Several of these acids are perhaps better known by their common names:

Common Name	Systematic Name
Formic acid	Methanoic acid
Acetic acid	Ethanoic acid
Butyric acid	Butanoic acid

A further discussion of organic nomenclature will be found in Chapters 10 and 11.

5.6 ATOMIC WEIGHTS

The electron configurations of magnesium and oxygen can be used to predict that the oxide of magnesium will be composed of equal

TABLE 5.5 Names of the First Ten Organic Acids

Number of Carbons	Name	Number of Carbons	Name
1	Methanoic acid	6	Hexanoic acid
2	Ethanoic acid	7	Heptanoic acid
3	Propanoic acid	8	Octanoic acid
4	Butanoic acid	9	Nonanoic acid
5	Pentanoic acid	10	Decanoic acid

numbers of magnesium and oxygen atoms. However, analysis shows that the *weight* of magnesium in the oxide far exceeds that of the oxygen. In 100 g of MgO there are 60.3 g of magnesium and only 39.7 g of oxygen. Since both of these weights represent the same number of atoms, and the weights are in the ratio 1.5:1, we must conclude that an atom of magnesium weighs approximately 1.5 times as much as one of oxygen.

Notice that it was not necessary to use the actual weight of either atom in arriving at this conclusion, and that we have created a relative weight scale on which the oxygen atom is the standard and the magnesium atom is 1.5 times as heavy.

It should be noted that the most abundant isotopes of magnesium and oxygen have mass numbers of 24 and 16, respectively, and so the same relative scale is produced from the ratio of their mass numbers (24:16 is the same as 1.5:1).

Since the weight of an individual atom is so small (in the order of 10^{-23} g), and reactions between single atoms cannot be observed, a relative weight is more useful in chemistry than the actual mass of an atom. The particular relative scale described above would be awkward, however, because the weights of many elements would be decimal fractions. For example, the most abundant isotopes of hydrogen and carbon have mass numbers of 1 and 12, respectively, and their weights relative to oxygen would be 0.063 and 0.75.

To overcome this difficulty, the ${}^{12}_{6}C$ isotope of carbon is used as the standard and is assigned a value of exactly 12.0000 atomic mass units. (One atomic mass unit equals 1.67×10^{-24} g.) The use of carbon as a standard results in values of 1.0000, 16.0000, and 24.0000 atomic mass units for the most common isotopes of hydrogen, oxygen, and magnesium, respectively.

It should be remembered, however, that there are other isotopes of these elements, and that in any measurable sample of an element they will be present. Hydrogen, for example, has three common isotopes whose mass numbers are 1.0000, 2.0000, and 3.0000. Recall from Chapter 3 that these isotopes are called protium, deuterium, and tritium. Their relative abundance is such that if 10,000 hydrogen atoms were sorted according to weight, there would be about two atoms of deuterium and 9,998 of protium. Tritium occurs to the extent of only about one atom per hundred billion. Since every atom of hydrogen has a mass number of at least one, the inclusion of the heavier isotopes means that the mass number of an "average" hydrogen atom is slightly more than 1.0000.

What is being described is a weighted average—that is, an average that takes into account the relative abundance of the different isotopes of an element. The weighted average of the mass numbers of the naturally occurring isotopes of an element is called the *atomic weight.* The atomic weights of the first 20 elements, accurate to three decimal places except in the case of calcium, are given in Table 5.6. For most purposes satisfactory results can be obtained by rounding off these values to the first decimal place.

TABLE 5.6 Atomic Weights of the First Twenty Elements

Hydrogen	1.008	Sodium	22.990
Helium	4.003	Magnesium	24.312
Lithium	6.939	Aluminum	26.982
Beryllium	9.012	Silicon	28.086
Boron	10.811	Phosphorus	30.974
Carbon	12.011	Sulfur	32.064
Nitrogen	14.007	Chlorine	35.453
Oxygen	15.999	Argon	39.948
Fluorine	18.998	Potassium	39.102
Neon	20.183	Calcium	40.08

5.7 THE MOLE CONCEPT

It was pointed out earlier that 60.3 g of magnesium contains the same number of atoms as 39.7 g of oxygen. These numbers are in the same ratio as the atomic weights of magnesium and oxygen (24.3 and 16.0). It is true, then, that a weight of magnesium equal to its atomic weight in grams (called its *gram atomic weight*) would contain the same number of atoms as there are in one gram atomic weight of oxygen. In other words, 24.3 g of magnesium and 16.0 g of oxygen contain the same number of atoms. This number has been calculated to be 6.023×10^{23} and is called *Avogadro's number,* or a *mole.* Thus the gram atomic weight of any element contains one mole of atoms. Since the carbon-12 isotope is the standard for atomic weights, the mole is more precisely defined as the number of atoms in 12.0000 g of carbon-12.

It has become an accepted practice to use the term mole to refer not only to a specific number of particles, but also to the weight of a substance containing that number of particles. For example, a mole of hydrogen atoms is 1.0 g of hydrogen, and a mole of magnesium is 24.3 g.

5.8 MOLECULAR AND FORMULA WEIGHTS

Atomic weights may also be used to determine how many molecules there are in a given weight of a compound, or to weigh out the same number of molecules of different compounds. As an example, the molecular formula of methane is CH_4. This tells us that one molecule of methane is composed of a single carbon atom and four atoms of hydrogen. A mole of methane molecules contains one mole of carbon atoms (weighing 12.0 g) and four moles of hydrogen (4.0 g). Thus one mole of methane weighs 16.0 g, two moles would weigh 32.0 g, etc. For a molecular substance the sum of the atomic weights of all the atoms in the molecule is called the *molecular weight.* The molecular weight of methane is 16.0 amu.

Since an ionic substance does not exist in molecular form, its formula shows the simplest ratio between its ions. The sum of the atomic weights of all the atoms in a formula is called the *formula weight.* For example, the formula weight of the ionic substance sodium carbonate is

$$\begin{aligned} 2\ \text{Na} \times 23.0\ \text{amu} &= 46.0\ \text{amu} \\ 1\ \text{C} \times 12.0\ \text{amu} &= 12.0\ \text{amu} \\ 3\ \text{O} \times 16.0\ \text{amu} &= 48.0\ \text{amu} \\ \hline \text{Formula weight} &= 106.0\ \text{amu} \end{aligned}$$

The distinction between molecular and formula weights is a matter of slight importance, and the calculation is performed in the same way regardless of the nature of the compound.

5.9 REACTIONS AND EQUATIONS

A chemical reaction is any change resulting in the formation and/or disruption of chemical bonds. A substance which is changed by the reaction is called a *reactant,* and any new substance appearing during the reaction is called a *product.* A chemical reaction may be described in words by naming the reactants and products as shown in the following example:

sulfur and oxygen react to form sulfur dioxide

Descriptive statements such as this are called *word equations.* This particular word equation tells us that sulfur and oxygen are the reactants and sulfur dioxide is the product. When the names of the reactants and products are replaced by their respective formulas, and the other words by appropriate symbols, the result is called a symbolic equation, or simply an *equation,* and is a more precise way of making the same chemical statement. The equation for the reaction between sulfur and oxygen is

$$S + O_2 \longrightarrow SO_2$$

In an equation the word "and" is represented by a plus sign, and the reactants (shown on the left) and products (on the right) are separated by an arrow pointing to the right. The use of two arrows pointing in opposite directions indicates that the reaction is *reversible,* i.e., that it reaches an *equilibrium* condition in which the original products are being used and replaced at the same rate. The reaction between water and carbon dioxide is an equilibrium reaction:

$$H_2O + CO_2 \rightleftharpoons H_2CO_3$$

Carbon dioxide, which is a waste product of cellular metabolism, combines with water and is transported in the blood as carbonic acid. The reverse reaction converts carbonic acid back to water and carbon dioxide in the lungs so that the CO_2 may be expelled. Other equilibrium reactions will be discussed in Chapters 6 and 8.

The plus sign and the arrow are essential parts of every equation. Other symbols, whose use is optional, are used to show the conditions necessary for reaction or to specify the physical state of a reactant or product. The most frequently used symbols are shown in Table 5.7.

TABLE 5.7 Symbols Used in Equations

Symbol	Meaning
$\longrightarrow$	Yields, forms, produces, etc.
$\rightleftharpoons$	Equilibrium reaction
Δ	Heat
(g)	A gas
(l)	A liquid
(s)	A solid
(aq)	An aqueous solution (a substance dissolved in water)

When the appropriate symbols are added to the preceding equations, they become

$$S_{(s)} + O_{2\,(g)} \xrightarrow{\Delta} SO_{2\,(g)}$$

$$H_2O_{(l)} + CO_{2\,(g)} \xrightleftharpoons{\text{carbonic anhydrase}} H_2CO_{3(aq)}$$

The first equation now shows that heat is required for the formation of gaseous sulfur dioxide from solid sulfur and gaseous oxygen, while the second equation tells us that a substance called carbonic anhydrase must be present or the reaction will not proceed in either direction. Carbonic anhydrase is written above the arrows because it is not incorporated into any of the products of the reaction. When a chemical affects the rate of a reaction but is not changed by it, it is called a *catalyst.* Catalysts found in living organisms are called enzymes. Enzymes will be discussed in Chapter 15.

Types of Reactions

The study of chemical reactions is simplified by the fact that most of them fall into easily recognized patterns. There are four basic patterns or types, some of which are known by two different names: they are synthesis (or combination), decomposition, substitution (single replacement), and metathesis (double replacement). Their patterns are shown in Table 5.8.

TABLE 5.8 Types of Chemical Reactions

Name	Pattern
Synthesis (combination)	$A + B \longrightarrow AB$
Decomposition	$AB \longrightarrow A + B$
Substitution (single replacement)	$A + BC \longrightarrow B + AC$
Metathesis (double replacement)	$AB + CD \longrightarrow AD + BC$

Synthesis

Two reactants combine to produce a single product. Examples:

$$S_{(s)} + O_{2(g)} \longrightarrow SO_{2(g)}$$

$$H_2O_{(l)} + CO_{2(g)} \xrightleftharpoons{\text{carbonic anhydrase}} H_2CO_{3(aq)}$$

$$C_2H_{4(g)} + Cl_{2(g)} \longrightarrow C_2H_2Cl_{2(g)}$$

Decomposition

A single reactant breaks down into two or more products. Examples:

$$CaCO_{3(s)} \longrightarrow CaO_{(s)} + CO_{2(g)}$$

$$KClO_{3(s)} \xrightarrow{MnO_2} KCl_{(s)} + O_{2(g)}$$

$$H_2SO_{4(aq)} \xrightarrow{300^\circ C} H_2O_{(g)} + SO_{2(g)} + O_{2(g)}$$

Substitution

A free element is substituted for one of the elements in a compound. Examples:

$$Al_{(s)} + HCl_{(aq)} \longrightarrow AlCl_{3(aq)} + H_{2(g)}$$

$$Zn_{(s)} + CuSO_{4(aq)} \longrightarrow ZnSO_{4(aq)} + Cu_{(s)}$$

$$Cl_{2(g)} + KI_{(aq)} \longrightarrow KCl_{(aq)} + I_{2(aq)}$$

Metathesis

A reaction which can be thought of as the exchange of the electropositive portions of two compounds. Examples:

$$HC_2H_3O_{2(aq)} + NaOH_{(aq)} \longrightarrow NaC_2H_3O_{2(aq)} + H_2O_{(l)}$$

$$CaCO_{3(s)} + HCl_{(aq)} \longrightarrow CaCl_{2(aq)} + H_2CO_{3(aq)}$$

$$AgNO_{3(aq)} + NaCl_{(aq)} \longrightarrow NaNO_{3(aq)} + AgCl_{(s)}$$

One value of classifying reactions by type is that it makes it possible in many cases to predict the products when the reactants are given. It should not be assumed, however, that every equation fits into one of these four categories. Many of they are rather unique, as the following examples will show:

$$C_{(s)} + HNO_{3(aq)} \longrightarrow NO_{2\,(g)} + CO_{2\,(g)} + H_2O_{(g)}$$

$$SO_{2\,(g)} + H_2S_{(g)} \longrightarrow S_{(s)} + H_2O_{(g)}$$

The study of equations like these is more appropriate in chemistry courses where the emphasis is on reaction mechanisms.

5.10 IONIC EQUATIONS

The formula for every compound used in the previous section was written in molecular form, which may give the impression that ions are not involved in chemical reactions. This is not true: the formulas were written in this way only for the sake of clarity, and several of the reactions actually involve ionic rather than molecular species. The reaction between aqueous silver nitrate and aqueous sodium chloride is an example. The ionic equation for the reaction is

$$Ag^+_{(aq)} + NO_3{}^-_{(aq)} + Na^+_{(aq)} + Cl^-_{(aq)} \longrightarrow AgCl_{(s)} + Na^+_{(aq)} + NO_3{}^-_{(aq)}$$

The essence of this reaction is the combination of aqueous ions of silver and chlorine to form insoluble (un-ionized) silver chloride. Free ions, such as those of sodium and nitrate, which are present at both the beginning and end of a reaction are called *spectator* ions. Since they have no active role in the changes that occur, the equation may be written without them. An equation from which spectator ions have been eliminated is referred to as a *net ionic equation.* The molecular, ionic, and net ionic forms of an equation are shown for comparison. Molecular form:

$$NaOH_{(aq)} + HCl_{(aq)} \longrightarrow NaCl_{(aq)} + H_2O_{(l)}$$

Ionic form:

$$Na^+_{(aq)} + OH^-_{(aq)} + H^+_{(aq)} + Cl^- \longrightarrow Na^+_{(aq)} + NO_3{}^-_{(aq)} + H_2O_{(l)}$$

Net ionic form:

$$OH^-_{(aq)} + H^+_{(aq)} \longrightarrow H_2O_{(l)}$$

Net ionic equations are especially useful for representing reactions between acids and bases (Chapter 8).

5.11 BALANCING EQUATIONS

Chemical reactions produce new associations of atoms, but they never produce new or different atoms. Thus the combined mass of the reactants will always be equal to the combined mass of the products. This unvarying characteristic of chemical reactions is the basis of the *Law of Conservation of Matter,* which states that matter is neither created nor destroyed in a chemical reaction. This law requires that every atom shown on the reactant side of an equation must also appear among the products. An equation conforming to the Law of Conservation of Matter is said to be *balanced.* The concept of a balanced equation is shown schematically in Figure 5.2. Water is produced by the reaction of two diatomic molecules, H_2 and O_2, which combine in a 2:1 ratio. The total number (and weight) of hydrogen and oxygen atoms in the reacting molecules must equal the number (and weight) of hydrogen and oxygen atoms in the product.

Balancing an equation is accomplished by simple bookkeeping and does not change the facts of the equation. Some of the equations in the preceding section are already balanced, while others are not. Balancing an equation, like working a jigsaw puzzle, is essentially a process of trial and error, and the approach varies with the individual. The following suggestions may be useful.

1. Be certain that the formulas of the reactants and products are correctly written and that they are not subsequently changed in any way.
2. Make a tally sheet that lists the symbol of each element appearing in the equation.
3. On the left side of the symbol show the total number of atoms of that element found on the reactant side of the equation. On the right side of the symbol show the same information for the products.
4. If the two numbers agree for every element, the equation is balanced. If any two numbers do not agree, the coefficients of the substances which contain that element are adjusted until they do. Coefficients of other substances may also have to be changed.
   As in mathematics, a formula written without a coefficient is understood to have a coefficient of one. The effect of a coefficient is to multiply every atom in the formula it precedes by the value of the coefficient. For example, $3H_2SO_4$ means six atoms of hydrogen, three of sulfur, and twelve of oxygen.
5. The coefficients of the balanced equation should be reduced to the lowest possible whole numbers. Thus

$$2\ H_{2(g)} + O_{2(g)} \longrightarrow 2\ H_2O_{(g)} \text{ is preferred to}$$

$$4\ H_{2(g)} + 2\ O_{2(g)} \longrightarrow 4\ H_2O_{(g)}$$

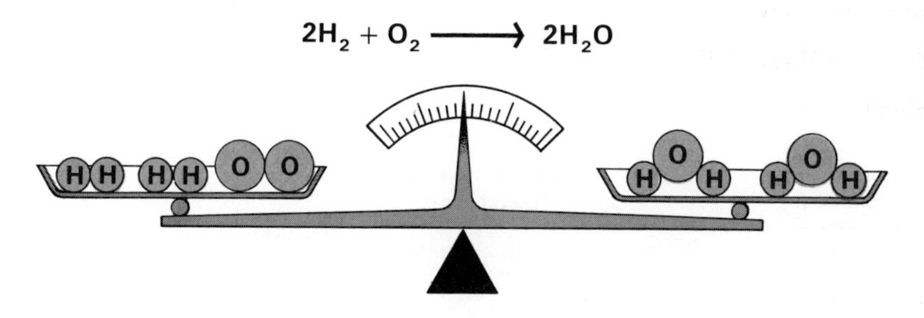

Figure 5.2
Schematic representation of a balanced equation.

Sample problem 5.11 Balance the equation

$$CaCO_{3(s)} \longrightarrow CaO_{(s)} + CO_{2\,(g)}$$

Solution The tally sheet should look like this:

Total # of Atoms on Reactant Side	Symbol of Element	Total # of Atoms on Product Side
1	Ca	1
1	C	1
3	O	3

Since the two numbers agree for every element, the equation is balanced as written.

Sample problem 5.12 Balance the equation

$$CaCO_{3(s)} + HCl_{(aq)} \longrightarrow CaCl_{2(aq)} + H_2CO_{3(aq)}$$

Solution First prepare the tally sheet:

Total # of Atoms on Reactant Side	Symbol of Element	Total # of Atoms on Product Side
1	Ca	1
1	C	1
3	O	3
1	H	2
1	Cl	2

This equation is not balanced because there are twice as many atoms of hydrogen and chlorine among the products as are present among the reactants. Since hydrogen and chlorine are both contained in the HCl molecule, doubling the number of these molecules will provide the required number of both kinds of atoms. This is accomplished by placing a coefficient of 2 before the HCl. The balanced equation is

$$CaCO_{3(s)} + 2\ HCl_{(aq)} \longrightarrow CaCl_{2(aq)} + H_2CO_{3(aq)}$$

Sample problem 5.13 Balance the equation

$$Al_{(s)} + HCl_{(aq)} \longrightarrow AlCl_{3(aq)} + H_{2\,(g)}$$

Solution Prepare the tally sheet:

Total # of Atoms on Reactant Side	Symbol of Element	Total # of Atoms on Product Side
1	Al	1
1	H	2
1	Cl	3

Additional atoms of hydrogen and chlorine are needed on the reactant side. Since they appear in a 1:1 ratio in the same molecule, the number of hydrogen atoms must always equal the number of chlorine atoms—yet the products show a 2:3 ratio between hydrogen and chlorine in the unbalanced equation. By selecting the least common multiple of 2 and 3, it is possible to satisfy both ratios. The least common multiple of 2 and 3 is 6. This is the required number of hydrogen atoms, as well as the correct number of chlorine atoms, and is obtained by placing a coefficient of 6 before HCl:

$$Al + 6\ HCl \longrightarrow AlCl_3 + H_2$$

The tally sheet shows that the coefficients on the right side must now be increased to show 6 hydrogen atoms and 6 chlorine atoms among the products. Since each hydrogen molecule contains two hydrogen atoms, a coefficient of 3 will result in a total of 6 hydrogen atoms, and 2 moles of $AlCl_3$ will contain 6 chlorine atoms:

$$Al + 6\ HCl \longrightarrow 2\ AlCl_3 + 3\ H_2$$

The tally sheet shows a need for two atoms of aluminum on the left side. A coefficient of 2 for aluminum balances the equation:

$$2\ Al_{(s)} + 6\ HCl_{(aq)} \longrightarrow 2\ AlCl_{3(aq)} + 3\ H_{2\,(g)}$$

Some slightly more difficult examples of balanced equations are

$$Mg(NO_3)_{2(aq)} + 2\ H_2SO_{4(aq)} \longrightarrow 2\ HNO_{3(aq)} + Mg(HSO_4)_{2(aq)}$$

$$C_3H_{8\,(g)} + 5\ O_{2\,(g)} \longrightarrow 3\ CO_{2\,(g)} + 4\ H_2O_{(g)}$$

$$C_6H_{12}O_{6(s)} + 6\ O_{2\,(g)} \longrightarrow 6\ CO_{2\,(g)} + 6\ H_2O_{(g)}$$

$$4\ Fe_{(s)} + 3\ O_{2\,(g)} \longrightarrow 2\ Fe_2O_{3(s)}$$

5.12 QUANTITATIVE ASPECTS OF REACTIONS

There are two very good reasons for balancing an equation: the equation is made to conform to the Law of Conservation of Matter, and it also becomes an instrument that can be used to determine mass relationships in reactions. This means, for example, that we can calculate the quantity of a reactant required for the production of a specific amount of product. The reaction between propane and oxygen will serve as an illustration. The balanced equation is

$$C_3H_{8\,(g)} + 5\ O_{2\,(g)} \longrightarrow 3\ CO_{2\,(g)} + 4\ H_2O_{(g)}$$

If the coefficients are used as numbers of moles, the equation states that one mole of propane requires five moles of oxygen for its complete combustion, and is converted into three moles of carbon dioxide and four of water in the process. Remember that a mole of a substance represents a number of atomic mass units as well as their equivalent in grams: thus the molar relationships in the balanced equation are also relationships of weight. For the burning of propane the weight relationships are obtained by multiplying each molecular weight by its coefficient:

Participants:

$$C_3H_8 + 5\ O_2 \longrightarrow 3\ CO_2 + 4\ H_2O$$

Molecular weights:

44 32 44 18

Molecular weights multiplied by coefficients:

44 g + 160 g + 132 g + 72 g

Therefore, complete combustion of 44 g of propane requires 160 g of oxygen and produces 132 g of carbon dioxide and 72 g of water.

For smaller or larger quantities of propane the weights of the other substances also vary, but maintain the same proportions. For instance, if only half as much propane is burned, the weight of each of the other three compounds will also be reduced by half.

The realization that a constant ratio exists between the weights of any two substances in a reaction gives us a powerful tool for answering questions of the type, "How much. . . ?" The following problem will illustrate how this is done. If television ads can be believed, conditions ranging from "nervous indigestion" to the painful aftermath of gluttonous eating can all be corrected by neutralizing some of the hydrochloric acid in the stomach. As a result, the American television audience spends millions of dollars each year for Tums,® Rolaids,® DiGel,® BiSoDol,® and similar preparations. The active ingredient in every antacid is a compound that neutralizes acids. Some of the most effective compounds are magnesium carbonate, magnesium hydroxide (milk of magnesia), and calcium carbonate (precipitated chalk).

Sample problem 5.14 Suppose that the recommended dose of an antacid contained 1 gram of $Mg(OH)_2$. How many grams of HCl would this neutralize? The equation for the reaction is

$$Mg(OH)_{2(aq)} + HCl_{(aq)} \longrightarrow MgCl_{2(aq)} + H_2O_{(l)}$$

Solution Since the weight relationships depend upon the molar coefficients, the equation must first be balanced.

$$Mg(OH)_{2(aq)} + 2\ HCl_{(aq)} \longrightarrow MgCl_{2(aq)} + H_2O_{(l)}$$

The balanced equation shows that one mole of $Mg(OH)_2$ can neutralize two moles of HCl. Since one mole of

$Mg(OH)_2$ is 58.3 g, and two moles of HCl is 73.0 g, the antacid can neutralize 73.0/58.3, or 1.25, times its own weight of acid. Thus one gram of $Mg(OH)_2$ can effectively remove 1.25 g of acid.

Problems of this sort can also be solved by the use of proportions involving two of the participants in the reaction. In the example under discussion, the participants are $Mg(OH)_2$ and HCl. From the balanced equation we learned that 1 mole of $Mg(OH)_2$ (or 58.3 g) is chemically equivalent to 2 moles of HCl (or 73.0 g). We also know the actual quantity of $Mg(OH)_2$ to be used in the reaction. The quantitative relationship between the two participants is expressed in the following proportion:

$$\frac{\text{Actual quantity of } Mg(OH)_2 \text{ used}}{\text{Chemical equivalent of } Mg(OH)_2} = \frac{\text{Actual quantity of HCl required}}{\text{Chemical equivalent of HCl}}$$

When the numerical values are inserted in the proportion, we get

$$\frac{1 \text{ gram}}{58.3 \text{ g}} = \frac{\text{Actual quantity of HCl required}}{73.0 \text{ g}}$$

This method is preferred by many people because it simplifies the task of setting up the problem.

APPLICATION OF PRINCIPLES

1. Draw the structural formula for carbon dioxide and the electronic formula for formic (methanoic) acid.
2. Write the empirical formula for $C_2H_4Cl_2$.
3. Write the empirical formula for the ionic compound of calcium and chlorate ions.
4. Write names to correspond to the following formulas. If the compound can be named in two different ways, give both names.
   (a) PCl_3
   (b) Fe_2O_3
   (c) $Al(OH)_3$
   (d) KCN
   (e) BF_3
   (f) Cu_2S
   (g) H_3PO_4
   (h) $Mg(NO_3)_2$
5. Write formulas to correspond to these names:
   (a) ammonium sulfate
   (b) mercurous bromide
   (c) lithium sulfide
   (d) ferric phosphate
   (e) calcium nitride
   (f) lead(II) chromate
   (g) sulfur hexafluoride
   (h) zinc acetate
6. Calculate the molecular weights of the following compounds:
   (a) $Pb(C_2H_3O_2)_2$
   (b) $(NH_4)_2SO_4$
   (c) CH_3CH_2OH
   (d) $C_6H_{12}O_6$

7. Calculate the percentage composition of each of the following compounds:
   (a) KSCN
   (b) CH_2Cl_2
   (c) $Al_2(SO_4)_3$
   (d) CH_2NH_2COOH
8. How many moles are contained in (a) 100 g of calcium hydroxide (b) 25 g of NH_3 (c) 0.90 g of NaCl?
9. Identify the type of reaction represented by each of the following equations:
   (a) $(NH_4)_2CO_3 \rightarrow NH_3 + H_2O + CO_2$
   (b) $P_4H_{10} + H_2O \rightarrow H_3PO_4$
   (c) $MnO_2 + HCl \rightarrow MnCl_2 + Cl_2 + H_2O$
   (d) $AgNO_3 + H_2S \rightarrow Ag_2S + HNO_3$
10. Write formulas for the products of the following substitution and double replacement reactions:
   (a) $AlCl_3 + AgNO_3$
   (b) $Mg + HgCl_2$
   (c) $NH_4I + Cl_2$
   (d) $H_3PO_4 + Ca(OH)_2$
11. Write a net ionic equation for the following reaction:

$$Pb(NO_3)_{2(aq)} + Na_2CrO_{4(aq)} \longrightarrow PbCrO_{4(s)} + 2\ NaNO_{3(aq)}$$

12. Balance the following equations:
   (a) $Mg_3N_2 + H_2O \rightarrow MgO + NH_3$
   (b) $Na_2O_2 + H_2O \rightarrow NaOH + O_2$
   (c) $SO_2 + H_2S \rightarrow S + H_2O$
   (d) $H_2SO_4 \rightarrow H_2O + SO_2 + O_2$
   (e) $NaHCO_3 + HCl \rightarrow NaCl + H_2O + CO_2$
   (f) $Mg(OH)_2 + H_3PO_4 \rightarrow Mg_3(PO_4)_2 + H_2O$
13. Sea shells are mostly calcium carbonate. When heated, calcium carbonate decomposes into calcium oxide (lime) and carbon dioxide. (a) Write a balanced equation for the reaction. (b) Calculate the number of pounds of calcium carbonate needed to produce 10 lb of lime.
14. An explorer, lost in the desert, stumbled on an abandoned automobile whose gas tank still held 5 liters of fuel. Gasoline has a density of 0.67 g/cc. Assuming gasoline to be composed of C_7H_{16} molecules, how many liters of water could he produce by burning the fuel and trapping the products? The reaction is

$$C_7H_{16(l)} + O_{2\,(g)} \longrightarrow CO_{2\,(g)} + H_2O_{(g)}$$

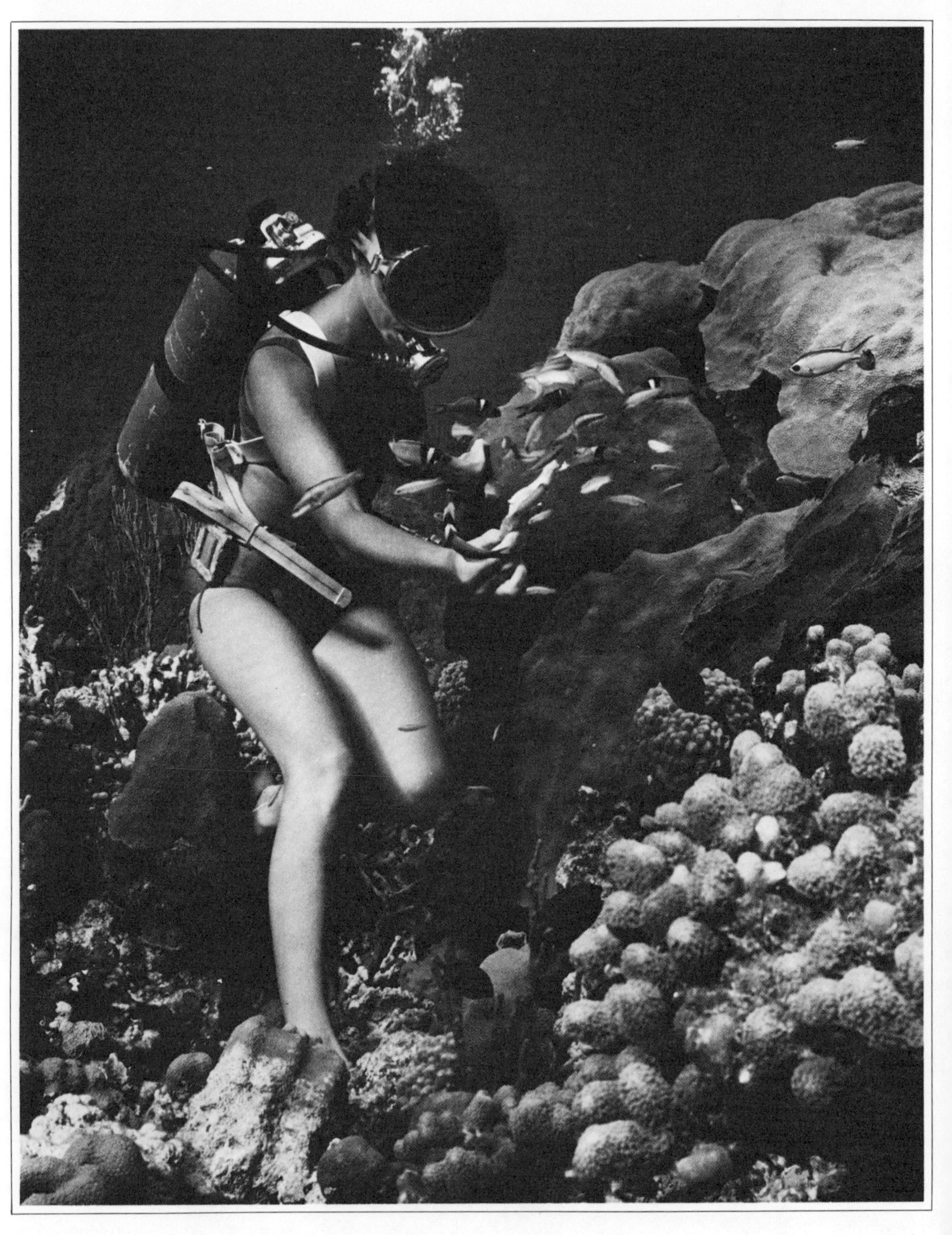

The undersea world. After centuries of speculation, man is at last beginning to learn something about the creatures who live in the seas. (Photograph by Paul J. Tzimoulis, Hollywood, California)

6

SOLUTIONS, DISPERSIONS, AND WATER

LEARNING OBJECTIVES

1. Describe the interaction that takes place between solute and solvent as a substance dissolves in water.
2. Describe the events that occur during the formation of an ionic solution.
3. Explain how solution formation is affected by such factors as the nature of the solute and the solvent, the temperature, the pressure, and the surface area of the solute.
4. Describe several different ways of expressing the concentration of a solution.
5. Compare a solution with a colloidal dispersion.
6. Explain how a solution of a specific concentration can be prepared by dilution.
7. Summarize the chemical properties of water.
8. Explain how the presence and concentration of a solute affect the properties of a solvent.
9. Describe and explain the events that take place when two solutions are separated by a differentially permeable membrane.
10. Describe the various processes that are employed in the production of potable water.
11. Distinguish between permanent and temporary hardness in water and explain how each type of hardness can be reduced.

KEY TERMS AND CONCEPTS

activity series
anhydrous
boiling point
colligative properties
colloidal dispersion
deliquescent
desiccator
dialysis
differentially permeable membrane
diffusion
edema
efflorescent
hard water
hydrate
hydrolysis
hygroscopic
hypertonic
hypotonic
ion exchange
isotonic
milligram percentage
molarity
osmosis
osmotic pressure
plasmolysis
potable water
saturated solution
serial dilution
standard solution
temporary hardness
Tyndall effect
vapor pressure

Before a chemical reaction can take place between two substances, their individual atoms, ions, or molecules must make contact with each other. Reactions between solids are usually quite slow because contacts can occur only at their surfaces. The surface area of a solid can be increased by crushing or grinding it or, for most chemicals, by dissolving it in a liquid.

Dissolving should not be confused with melting. When a substance dissolves, molecules or ions of it are dispersed among molecules of solvent. Melting occurs when the kinetic energy of a solid is increased to the point that its ions or molecules begin to move among each other.

6.1 CHARACTERISTICS OF SOLUTIONS

A mixture formed by dissolving one substance in another is called a *solution.* The dissolved material is the *solute* and the substance in which it is dissolved is the *solvent.* Among the various types of solutions are solid in liquid, liquid in liquid, and gas in liquid. A solution consists of individual atoms, ions, or molecules separated by particles of solvent, and has the following characteristics:

1. It is a homogeneous mixture.
2. It is clear but not necessarily colorless.
3. The individual particles of solute are smaller than 10 Å in diameter.
4. The solute does not settle out on standing.
5. The solute passes through filters, and many solutes pass through membranes as well.

Since a solution is a mixture, its components may be separated by physical processes such as evaporation or distillation.

Solutions containing water as the solvent are said to be *aqueous*. Although aqueous solutions are extremely important from a biological standpoint, solvents less polar than water must be used for dissolving nonpolar compounds. Ethyl alcohol is a slightly polar solvent whose solutions are called *tinctures* (e.g., tincture of iodine), *extracts* (extract of vanilla), and *elixirs* (elixir of terpin hydrate). Common nonpolar solvents are diethyl ether (commonly called ether) and benzene.

Kinds of Solutions

gas in liquid	champagne
liquid in liquid	vinegar
solid in liquid	sugar in water
gas in gas	dry air
liquid in gas	humid air
solid in gas	vapor from mothballs in air
gas in solid	no common example
liquid in solid	mercury in cadmium (dental fillings)
solid in solid	silver in gold (jewelry alloys)

6.2 SOLUTION FORMATION

Solution formation is a process in which individual particles of solute are first separated from each other and then distributed uniformly between particles of solvent. Thus two energy-consuming events must take place when a substance dissolves: (a) the bonds between solute particles must be broken, and (b) particles of solvent must be moved apart to make room for the solute. Energy is released and becomes available for these two processes by the formation of weak bonds between solvent and solute. However, if the energy released by interaction of solvent and solute is less than the amount required for the separation of the solute and the disruption of the solvent, the solvent does not dissolve readily (Figure 6.1). The formation of the aqueous solution of an ionic solid will serve as an example of the process.

The particles of an ionic solid attract each other electrostatically. This attractive force must be overcome before the individual ions can be separated. However, the ions on the surface of the crystal attract the polar water molecules and form numerous weak electrostatic bonds with them (Figure 6.2). The positive ions interact with the partial negative charge on the oxygen side of the dipole, while the nega-

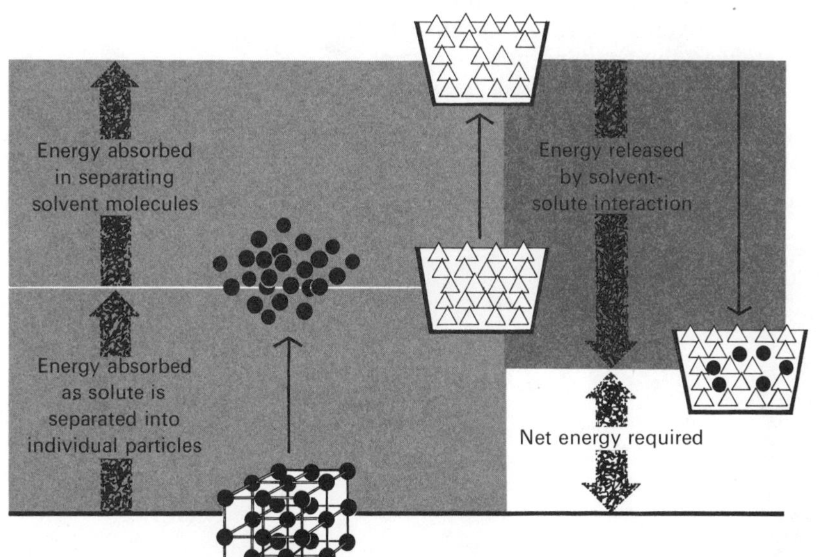

Figure 6.1
An example of an insoluble substance. The energy required for the process of dissolving is less than the available energy.

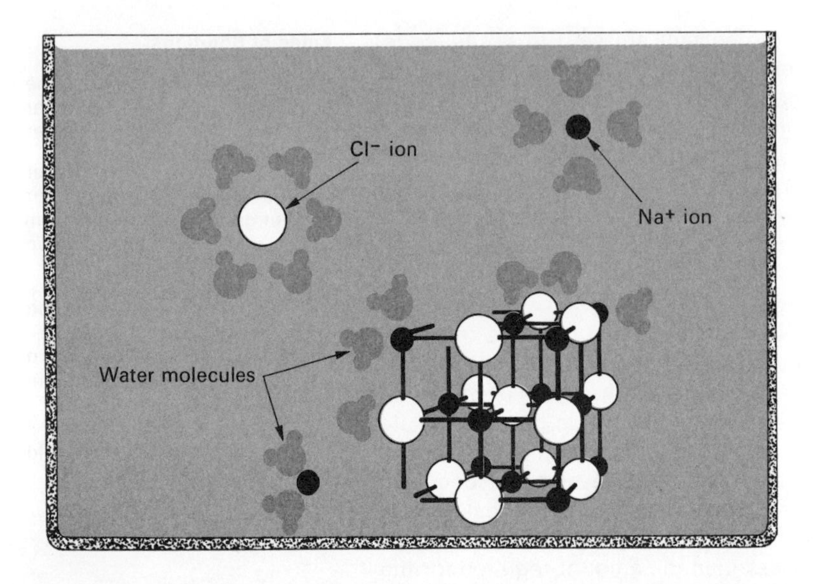

Figure 6.2
An ionic solid dissolving in water. Water molecules form weak electrostatic bonds with ions on the surface of the crystal.

tive ions and the partial positive charge attract one another. The formation of bonds with the solvent releases not only enough energy to separate the ion from the crystal, but also the energy needed for shifting molecules of solvent so that the ion can be accommodated.

Nonpolar substances do not dissolve very well in polar solvents. The molecules of a polar solvent are strongly attracted to each other but have very little attraction for nonpolar molecules. Thus the energy released by the formation of solvent-solute bonds is not enough to overcome the solvent-solvent attractions (Figure 6.3). Nonpolar substances readily dissolve in nonpolar solvents, however, because the solvent-solvent attractions are quite weak and shifting of solvent molecules to make room for the solute requires only a small amount of energy (Figure 6.4).

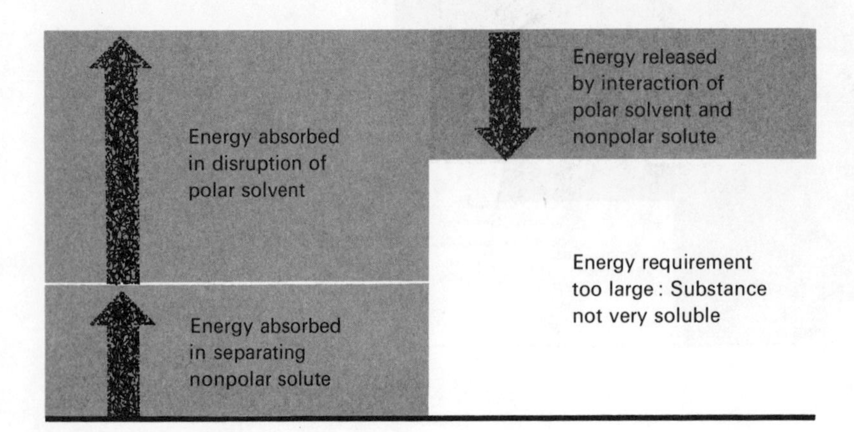

Figure 6.3
The interaction between a polar solvent and a nonpolar solute. A nonpolar substance does not dissolve very well in a polar solvent because too much energy is required to overcome attractions between polar solvent molecules.

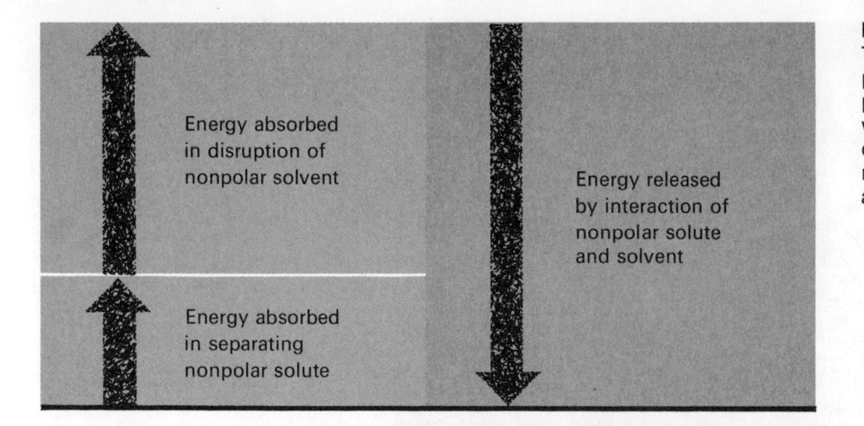

Figure 6.4
The interaction between nonpolar solvent and solute. Nonpolar substances dissolve quite well in nonpolar solvents because very little energy is required to separate the solute and disrupt the solvent.

6.3 FACTORS AFFECTING SOLUTION FORMATION

The quantity of a given solute that will dissolve in a specified amount of solvent depends upon three factors: the nature of the solvent and solute, the temperature, and the pressure.

The Nature of the Solvent and Solute

The relationship between chemical character and solubility can be summed up in the words "like dissolves like." This means that the best solvent for a nonpolar solute is another nonpolar substance, while polar and ionic compounds tend to be most soluble in polar solvents. Thus the nonpolar oxygen molecule has an extremely low solubility in water, and ionic NaCl is almost insoluble in nonpolar benzene.

Temperature

Temperature has a marked effect on the solubility of many solutes, but it is impossible to predict, except in the case of gases, whether an increase in temperature will cause an increase or a decrease in the solubility of a given substance. Experimental evidence (and subsequently our molecular model) suggests that the solubility of a gas should decrease with an increase in temperature. This is because increasing the temperature causes both the molecules of gas and the particles of solvent to move more rapidly and imposes greater strains on any solvent-solute bonds. The net result is that molecules of a gas escape from the solution at a higher rate as the temperature increases.

The transfer of oxygen from the alveoli of the lungs into the blood stream is aided by the fact that the temperature of the blood leaving the lungs is slightly lower than that of blood coming into the lungs; thus oxygen passes more readily across the alveolar membranes into

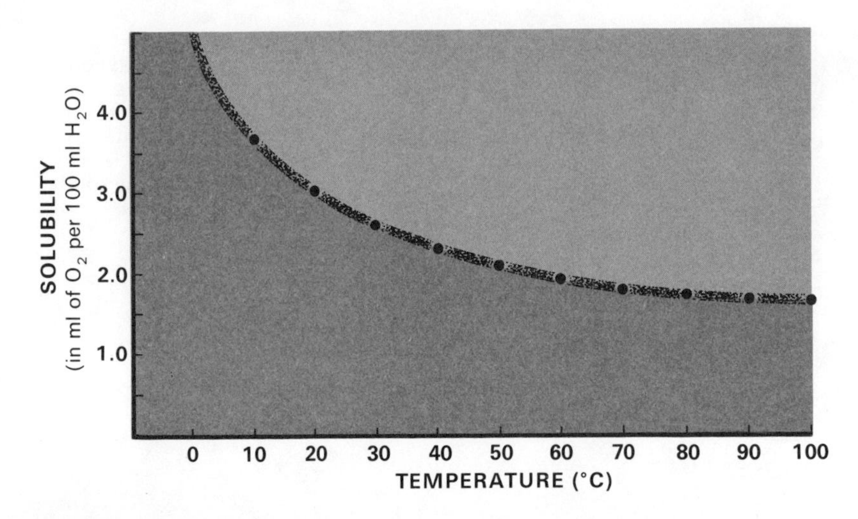

Figure 6.5
The effect of temperature on the solubility of oxygen gas in water.

the cooler blood. The effect of temperature on the solubility of oxygen in water is shown in Figure 6.5.

For nongaseous solutes the general observation is that raising the temperature increases their solubility, but there are many exceptions to this statement. The variation in solubility with changes in temperatures is shown for several common ionic and molecular solutes in Figure 6.6.

Pressure

Pressure changes have an effect upon solubility only when the solute is a gas. Let us consider a closed container in which a liquid is in con-

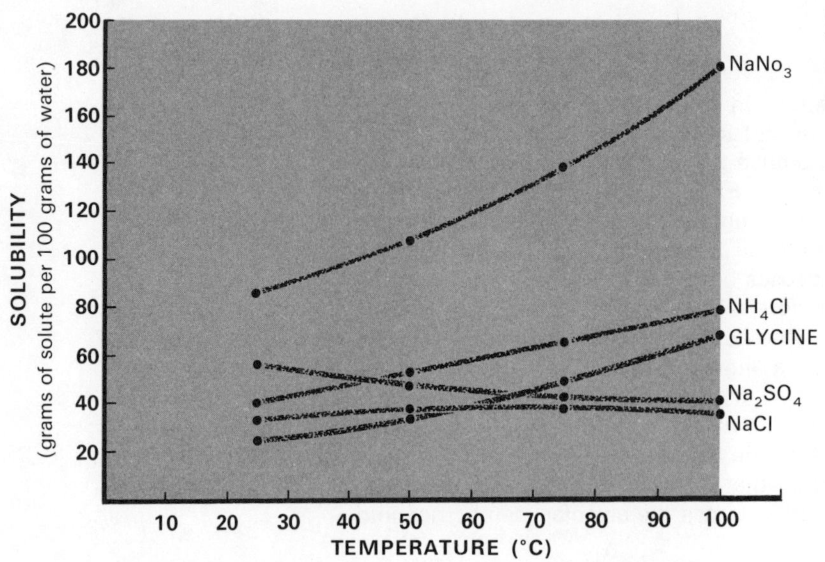

Figure 6.6
The effect of temperature on the solubility of various ionic and molecular solutes.

tact with carbon dioxide gas. Because the molecules of the gas are in constant random motion and cannot escape from the container, some of them penetrate the surface by chance and dissolve in the liquid. Molecules of carbon dioxide also leave the liquid and reenter the gaseous phase. A conditon of dynamic equilibrium is reached when molecules of the gas are going into and coming out of solution at the same rate (Figure 6.7). A solution in this state of equilibrium is said to be *saturated* with respect to the solute: that is, the amount of dissolved solute remains the same unless the conditions are changed.

Increasing the pressure on the gas crowds the molecules closer together. This disturbs the equilibrium because it forces additional molecules to move into the liquid. Eventually a new equilibrium condition is reached in which the solution contains a greater amount of dissolved carbon dioxide than before (Figure 6.8).

Champagne, beer, and soft drinks contain carbon dioxide under pressure. Opening the container reduces the pressure, the gas eventually escapes, and the beverage is said to taste "flat."

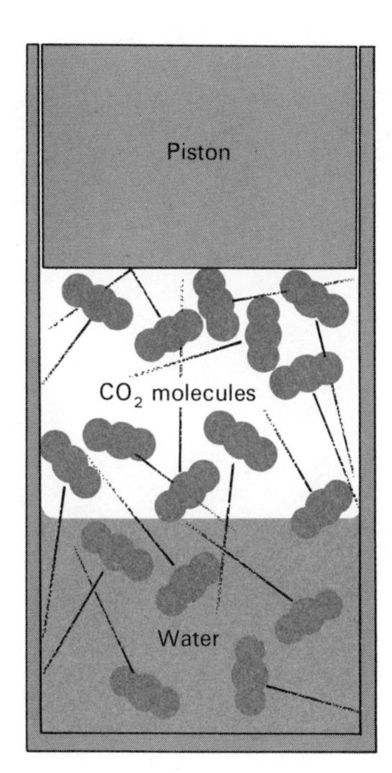

Figure 6.7
Dynamic equilibrium. Molecules of carbon dioxide are entering and leaving the water at the same rate. A solution in a state of dynamic equilibrium is said to be saturated.

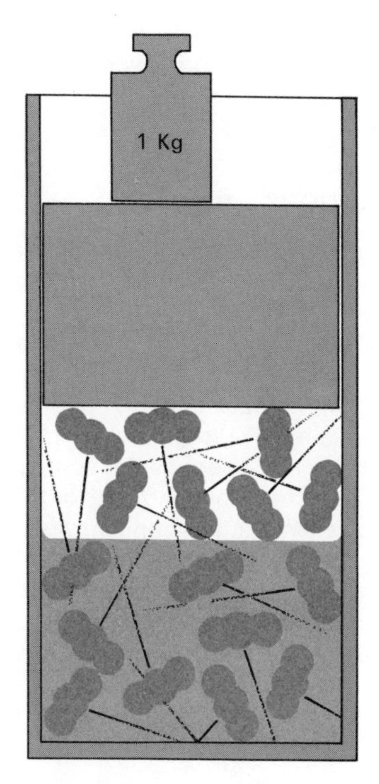

Figure 6.8
The effect of increased pressure on a system at equilibrium. Increasing the pressure forces more of the gas into solution. A new equilibrium is established at the higher pressure.

6.4 FACTORS AFFECTING THE RATE OF DISSOLVING

In addition to the three factors affecting the concentration of solute in a solution, there are two factors which affect the rate at which a solute goes into solution: the surface area of the solute and agitation.

Surface area

Since dissolving takes place only where the solute contacts the solvent, the rate of dissolving can be increased by increasing the surface area of the solute through grinding or crushing. Although this makes the solute dissolve more quickly, it does not increase the quantity of solute that will dissolve.

Agitation

If a lump of solute is allowed to stand undisturbed in a solvent, the solution in the immediate vicinity of the lump soon contains a large number of solute particles, while the solvent at a greater distance contains relatively little solute. Stirring or shaking speeds up the solution process because it brings the undissolved solute into contact with the less concentrated solution.

6.5 CONCENTRATION OF SOLUTIONS

A reexamination of Figure 6.6 provides us with some interesting details concerning solutions. For instance, a sodium chloride solution at 20°C is saturated when it contains 36 g of NaCl in 100 ml of solution, but a saturated solution of sodium sulfate at the same temperature contains 60 g per 100 ml. Furthermore, the solubility of NaCl increases slightly at a higher temperature, while that of Na_2SO_4 decreases. It is apparent from this that saturation is relative to the nature of the solute and the conditions and, therefore, is not a very useful concept.

Solutions are also classified as either concentrated or dilute, but these, too, are relative terms. For example, the concentrated solution of sulfuric acid used in the laboratory is almost pure H_2SO_4, while concentrated hydrochloric acid is only about one-third HCl. *Concentrated* simply implies that the solute constitutes a greater proportion of the solution, but does not indicate just what the proportion is. In solutions where the solute is the active ingredient (as in disinfectants, antibiotics, and solutions for injection), too much or too little solute may have undesirable effects, and so the terms *concentrated* and *dilute* are not precise enough.

Chemists refer to the makeup of a solution as its concentration. The *concentration* of a solution may be defined as the quantity of solute contained in a given amount of solution. Some of the ways of expressing concentration are percentage, milligram percentage, ratio, and molarity. A solution whose concentration is precisely known is called a *standard* solution.

Percentage

Percent means parts in a hundred. The *percentage* concentration of a solution is usually expressed as the number of grams of solute in 100 ml of solution. Physiological saline, for example, is a 0.92% solution of NaCl in water—that is, there are 0.92 g of NaCl in 100 ml of solution. This is prepared by placing 0.92 g of solid NaCl in a calibrated container and adding enough distilled water to make 100 ml of solution. For liquid solutes the percentage concentration tells the number of milliliters of solute per 100 ml of solution.

Milligram percentage

The concentrations of substances in body fluids such as urine or blood are rather low, and for this reason they are generally expressed in milligram percent (mg%). *Milligram percent* is the number of milligrams of solute in 100 ml of body fluid. For example, the normal fasting level of glucose in human blood is about 80-120 mg per 100 ml, or 80-120 mg%. This corresponds to a true concentration of 0.080-0.120%, and so the use of mg% eliminates an awkward decimal fraction.

Ratio

Chemicals such as antiseptics and disinfectants are sometimes furnished as concentrated solutions and must be diluted to a lower concentration prior to use. The concentration of a dilute solution is often expressed as a ratio between the parts of solute and the total parts of solution. For instance, a 1:1000 concentration means that one part of solute is present in each 1000 parts of solution. Sometimes the concentration of an extremely dilute solution is expressed in parts per million (ppm), which is also a ratio, but it is written in such a way that it does not require the use of as many figures as either the straight ratio or the percentage.

To show how the ppm designation is used, let us consider the concentration of chlorine in a swimming pool. The spread of disease organisms is effectively prevented when the concentration of chlorine falls in the range of 0.5 to 1.0 ppm. This corresponds to a range of 0.00005% to 0.0001%, and to dilutions of 1:2,000,000 and 1:1,000,000, respectively. The concentrations of pollutants in air and water are normally given in ppm. Table 6.1 shows the maximum concentrations of various chemicals in drinking water that are permitted by the U.S. Public Health Service.

Molarity

The *molarity* of a solution is the number of moles of solute per liter of solution. A *molar solution* (*M*) contains one gram molecular weight of solute per liter of solution. Solutions of known molarity can be prepared with great accuracy, and provide a convenient way of mea-

TABLE 6.1 Chemical Standards for Potable Water

Species	Maximum Permissible Concentration (parts per million)
Arsenic (3+ and 4+ oxidation states)	0.05
Barium ion	1.0
Cadmium ion	0.01
Chloride ion	250
Chromium (6+ oxidation state)	0.05
Copper (2+ oxidation state)	1
Cyanide ion	0.2
Fluoride ion	2.0
Iron (3+ oxidation state)	0.3
Lead (2+ oxidation state)	0.05
Detergent	0.5
Manganese (2+ oxidation state)	0.05
Nitrate ion	45
Organic contaminants	0.2
Selenium (3+ and 4+ oxidation states)	0.01
Silver ion	0.05
Sulfate ion	250
Zinc ion	5

suring precise multiples or fractions of moles. For example, 1 ml of any 1 *M* solution contains exactly 1/1000 mole of solute; or 10 ml of a 0.1 *M* solution contains exactly 1/1000 mole. Most standard solutions used in the laboratory are prepared to a given molar concentration.

6.6 PREPARING DILUTIONS

Laboratory procedures in chemistry, physiology, and microbiology often involve the use of several concentrations of the same material. For example, a study of digestive processes might require that saliva be collected and diluted to give concentrations of 1:5 and 1:50. Let us consider how these dilutions might be prepared.

Remember that the first number in the ratio gives the number of parts of solute and the second number the total parts of solution. Thus a 1:5 dilution contains one volume of saliva with four of distilled water. Any quantity of solution can be prepared as long as four volumes of solvent are used for every volume of solute.

The obvious way to prepare the 1:50 dilution is to combine 1 volume of saliva with 49 volumes of water. However, this makes 50 volumes of solution, which may be much more than is needed. Another way of arriving at a 1:50 concentration is by diluting the 1:5 solution until it is one-tenth as strong: in other words, by making a 1:10 dilution of it. To do this, we simply mix one volume of the 1:5 concentration with nine volumes of water. Figure 6.9 illustrates the procedure. This method of preparing several concentrations of a material is called *serial dilution.*

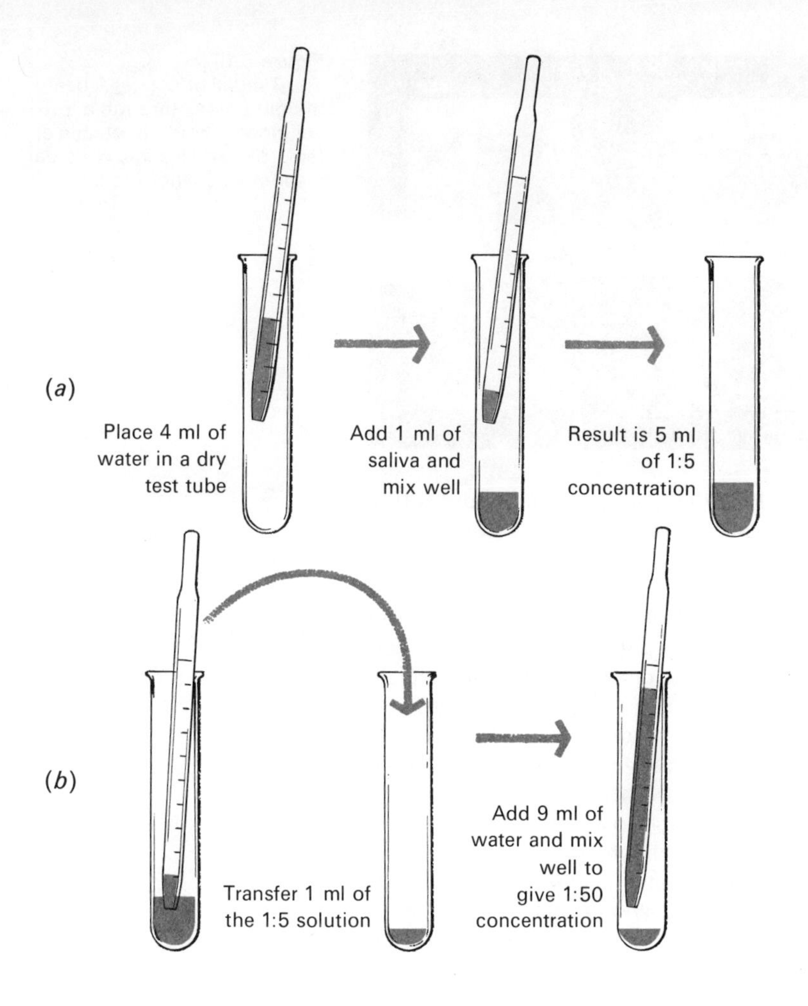
(a)
Place 4 ml of water in a dry test tube
Add 1 ml of saliva and mix well
Result is 5 ml of 1:5 concentration
(b)
Transfer 1 ml of the 1:5 solution
Add 9 ml of water and mix well to give 1:50 concentration

Figure 6.9
Preparing dilutions of saliva in distilled water. (a) Preparing a 1:5 dilution. (b) Using the 1:5 dilution to prepare a 1:50 dilution.

6.7 COLLOIDAL DISPERSIONS

A *colloidal dispersion* is a heterogeneous mixture from which the components do not separate. As is the case with true solutions, a colloidal dispersion may consist of two liquids, a liquid and a solid, a liquid and a gas, or any of the other possible combinations of physical states. However, biologically important colloids are most often liquids or solids dispersed in water. The protoplasm of a cell and the fluid portion of the blood are examples of biological colloids. Whipped cream is a colloidal dispersion of air in a liquid.

A colloidal dispersion differs from a true solution in that the individual particles of at least one of its components are actually aggregates of ions or molecules ranging in size from about 10 Å to 1000 Å. These particles may pass readily through most filters but they are retained by membranes. Colloidal dispersions may be further distinguished from solutions by their behavior in a strong beam of light. Light rays pass

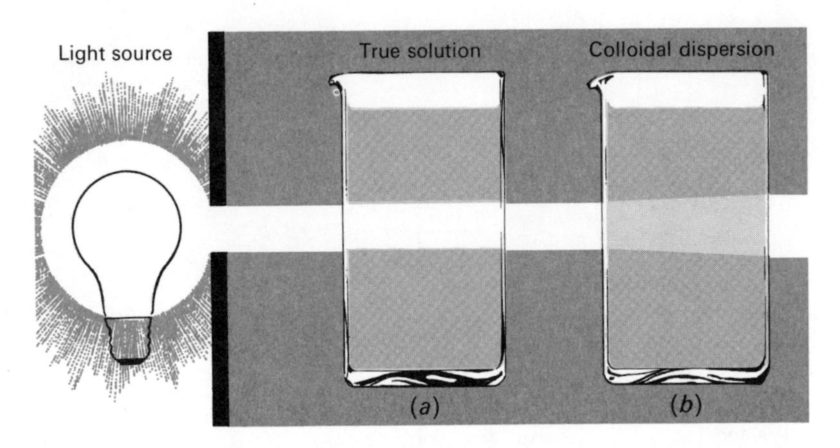

Figure 6.10
The Tyndall effect. (a) A beam of light passes through a true solution without noticeable effect. (b) Particles in a colloidal dispersion scatter the light.

through a solution without noticeable effect, as shown in Figure 6.10a, but colloids scatter the light and the beam is well-defined (Figure 6.10b). The light-scattering property of colloids is called the *Tyndall effect.*

Colloidal dispersions may be prepared from a small number of true solutions by reactions which cause solute particles to stick to each other and form submicroscopic invisible clusters. High-molecular-weight substances such as starch, glycogen, and most proteins are relatively insoluble in water and they, too, form colloidal dispersions. Some materials can be reduced to colloidal dimensions by grinding.

Colloidal particles tend to adsorb ions from the medium in which they are dispersed, and become either positively or negatively charged. All of the particles in a given dispersion have the same charge, and it is their mutual repulsion that prevents them from settling. The addition of an electrolyte, especially one that produces multiply charged ions in solution, disrupts the dispersion by linking the particles together.

6.8 PROPERTIES OF WATER

Physical Properties

Water is an important compound in many ways. Without water, life on this planet could not exist. Water is not only the chief solvent for biological substances, but it plays a key role in temperature regulation, and influences climatic conditions as well. (See discussion in Chapter 2.) Recall that the Celsius temperature scale is derived from the melting and boiling points of water, and that the calorie and the gram are both based on physical properties of water. The physical properties of water are summarized in Table 6.2.

Chemical Properties

Water is very little affected by high temperatures. If water is dropped on a surface that has been heated to a temperature of 2000°C, less

TABLE 6.2 Physical Properties of Water

1. Water is colorless, odorless, tasteless, and an extremely poor conductor of electricity.
2. Water is an excellent solvent for ionic and polar substances.
3. At a pressure of 760 torr it has a melting point of 0°C and a boiling point of 100°C.
4. Its density at 4°C is 1 g/cc. Above or below this temperature the density decreases.
5. The heat of fusion is 80 cal/g and the heat of vaporization is 540 cal/g.
6. The specific heat of water is 1 cal/g/°C.

than 2% of the molecules will be decomposed into hydrogen and oxygen. It can, however, be broken down by the passage of electric current after a suitable conductor has been added (see electrolysis, Chapter 2). Despite its extreme stability at high temperatures, water reacts chemically in the following ways:

Reactions between Water Molecules

Water molecules react with each other to form hydroxyl (OH^-) and hydronium (H_3O^+) ions. The reversible reaction is

$$2\ H_2O \rightleftharpoons H_3O^+ + OH^-$$

The unequal length of the arrows indicates that the reaction which produces water is favored, so the concentration of hydroxyl and hydronium ions at any instant is quite low—approximately one ten-millionth of a mole (10^{-7} mole) per liter.

The ionization of pure water always produces equal numbers of OH^- and H_3O^+ ions. Some substances cause a change in this 1:1 ratio when they are added to, or react with, water. If the H_3O^+ ions outnumber the OH^- ions, the added substance is called an *acid;* and if the OH^- ions are most numerous, it is called a *base.* Acids and bases are discussed in detail in Chapter 8.

The formation of hydrogen bonds between water molecules was discussed in Chapter 4. Water is also capable of hydrogen bonding with other molecules containing a highly electronegative element.

Reactions with Metals

Certain active metals react with cold water to produce hydrogen gas, as shown by the examples of sodium and potassium.

$$2\ K_{(s)} + 2\ H_2O_{(l)} \longrightarrow H_{2\,(g)} + 2\ KOH_{(aq)}$$

$$2\ Na_{(s)} + 2\ H_2O_{(l)} \longrightarrow H_{2\,(g)} + 2\ NaOH_{(aq)}$$

These reactions generate so much heat that the hydrogen is usually explosively ignited.

Less active metals such as aluminum, zinc, and iron react with steam, but not with cold water, and the products are hydrogen gas and a metallic oxide.

Production of hydrogen gas by the reaction between water and potassium metal. The reaction is so vigorous that the hydrogen gas is ignited. (Photograph by J. Asdrubal Rivera)

$$2\ Al_{(s)} + 3\ H_2O_{(g)} \longrightarrow Al_2O_{3(s)} + H_{2\,(g)}$$

$$Zn_{(s)} + H_2O_{(g)} \longrightarrow ZnO_{(g)} + H_{2\,(g)}$$

Metals such as copper, silver, and gold do not react with water.

When the metals are arranged in order according to their relative activities, the resulting list is called the activity series. The relative activities of the more common metals are shown in Table 6.3. The most active metal is at the top of the list and the least active at the bottom. Note the inclusion of hydrogen. Although hydrogen is not a metal, it divides the metals into two groups. Elements below hydrogen in the activity series do not react with water, nor do they react with any other compound to produce hydrogen gas.

Reactions with Nonmetals

Reactions between nonmetals and water are not very common. One of the most important occurs when chlorine is added to water, producing hydrochloric and hypochlorous acids.

$$Cl_{2\,(g)} + H_2O_{(l)} \longrightarrow HCl_{(aq)} + HCLO_{(aq)}$$

Because hypochlorous acid is a powerful germicide, chlorine is used in swimming pools, and in treatment of drinking water and sewage.

TABLE 6.3 The Activity Series of Metals

Metal	Reaction
Potassium Barium Strontium Calcium Sodium Magnesium	React with cold water to form H_2 and a hydroxide
Aluminum Manganese Zinc Chromium Iron Cobalt Nickel Tin Lead	React with steam to form H_2 and an oxide
Hydrogen	
Copper Mercury Silver Platinum Gold	No reaction with steam

Reactions with Oxides

When a reaction takes place between water and an oxide, the product depends upon the nature of the oxide. Two general patterns are found: nonmetallic oxides produce sour-tasting solutions which have high concentrations of H_3O^+ ion, while solutions of metallic oxides taste bitter, have a soapy feel, and contain greater concentrations of OH^- ion. In other words, oxides of nonmetals produce acidic solutions, and oxides of metals form basic solutions. The reaction in each case involves a combination with water followed by splitting of the product into two ions. Typical examples of reactions with *nonmetallic oxides* are

(a) $SO_{2\,(g)} + H_2O_{(l)} \longrightarrow H_2SO_{3(aq)}$

$H_2SO_{3(aq)} \longrightarrow H_3O^+_{(aq)} + HSO_3^-{}_{(aq)}$

(b) $CO_{2\,(g)} + H_2O_{(l)} \longrightarrow H_2CO_{3(aq)}$

$H_2CO_{3(aq)} \longrightarrow H_3O^+_{(aq)} + HCO_3^-{}_{(aq)}$

Some typical examples of reactions with *metallic oxides* are

(a) $MgO_{(s)} + H_2O_{(l)} \longrightarrow Mg(OH)_{2(aq)}$

$Mg(OH)_{2(aq)} \longrightarrow Mg^{2+}_{(aq)} + 2\ OH^-_{(aq)}$

(b) $CaO_{(s)} + H_2O_{(l)} \longrightarrow Ca(OH)_{2(aq)}$

$Ca(OH)_{2(aq)} \longrightarrow Ca^{2+}_{(aq)} + 2\ OH^-_{(aq)}$

Hydrolysis

Hydrolysis is a composite word formed from *hydro,* which refers to water, and *lysis,* which means a loosening. It can be interpreted literally as "breaking down by means of water." *Hydrolysis,* then, is a chemical reaction that takes place between water and a second compound with the result that the other compound is split. Many different kinds of substances undergo hydrolysis, especially molecules of biological origin. The digestive process, for instance, breaks down large molecules, such as fats and proteins, into smaller units that can pass through the intestinal wall. Although the digestion of a fat or protein involves a number of hydrolytic steps, the overall reactions can be summarized as follows:

$$\text{Fats} \xrightarrow{\text{catalysts}} \text{Fatty acids} + \text{Glycerol}$$

$$\text{Proteins} \xrightarrow{\text{catalysts}} \text{Amino acids}$$

Note that hydrolytic reactions in a living organism require the presence of a catalyst. As mentioned previously, the catalysts involved in biological reactions are called enzymes.

Hydrolysis of polar molecules such as HCl and $HC_2H_3O_2$ takes place in aqueous solutions without a catalyst. Hydrolysis of a polar compound produces positive and negative ions, as shown by the following examples:

$$H_2O_{(l)} + HCl_{(g)} \rightleftharpoons H_3O^+_{(aq)} + Cl^-_{(aq)}$$

$$H_2O_{(l)} + HC_2H_3O_{2(l)} \rightleftharpoons H_3O^+_{(aq)} + C_2H_3O_2^-{}_{(aq)}$$

Note the relative lengths of the two arrows. They indicate that in a dilute solution of HCl virtually all of the molecules have been hydrolyzed and exist in the ionic form. A schematic representation of the hydrolysis of HCl is shown in Figure 6.11. The bonding electrons are more strongly attracted to the chlorine atom because of its higher electronegativity. Thus a strong dipole is created, with the chlorine acquiring a slight negative charge and the hydrogen a slight positive charge. Polar water molecules interact with this dipole and cause the hydrogen and chlorine to separate. The shared electrons stay with the chlorine, and the hydrogen forms a coordinate covalent bond with a molecule of water.

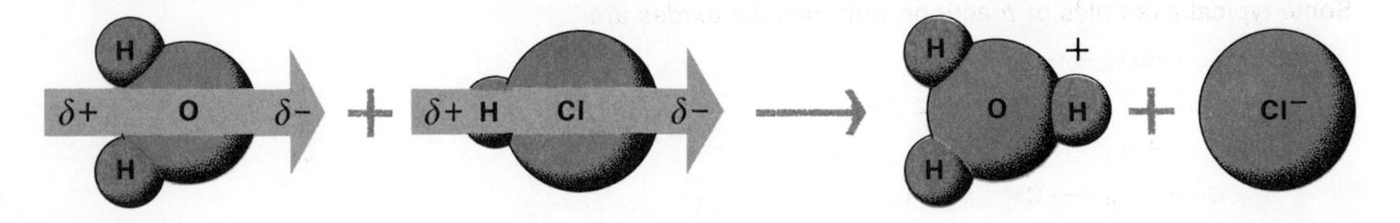

Figure 6.11
A schematic representation of the hydrolysis of hydrogen chloride.

The less polar acetic acid molecules exist predominantly in the molecular rather than the ionic form in aqueous solutions. The acetate ions and water molecules are competing for the same hydrogen atoms in the acetic acid molecules, and the acetate ions have a stronger pull.

Hydrate Formation

Water molecules are incorporated into the lattice structures of some ionic compounds as they crystallize from solution. Compounds in which water forms a part of the crystalline structure are called *hydrates.* The water in a hydrate is accommodated in spaces between the ions and is held in place by attractions between the positive ions and the lone pairs of electrons on the oxygen atom, and by hydrogen bonding with the negative ion (Figure 6.12). The ratio between formula units and water molecules in a hydrate is indicated by a dot in the formula, as in $Na_2CO_3 \cdot 10\ H_2O$ (washing soda) and $CuSO_4 \cdot 5\ H_2O$. Salts such as NaCl and KNO_3 do not form hydrates because there is not enough room between their ions for water molecules.

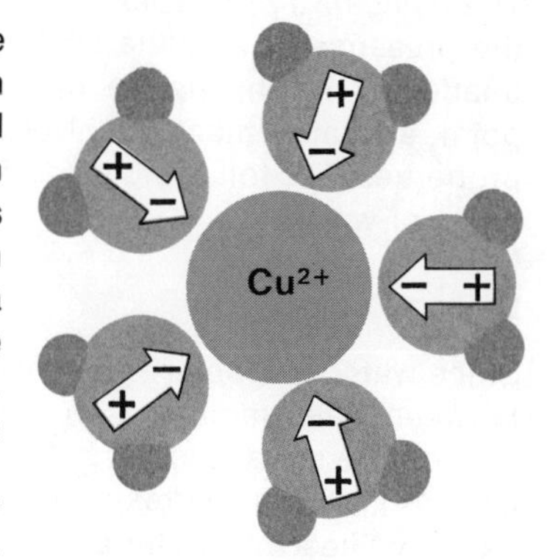

Figure 6.12
Weak bonds hold the water in a crystalline hydrate.

Vigorous heating of a hydrate breaks the weak bonds and drives off the water, and the solid that remains is said to be *anhydrous* (without water). In the presence of moisture the anhydrous salt gradually returns to the hydrated form.

One interesting observation is that the hydrated crystal is often not the same color as the anhydrous salt. For example, $CuSO_4 \cdot 5\ H_2O$ is deep blue, while $CuSO_4$ is almost white. Because the copper(II) sulfate crystals change color as they are hydrated, they can be used to detect the presence of moisture.

Another important use of hydrates involves the change of calcium sulfate from a partially hydrated to the fully hydrated form. Partially hydrated calcium sulfate [$(CaSO_4)_2 \cdot H_2O$] is called plaster of paris, and the fully hydrated form ($CaSO_4 \cdot 2\ H_2O$) is gypsum. When powdered plaster of paris is mixed with water to form a paste, the following reaction occurs:

$$(CaSO_4)_2 \cdot H_2O_{(s)} + 3\ H_2O_{(l)} \longrightarrow 2\ CaSO_4 \cdot 2\ H_2O_{(s)}$$

and the paste solidifies. Hardening is accompanied by a slight expansion; thus plaster of paris is useful for making molds and impressions. It is also used to immobilize broken bones. A person who applies a cast should remember to allow for expansion so that the cast is not too tight.

Some hydrates spontaneously lose water of hydration when exposed to dry air, and are said to be *efflorescent.* A number of other compounds readily absorb moisture from the air and are called *hygroscopic.* A few crystalline substances are capable of absorbing so much moisture from the air that their surfaces become wet and the crystal begins to dissolve. The term *deliquescent* describes this behavior. Hygroscopic and deliquescent substances are useful as drying agents since they remove moisture from the surrounding atmosphere or from neighboring chemicals. Figure 6.13 shows a *desiccator,* a vessel for drying chemicals or keeping them dry. A hygroscopic substance such as $CaCl_2$ is put in the bottom of the container and the other chemicals are placed on the perforated shelf.

Ground-glass joint

Drying agent

Figure 6.13
A desiccator. The hygroscopic material in the lower chamber absorbs moisture.

Colligative Properties

The *colligative properties* of a solvent are those which are affected by the presence of a solute and by its concentration, but are relatively unaffected by the *nature* of the solute. The boiling point, freezing point, and movement of solvent through a membrane are colligative properties. The following discussion will consider the colligative properties of water.

Boiling Point

Since water is a polar substance, rather strong attractive forces exist between water molecules in the liquid state. However, a water molecule on the surface is not held as tightly as one inside the liquid (Figure 6.14), and has a greater freedom of motion. If a closed container is partially filled with water at room temperature, some of the molecules at the surface will possess enough kinetic energy to overcome the intermolecular attractive forces and escape from the liquid. These molecules of water vapor collide with each other, with the molecules of air, and with the walls of the container, and some of them return to the liquid from which they have escaped. Eventually an equilibrium condition is reached where the rate of evaporation equals the rate of condensation, and the number of molecules in the vapor state remains almost constant. Pressure is generated by molecules of water vapor striking the walls of the container. This pressure is called the *vapor pressure* of the liquid.

Raising the temperature increases the average kinetic energy of the water molecules and disturbs the equilibrium. At the higher temperature a new equilibrium is established in which larger numbers of molecules are found in the vapor phase and the vapor pressure is correspondingly higher. Figure 6.15 shows the relationship between the temperature of water and its vapor pressure.

If water were placed in an open container with air pressure of

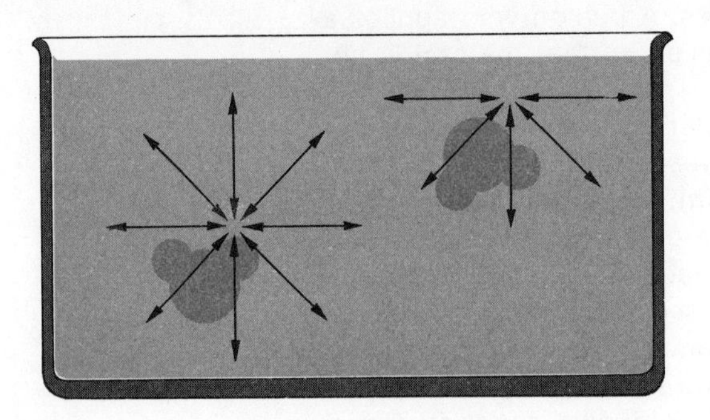

Figure 6.14
Attractions between molecules of water. Molecules within the liquid experience greater attractive forces than those on the surface.

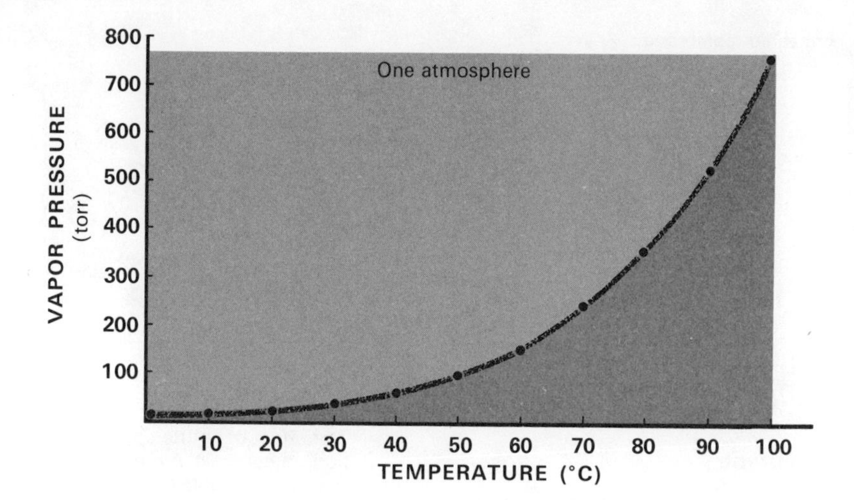

Figure 6.15
Variation of the vapor pressure of water with changing temperature.

760 torr pressing down on its surface, the vapor pressure would equal the atmospheric pressure when the temperature reached 100°C, and the water would boil. The temperature at which the vapor pressure is equal to atmospheric pressure is called the *boiling point* of the liquid. It is apparent from Figure 6.15 that if the atmospheric pressure were only 290 torr, water would boil at 75°C. Likewise, increasing the pressure increases the boiling point. If water is heated in a closed container, its temperature, and the temperature of the water vapor, will exceed 100°C. This principle is used to shorten the cooking time of foods in a pressure cooker. High-temperature steam (1520 torr and 121.6°C) is used in autoclaves to sterilize containers, instruments, surgical dressings, and some chemical solutions.

The Effect of Altitude on the Boiling Point of Water

Mt. Everest	28,028 ft.	87.6°C
Lake Titicaca, Peru	12,507 ft.	91.5°C
Denver, Colorado	5,280 ft.	95.0°C
San Francisco, California	sea level	100°C

Particles of a nonvolatile solute such as sugar or salt raise the boiling point of water because they interfere with the escape of water molecules: thus a higher temperature must be reached before the vapor pressure equals that of the atmosphere. One mole of molecules or ions dissolved in 1000 g of water raises its boiling point by about 0.5°C.

Freezing Point

A solution freezes at a lower temperature than the pure solvent because the solute interferes with the formation of a regular solid structure, and the energy level of the solvent molecules must be reduced before the solute can be accommodated. One mole of ions or molecules dissolved in 1000 g of water lowers its freezing point by approximately 1.9°C.

A water-soluble, low-molecular-weight substance such as methyl alcohol (CH_3OH) or ethylene glycol [$C_2H_4(OH)_2$] is added to the radiator of an automobile so that the water in the cooling system will not freeze until the temperature becomes extremely low.

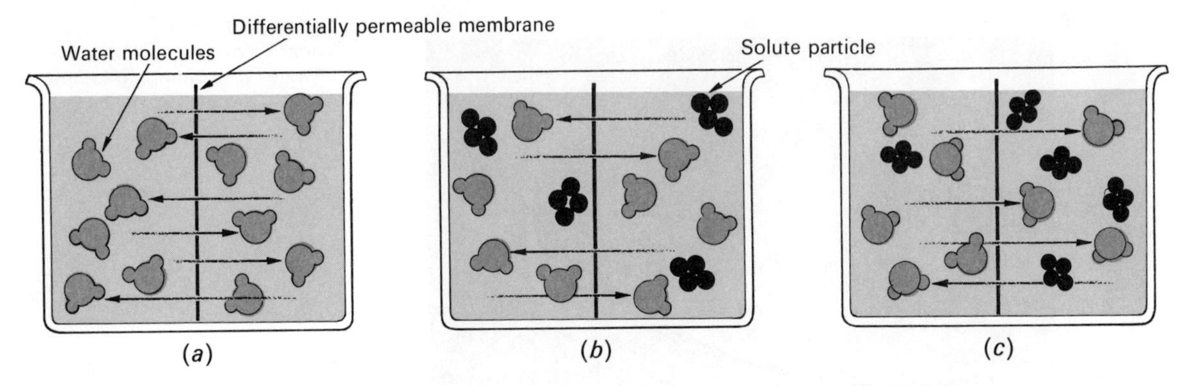

Figure 6.16
The effect of solute concentration on osmosis. A membrane that is permeable to water molecules but not to particles of solute divides the container into two compartments. In both (a) and (b) the concentration of water is the same in both compartments, and water molecules move at the same rate in both directions. In (c) the concentration of water is greater on the left side, and water molecules move from left to right at a greater rate.

Passage of Solvent through a Membrane

Water passes easily through cell membranes and through barriers such as the walls of blood vessels and the lining of the digestive tract. However, larger molecules and many ions pass through these same membranes with difficulty. Since living membranes are freely permeable to water but retain the majority of solutes, they are said to be *differentially permeable,* or semi-permeable. The movement of water through a differentially permeable membrane is called *osmosis.* Osmosis is a colligative property because it is affected by solute concentration, as shown in Figure 6.16. Osmosis occurs because of random motion of water molecules. With pure water on both sides of the membrane the rate of movement of water molecules from either side to the other is the same (Figure 6.16a). This is also true when the membrane separates two solutions of equal concentration (Figure 6.16b). However, an increase in the solute concentration on only one side of the membrane means that the concentration of water on that side must decrease, and we can expect that the rate of movement of water through the membrane from that side will also decrease (Figure 6.16c). Thus the net flow of water will be from the solution having the greater water concentration (i.e., the more dilute solution) to the solution with the lower water concentration.

The net movement of a substance from an area of higher to an area of lower concentration, whether or not it involves passage through a membrane, is called *diffusion.* Viewed in this light, osmosis is simply the diffusion of water through a differentially permeable membrane.

6.9 MEMBRANES AND THE DIFFUSION OF SOLUTES

The selective permeability of membranes was explained in a model proposed by J. F. Danielli in 1940, and supported by the experimental results of numerous other workers. Danielli suggested that a typical membrane consists of a double layer of fatty substance sandwiched between two layers of protein and pierced by many small pores (Fig-

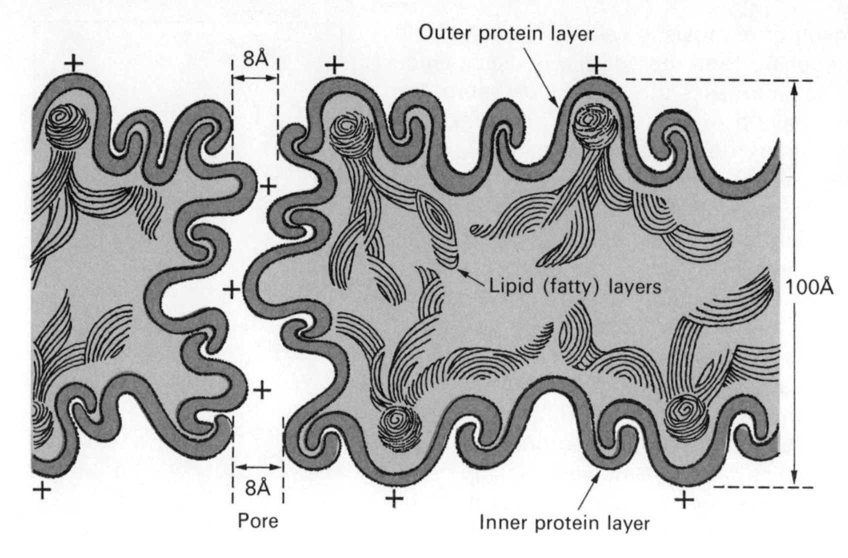

Figure 6.17
The Danielli model of the plasma membrane. It consists of a double layer of fatty substances sandwiched between two layers of protein. Polar protein molecules line the pores and their positively charged portions project into the passage.

ure 6.17). Calculations indicate that the pores have an average diameter of about 8 Å. It also appears that the polar protein molecules that form the lining of the pores are arranged so that positively charged segments project into the passage. The size of the pores means that large molecules will be retained, and the structure of the passage means that positively charged ions, especially those having a large effective charge diameter, will diffuse through the membrane more slowly than negative ions. This hypothesis is confirmed by studies of the diffusion rates of hydrated K^+, Na^+, and Cl^- ions into red blood cells.

The charge on a sodium ion can be felt at a distance of 2.56 Å, giving it an effective charge diameter of 5.12 Å. The effective charge diameter of the potassium ion is only slightly less—3.96 Å. However, diffusion of the potassium ion is 100 times as fast as that of sodium. In comparison, the effective charge diameter of the chloride ion is 3.86 Å, but the diffusion rate is almost as high as that of water. It should be remembered that these rates pertain to the plasma membrane of the red blood cell, and that different membranes may not behave in the same way.

Osmotic Pressure

A living cell can be regarded as a selectively permeable bag filled with a dilute solution of ions and variously sized molecules. Most cells are bathed in a liquid which is similar to their contents; therefore, water diffuses into and out of cells at the same rate. However, if a cell is removed from its normal environment and placed in a solution having a higher concentration of water, the water will enter the cell at a higher rate than it leaves, and the cell will become swollen and may eventually burst.

The rupture of a cell as a result of osmosis is called *plasmolysis.* In the case of red blood cells a specific term, *hemolysis,* is used. Since the diffusion of water into the cell increases the internal pressure, the contents of the cell are said to have a *higher osmotic pressure* than the external solution. If two solutions do not have the same osmotic pressure, the solution having the greater pressure is said to be *hyperosmotic,* or *hypertonic,* with respect to the other. The one with the lower osmotic pressure is *hypotonic.* The net diffusion of water always takes place from a hypotonic to a hypertonic solution. When the net diffusion of water between two solutions is zero, as between a red blood cell and the fluid of the blood, the solutions are said to be *isotonic.*

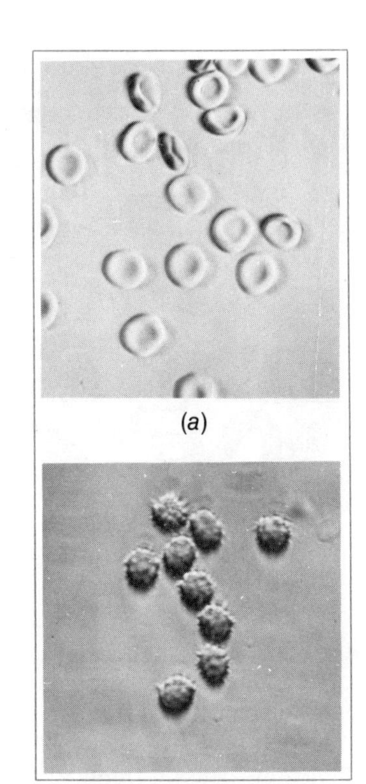

(a) Normal human red blood cells. (b) Red blood cells after exposure to a hypertonic solution. (Photographs by A. M. Winchester, University of Northern Colorado)

Red blood cells may safely be placed in physiological saline solution because the saline and the cell contents are isotonic. Recall that physiological saline contains sodium ions and that the membrane of a red blood cell is relatively impermeable to cations. Since the plasma membrane *is* readily permeable to chloride ions, the sodium ions must be responsible for the osmotic pressure of the solution. A 5.5% solution of glucose is also isotonic with red blood cells.

Not all membranes are as selective toward the passage of small molecules and electrolytes as the plasma membrane. The membranes which are the walls of blood vessels readily pass glucose and Na^+, Cl^-, K^+, and HCO_3' ions, while retaining larger molecules such as albumin and globulin. Thus the osmotic pressure of the blood—approximately 25 torr—is due primarily to its protein constituents.

The membranes that form the filtering units of the kidneys (glomeruli) are especially permeable to ions and small molecules like glucose and urea [$CO(NH_2)_2$]. As blood passes through the glomerulus, simple hydrostatic pressure forces water and solutes through the glomerular membrane. This filtrate contains glucose, amino acids, electrolytes, vitamins, and waste products. Then, as the filtrate flows through the tubules of the kidney, all of the solutes, except the waste products and some of the electrolytes, are actively reabsorbed; and most of the water follows by osmosis. The fluid remaining in the tubules is called urine.

If the kidney is damaged, albumin from the blood escapes through the glomerular membrane into the urine. The loss of protein decreases the osmotic pressure of the blood, and water moves from the blood into the tissues. The accumulation of water in the tissues is called *edema.* Conversely, if the osmotic pressure of the blood rises above that of the tissues, water leaves the tissues and enters the blood, increasing the volume of urine and causing dehydration.

Dialysis

The metabolic waste products normally removed from the blood by the kidney are extremely toxic. Failure to excrete these wastes leads, in about one week, to uremic coma, and their retention for longer periods is usually fatal. However, thousands of people with severely

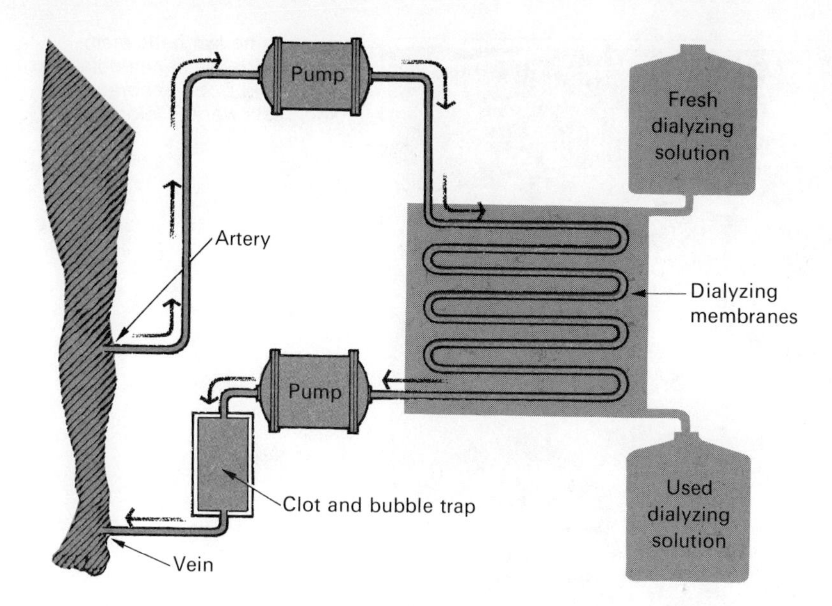

Figure 6.18
An artificial kidney. Metabolic wastes in the blood are removed by dialysis.

impaired kidney function continue to enjoy relatively good health because their blood is periodically cleansed by dialysis. *Dialysis* is the removal of low-molecular-weight solutes from those of higher molecular weight by passage through a synthetic membrane. Thus the membrane acts as a sieve: it holds back the large protein molecules, but is freely permeable to other solutes in the blood. Since the metabolic waste products are relatively small molecules, they are removed by dialysis. A schematic diagram of the dialyzing apparatus—or artificial kidney—is shown in Figure 6.18.

The patient's blood is pumped through a network of tiny tubes made of cellophane which are immersed in the dialyzing fluid. Low-molecular-weight solutes can move readily across the cellophane membrane in both directions. The dialyzing fluid contains the same concentration of glucose, amino acids, and electrolytes as the blood, but none of the metabolic wastes. Thus the wastes diffuse rapidly into the dialyzing fluid, but the other constituents of the blood are not removed because their net diffusion rate is zero.

Dialyzing membranes are also used to remove sodium chloride in the preparation of low-salt foods, and for desalting sea water so that it can be used for drinking. The latter process, called reverse osmosis, is discussed in Section 6.11.

A rather noteworthy example of the importance of osmotic pressure is the secondary effect which follows drowning in fresh water. When fresh water is inhaled into the lungs, the higher osmotic pressure of the blood causes water to enter the pulmonary capillaries and dilute the blood. The red blood cells are hemolyzed because of the hypotonic solution that surrounds them, and release K^+ ion into the plasma. The unusually high concentration of potassium ions causes erratic heartbeat and ventricular fibrillation. For this reason a victim of

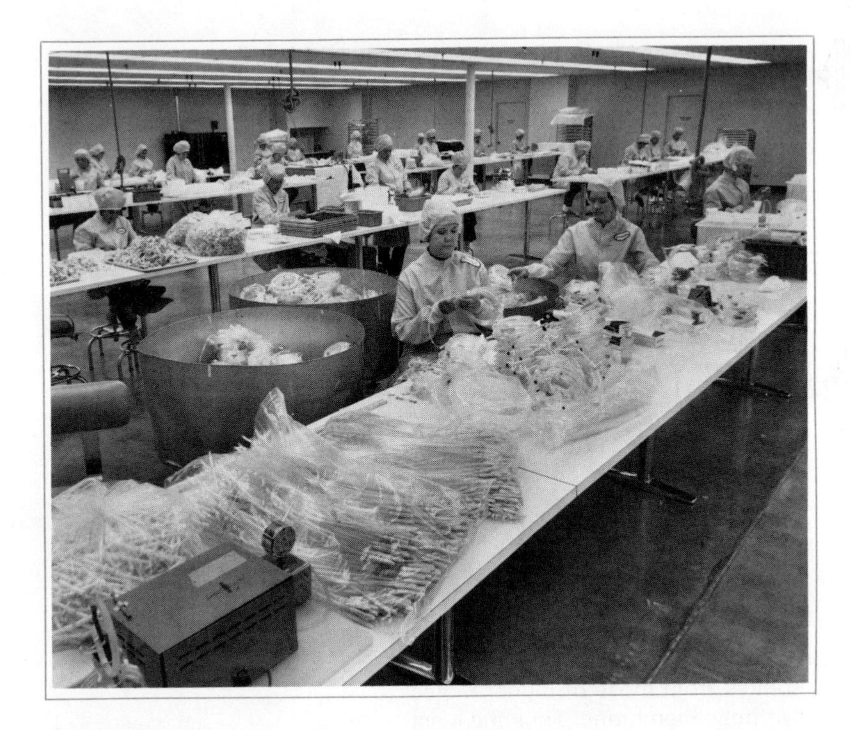

Packaging synthetic membranes for use in hemodialysis. (Courtesy Cobe Laboratories, Inc., Lakewood, Colorado)

fresh-water drowning may require closed-chest cardiac compression to keep the blood flowing after breathing is restored.

6.10 THE PHYSIOLOGICAL IMPORTANCE OF WATER

Water accounts for about 70% of the body weight of an adult human and slightly more for infants. Blood, lymph, saliva, sweat, and other body fluids are aqueous solutions. Thus water performs a number of important functions in man which can be broadly described under the headings of transport, lubrication, temperature regulation, and hydrolysis. The following paragraphs give representative examples of these functions.

Transport

The blood takes digested food to the tissues and takes cellular wastes from the tissues to the excretory organs. It also distributes internal secretions such as hormones and is the carrier for oxygen and carbon dioxide.

Lubrication

Painful friction between moving body parts is eliminated by secretions that coat their surfaces. Each of the major joints is enclosed in a sac of synovial fluid, and the eyeball swivels smoothly in its socket because it is bathed by a similar liquid. Food is more easily swallowed after being moistened with saliva, and the lungs slide painlessly against the walls of the thorax because of their slippery coating.

Temperature Regulation

The heat generated by muscle activity is dissipated by the evaporation of water from the skin and the lungs. The high specific heat of water also enables the body to be exposed to hot or cold environments without marked changes in body temperature.

Hydrolysis

As mentioned earlier, the digestive process is the chemical addition of water to foodstuffs, which splits the larger molecules into fragments that can be absorbed into the bloodstream.

6.11 WATER USE AND REUSE

It is bitterly ironic that man has designated a number of other creatures as endangered species when his peril is equal to theirs. In the process of creating a technological society, he has wasted his resources, poisoned his air, and fouled his water, and his numbers have increased beyond reason. Although a general discussion of these problems is beyond the scope of this book, a few words about the use and reuse of water are certainly appropriate.

The demand for water parallels the standard of living. Table 6.4 shows the amount of water used in the production of various kinds of consumer goods. In comparison, an individual uses only about 150 gallons per day for drinking, cooking, washing, disposal of wastes, and watering the lawn.

TABLE 6.4 Water Consumption in Manufacturing Processes

Product	Water
1 gallon of gasoline	500 gallons of water
1 yard of wool cloth	510 gallons of water
1 ton of paper	90,000 gallons of water
1 ton of steel	110,000 gallons of water

Strange as it may seem, water used in many industrial processes must meet higher standards than water for drinking. Electroplating, the manufacture and processing of photographic film, and the production of vaccines and antibiotics all require water of exceptional purity: a few parts per million of a specific solute may be unacceptable. For some of these applications, such as the production of measles vaccine, even water that has been distilled in glass is not pure enough, and an additional treatment is needed to remove the extremely small amount of glass that dissolves in hot water.

Such exacting standards do not apply to municipal water supplies, however, where the aim of purification is to make the water potable (that is, safe for drinking) and aesthetically pleasing. Water is said to be *potable* when it is free of toxic substances and disease-producing organisms. Making water aesthetically pleasing involves the elimination of objectionable odor, taste, color, and particulate matter. The nature and extent of the treatment depends upon the initial quality of the water. Water from deep artesian wells can often be used without treatment, but few communities are fortunate enough to enjoy such high-quality sources. More often the source is a lake or a stream, and the water contains one or more of the following:

1. Dissolved gases, such as NH_3 or H_2S, which have an unpleasant odor
2. Dissolved organic compounds, including detergents and pesticides
3. Dissolved minerals which have been leached from the soil
4. Insoluble minerals such as clay
5. Leaves, twigs, insect bodies, and other organic particulates
6. Living organisms—algae, molds, bacteria, and viruses.

Water that contains cyanide, lead, mercury, arsenic, or other toxic ions presents a difficult and expensive treatment problem which is not normally attempted.

Water Purification

The process that will be described pertains to water from a lake or pond, or flowing water that has been impounded in a reservoir.

Addition of Copper(II) Sulfate

Minute quantities of copper(II) sulfate are dusted on the surface of the water to prevent the growth of algae. Although algae are not poisonous, they make the water appear murky and give it an unpleasant taste.

Screening

As water is pumped into the treatment plant, it passes through a screen to remove paper, leaves, and large particulates.

Sedimentation

Following screening, lime (CaO) and alum $[KAl(SO_4)_2 \cdot 12\ H_2O]$ are added to promote the settling of suspended particles. The water is let into a large shallow basin where the two chemicals react to form aluminum hydroxide—a gelatinous precipitate which traps and carries the suspended particles to the bottom as it sinks.

Filtration

The water passes through a bed of gravel overlying several layers of fine sand. In modern installations a bed of activated charcoal is also used. The sand filter removes any remaining suspended matter and helps to reduce the amount of cellular contamination. The charcoal adsorbs colored solutes and dissolved gases.

Aeration

The water is sprayed into the air where the combination of oxygen and exposure to ultraviolet light acts to destroy living cells and to chemically alter many unpleasant-tasting organic solutes.

Chlorination

This is the final step in the treatment process. Chlorine gas is metered into the water to leave a residual of about 0.5 ppm. The chlorine destroys any remaining disease-producing organisms and causes further changes in any organic solutes that remain. The chlorine residue is volatile and is lost during the passage of the water through the distribution mains.

Hardness of Water

Water from lakes, streams, and wells usually contains minerals that have been leached from the soil. Water that contains Mg^{2+} and/or Ca^{2+} ions (usually in combination with either SO_4^{2-} or Cl^-) is said to be *hard,* and may cause problems when it is used for washing. Most of us probably live in areas where the water is hard, and have probably noticed the waxy scum that forms when soap is used in hard water. The scum is precipitate formed from the stearate ion of soap and either calcium or magnesium ion from the water. The scum represents wasted soap since cleansing action cannot take place until enough soap has been added to precipitate the metal ions. In addition, the scum may be deposited on fabrics, making them stiff, or on the skin, making it feel rough and dry.

What is the source of the hardness in water?
Carbon dioxide from the air dissolves in water and produces carbonic acid:

$$H_2O_{(l)} + CO_{2(g)} \rightarrow H_2CO_{3(aq)}$$

Carbonic acid reacts with minerals—limestone ($CaCO_3$) for example—to form soluble calcium compounds:

$$H_2CO_{3(aq)} + CaCO_{3(s)} \rightarrow Ca(HCO_3)_{2(aq)}$$

There are two common methods for combating hardness: one involves the addition of a soluble carbonate or borate to precipitate the metal ions, and in the other method (called *ion exchange*) the troublesome ions are replaced by ions of sodium which do not form a scum with soap. Either sodium carbonate (washing soda) or sodium borate (borax) may be used to precipitate the metal ions. Typical reactions with washing soda are

$$2\ Na^+_{(aq)} + CO_3^{2-}{}_{(aq)} + Ca^{2+}{}_{(aq)} + SO_4^{2-}{}_{(aq)} \longrightarrow CaCO_{3(s)} + SO_4^{2-}{}_{(aq)}$$

$$2\ Na^+_{(aq)} + CO_3^{2-}{}_{(aq)} + Mg^{2+}{}_{(aq)} + SO_4^{2-}{}_{(aq)} \longrightarrow MgCO_{3(s)} + 2\ Na^+_{(aq)} + SO_4^{2-}{}_{(aq)}$$

The ion exchange method employs a complex substance called a zeolite which has been formed into beads a few tenths of a millimeter in diameter and which carries adsorbed Na^+ ions. As the hard water trickles through a bed of zeolite beads, calcium and magnesium ions in the water replace the sodium ions. The reaction is

$$\text{Zeolite} \cdot Na^+ + Ca^{2+} \longrightarrow \text{Zeolite} \cdot Ca^{2+} + Na^+$$

A water pipe that has been partially blocked by accumulated hard water scale. (Courtesy Culligan International Co., Chicago, Illinois)

The zeolite can be recharged with sodium by washing it with a solution of NaCl.

Water that has been softened by the zeolite process should not be used for drinking or cooking because it contains such a high concentration of sodium ions. Such water may be harmful to persons who have either heart trouble or hypertension.

Water that contains HCO_3^- in addition to calcium and/or magnesium is said to be *temporarily hard* because the metal ions may be removed by heating. The reaction is

$$Ca^{2+}{}_{(aq)} + 2\ HCO_3^-{}_{(aq)} \longrightarrow CaCO_{3(s)} + H_2O_{(l)} + CO_{2\,(g)}$$

The formation of insoluble calcium carbonate is responsible for the scale that builds up in teakettles and boilers.

Water Reclamation

Despite our apparent mastery of the atom and our limited conquest of space, it is unlikely that our efforts to control the weather will bear fruit before the end of this century. Since many of our population centers, especially in the Southwest, are in semi-arid areas, we must reconcile ourselves to the prospect of extensive reuse of available water.

The major barrier to substantial water reclamation is not technological, but psychological. The average person recoils from the idea of reusing waste water because he does not realize that modern methods of waste-water management can restore the water in sewage to its sparkling pure condition. Many cities are already reclaiming their waste water and using it for irrigation of parks and golf courses, while others allow it to percolate down through the soil to replace the water drawn off through wells. Many other cities could begin to recycle their water simply by adding an additional step to their normal two-stage process.

A less controversial, but more expensive, way of increasing their water supply is possible for seashore communities. The salt can be removed from sea water either by flash distillation employing heat from a nuclear power plant, or by reverse osmosis. In the latter process, sea water is forced under pressure through a membrane that is permeable to water molecules but not to the dissolved salt. Water produced by either process is still many times as costly as reclaimed water and should not be viewed as a substitute for reprocessing, but as a supplement to it.

APPLICATION OF PRINCIPLES

1. Water inside a sealed container does not boil. Explain this observation.
2. Consult Table 6.2 and write a balanced equation for the reaction between calcium and water.
3. Write a balanced equation for the reaction between water and the oxide of sodium.
4. A negatively charged colloidal dispersion of sulfur can be prepared by combining hydrochloric acid and sodium thiosulfate. Which of the following ionic substances would be most effective in causing the dispersed sulfur to settle: (a) NaCl, (b) $MgCl_2$, or (c) Na_2SO_4? Justify your choice.
5. How would you dilute a 10% solution of hydrogen peroxide to a concentration of 3%?
6. Describe a method for preparing 0.5 liter of physiological saline solution (0.92% NaCl).
7. Within recent years a number of infants died in a hospital because their formula was prepared with table salt instead of sugar. Explain how the intake of large amounts of salt might lead to death.
8. Describe a method for preparing 1 liter of 1:10,000 epinephrin from a stock that has a concentration of 1:50.
9. What is the molarity of a solution prepared by dissolving 100 g of glucose ($C_6H_{12}O_6$) in sufficient water to make 400 ml of solution?
10. Vegetables that have become flaccid or wilted can be restored to a firm, crisp condition by soaking in water. Explain how this works.
11. Anaesthetic ether boils at a temperature only slightly higher than that of the human body. Using this information, suggest how the vapor-pressure curve for ether differs from that of water.

12. Bacterial cells dispersed in physiological saline solution exhibit the Tyndall effect, but eventually settle to the bottom of their container. Can dispersions of bacterial cells be classed as colloidal dispersions? Explain.
13. The maximum safe concentration of fluoride ion in drinking water is 2.0 ppm. Express this concentration in milligram %.
14. To what volume should 10 ml of 6 *M* HCl be diluted to give 0.5 *M* HCl?
15. Describe a simple experimental procedure for measuring the osmotic pressure of a sugar solution.

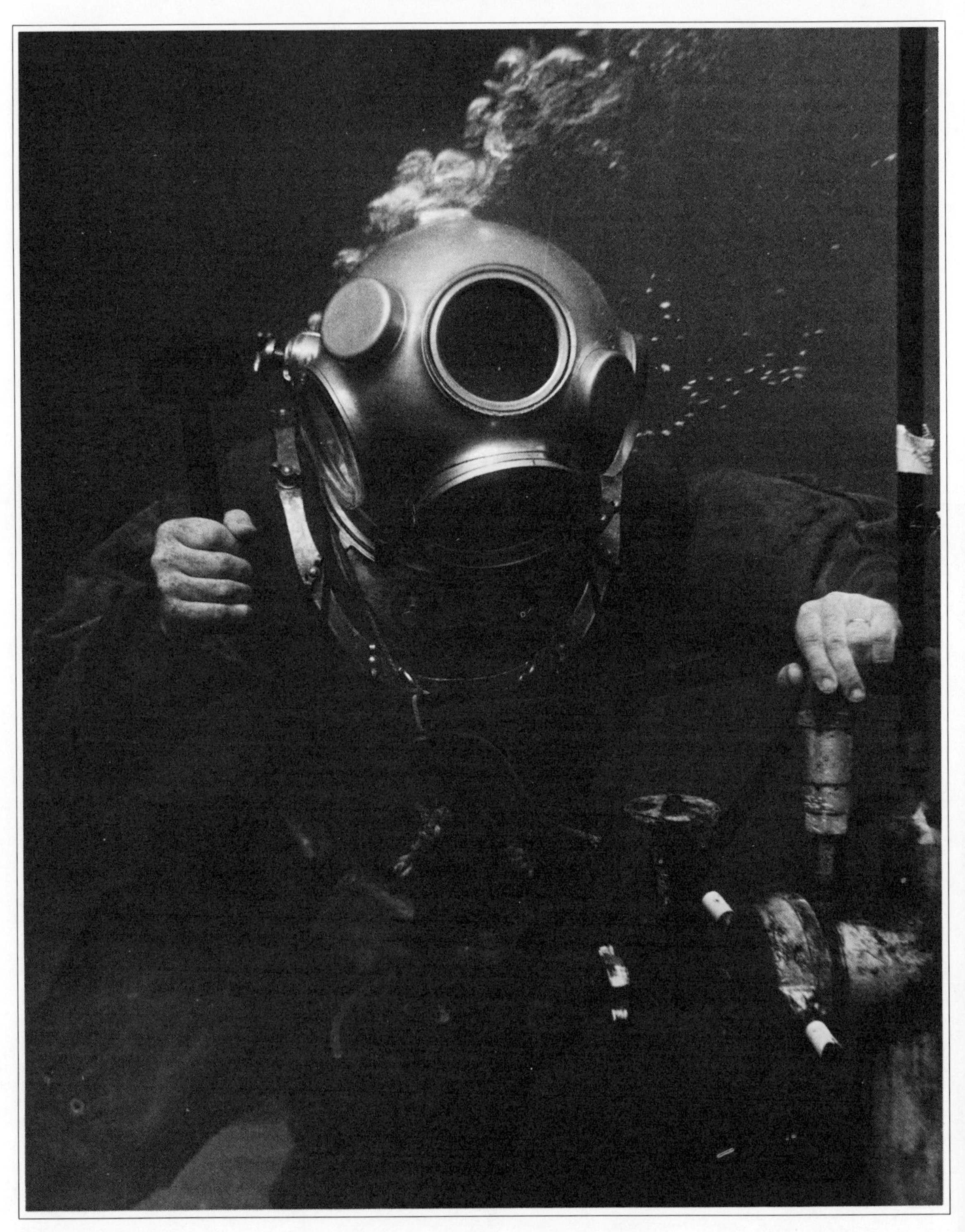

A practical application of the theory of gases. The diver is breathing a special mixture of gases designed to prevent decompression sickness (the bends). (Photograph by Rob Reilly, Santa Barbara City College)

7

GASES

LEARNING OBJECTIVES

1. Summarize the main points of the kinetic molecular theory.
2. Explain Charles', Boyle's, Henry's, and Dalton's Laws in terms of the kinetic molecular theory.
3. Use Boyle's and Charles' Laws to predict how the volume of a gas sample would change with changes in temperature and/or pressure.
4. In terms of partial pressures, explain the mechanics of breathing and the movement of respiratory gases between the lungs and the blood, and between the blood and the tissues.
5. Summarize the evidence supporting Avogadro's Hypothesis.
6. Use Dalton's Law to calculate the partial pressures of the individual gases in a mixture.
7. Describe the transport of oxygen and carbon dioxide in the blood and explain how these processes are affected by factors such as temperature, partial pressures, and the concentration of hydronium ions.
8. Summarize the main chemical and physical properties of oxygen and describe its role in medical therapy.
9. Distinguish between a real gas and an ideal gas.

KEY TERMS AND CONCEPTS

Avogadro's Hypothesis
Boyle's Law
Charles' Law
Dalton's Law
dependent variable
fire triangle
Henry's Law
hyperbaric chamber
ideal gas
independent variable
kinetic energy
kinetic molecular theory
molar gas volume
partial pressure
pneumothorax
STP
the bends

It is difficult to imagine a more intimate relationship than that which exists between man and his gaseous environment. We can exist for several weeks without food, and for a few days without water, but only for a few minutes without air. Thus our lives depend upon a material which we cannot feel, see, smell, taste, or hear. Considering that most of these characteristics apply to all gases, it is rather remarkable that we have been able to learn so much about their nature and behavior, for the model of the gaseous state is one of the most successful of all scientific theories.

7.1 THE GASEOUS STATE: A MODIFIED MODEL

According to the model that was proposed in the first chapter, a substance is said to be gaseous when it has neither a definite shape nor a definite volume, but adjusts to the size and shape of its container. This behavior was explained by assuming that gases are composed of rapidly moving, minute particles which are widely separated from each other. Thus if the size of the container is increased, the particles move farther apart; and if the gas is compressed into a smaller container, the particles are crowded closer together. The model also suggests that if we apply enough pressure, the particles will be crowded so closely together that the gas will liquefy—which is indeed the situation for most gases. The motion of the particles also provides a mechanism for the diffusion of gases, which was described in the previous chapter.

However, since there are other aspects of the behavior of gases that are not so readily explained by our model, it may be that the model is incomplete. For example, spray cans of the type used for paint, insecticide, hair conditioner, and other products contain a gas that acts as a propellant. These containers will usually burst if they are stored at temperatures in excess of 130°F (54°C). If our model is to account for this behavior, it must be modified, since in its present form it says nothing about temperature effects of gases.

The model proposes that the particles of a gas are in constant motion. If this is true, then collisions are inevitable: some particles will run into each other and some will strike the walls of the container. The continuous impact of particles with the container walls is responsible for the pressure exerted by the gas. Recall from Chapter 1 that the air exerts a pressure of 14.7 pounds per square inch at sea level.

If energy is lost by the particles as a result of these collisions, they will eventually slow down to the point where van der Waals forces of attraction become effective and the substance ceases to be a gas. Since gases have not been observed to revert spontaneously to the liquid state, we must conclude that whatever energy is lost by a particle as a result of a collision is somehow transferred to other particles, so that the net energy of the system remains the same. Thus as long as the energy of the system remains constant, the pressure in the aerosol can will also be constant. Putting the can in direct sunlight or in a warm place causes it (and the particles of gas in it) to absorb energy. The energy of a moving particle (called *kinetic energy*) is directly related to its mass and velocity. The relationship is

$$\text{kinetic energy} = \frac{(\text{mass})(\text{velocity})^2}{2}$$

Since the absorption of energy cannot affect the mass of the particle, it must be that the velocity of the particle is increased, causing it to strike the container more often and resulting in a higher pressure. At high temperatures the pressure may be great enough to rupture the container.

7.2 THE DISTRIBUTION OF MOLECULAR ENERGIES

Measurements have shown that there is a wide variation in velocity among the individual particles in a sample of gas. This variation may be represented by the distribution curve shown in Figure 7.1. The curve shows that at a given temperature most of the particles fall within a fairly narrow velocity range, but that the sample also contains

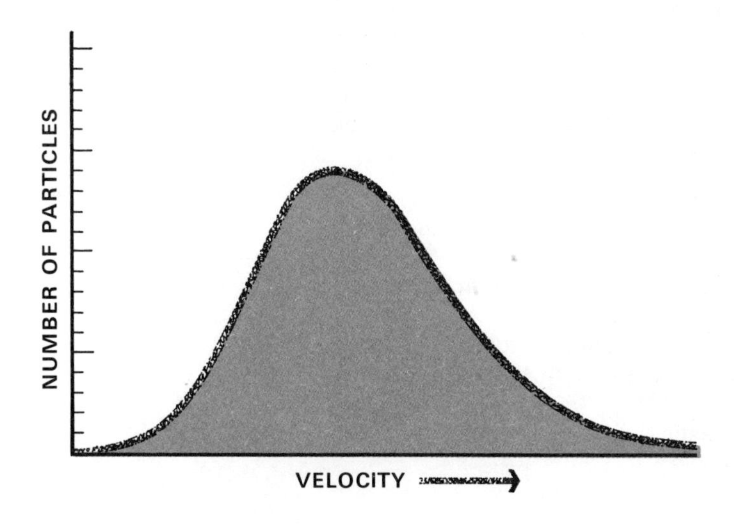

Figure 7.1
Molecules in a sample of gas vary widely in kinetic energy or velocity.

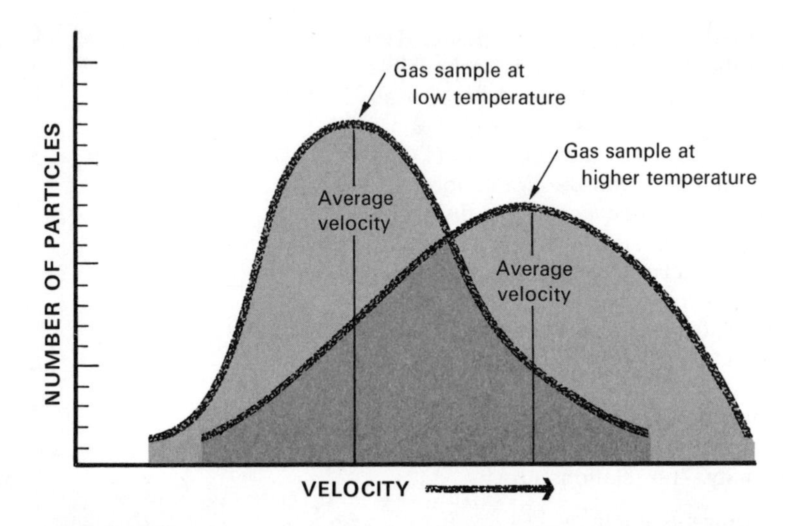

Figure 7.2
Distribution of velocities for gaseous particles at two different temperatures. The velocity range is roughly the same, but the particles at the higher temperature have a greater average velocity.

particles having extremely high or extremely low velocities. The effect of an increase in temperature is to shift the curve by increasing the average velocity, but it does not appreciably alter the velocity range of the particles (Figure 7.2). Numerous experiments have confirmed that an increase in absolute temperature causes a proportionate increase in the average velocity of the molecules in a sample of gas. Likewise, a decrease in temperature results in a decrease in the average velocity. It is interesting to note that if we follow the temperature-velocity relationship to its logical conclusion, it leads us to absolute zero—a temperature at which molecules have a kinetic energy of zero and molecular motion ceases.

7.3 THE KINETIC MOLECULAR THEORY

By adding to the initial assumptions of our model we have not only generated a plausible explanation for one of the properties of gases, but we have also produced a rather complete model of the gaseous state that is commonly referred to as the *kinetic molecular theory.* Although the precise wording of the theory varies from one textbook to another, the essential points are these:

1. A gas consists of tiny particles (atoms or molecules).
2. The volume of the particles themselves is negligible, and the space between them is very great in comparison to their size.
3. Because of the relatively large distances between particles, the attractions between them are extremely weak.
4. Particles of a gas are in constant, rapid, random motion.
5. The particles collide with each other and with the walls of their container.
6. These collisions are perfectly elastic, i.e., they do not reduce the net kinetic energy of the gas sample.

7. The kinetic energy is directly proportional to the absolute temperature of the gas.

7.4 TEMPERATURE-VOLUME RELATIONSHIPS (CHARLES' LAW)

The kinetic molecular theory is a very useful tool because it enables us to predict how the behavior of a gas will be affected by changes in temperature, pressure, container size, and the type and concentration of the particles themselves. Repeated observations have established that quantitative relationships exist among these variables—relationships so regular and unvarying that they are given the status of laws. The relationship between temperature and volume will be examined first.

Let us consider a container with rigid walls, such as an aerosol can. Let us further suppose that the gas within the can exerts a pressure of 1 atmosphere on the walls (Figure 7.3a). According to the kinetic molecular theory, an increase in the temperature of the gas by a fixed amount will cause a proportional increase in the velocity of the gas particles, and the pressure will become greater than 1 atmosphere (Figure 7.3b).

Now consider a container with movable walls such as a cylinder fitted with a piston. Assume that a gas within the cylinder exerts an initial pressure of 1 atmosphere and that the pressure of the air on the outside of the cylinder is also 1 atmosphere. The piston will remain stationary because the opposing pressures are the same (Figure 7.4a). If the temperature of the gas in the cylinder is raised by a fixed amount, the pressure on the inner walls will increase proportionately,

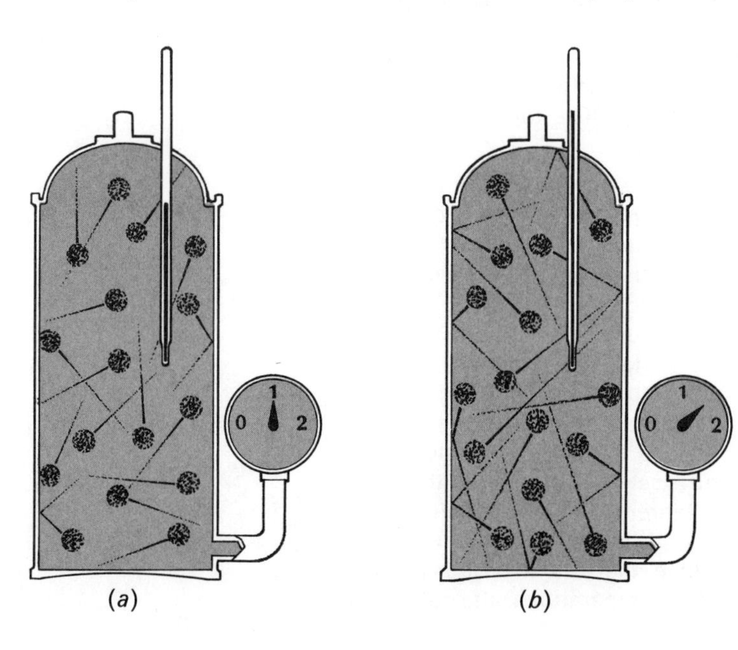

Figure 7.3
The effect of a temperature increase on the pressure of a confined gas. (a) The gas exerts a pressure of 1 atmosphere against the walls of its container. (b) Raising the temperature of the gas increases the velocity of its particles. They strike the wall more often, causing an increase in pressure.

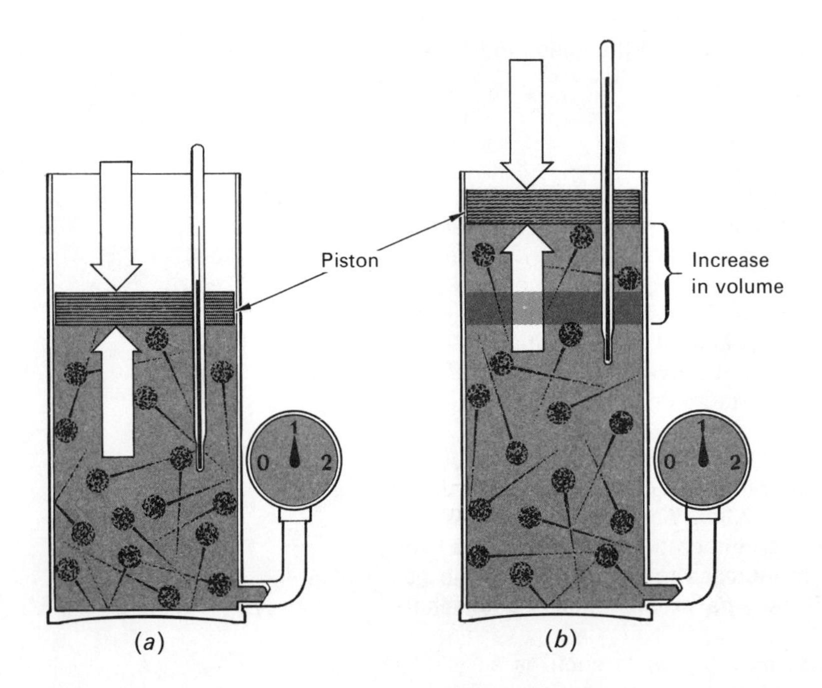

Figure 7.4
The effect of a temperature increase on the volume of a gas in a container fitted with a movable piston. (a) The volume remains constant as long as the pressure on both sides of the piston is the same. (b) Raising the temperature of the gas increases its pressure. The piston rises until the pressures are once again equal. The result is an increase in volume.

and the piston will be pushed outward. As the space within the cylinder increases, the particles will have to travel greater distances to strike the walls. The number of collisions will decrease and so will the pressure. The piston will stop moving when the pressures on it are once again equal (Figure 7.4b). The increase in the volume of the gas will be directly proportional to the increase in its absolute temperature. This relationship is the basis of *Charles' Law,* which states that the volume of a confined gas is directly proportional to its absolute temperature if the pressure remains constant. The following sample problems may help to clarify the relationship.

Sample problem 7.1 The temperature of the outside air is 0°C. The air is inhaled and its temperature rises to 37°C as it moves through the respiratory passages. If the volume of outside air measured 500 ml, what would be its volume after warming?

The temperature of the air is being raised from 0°C to 37°C, or from 273°K to 310°K. According to Charles' Law, the volume of the air should increase by a proportionate amount. Two alternative methods will be given for the solution of this problem. You may find one more attractive than the other, or you may prefer a different method altogether.

Solution The Proportion Method. Since there is a proportionality between the volumes and the temperatures, the following relationship is true:

$$\frac{\text{Initial temperature}}{\text{Final temperature}} = \frac{\text{Initial volume}}{\text{Final volume}}$$

Next we simply insert the appropriate numbers and solve the equation.

$$\frac{273°\text{K}}{310°\text{K}} = \frac{500 \text{ ml}}{\text{Final volume}}$$

$$(273°\text{K})(\text{Final volume}) = (310°\text{K})(500 \text{ ml})$$

$$\text{Final volume} = \frac{(310°\text{K})(500 \text{ ml})}{(273°\text{K})}$$

$$\text{Final volume} = 568 \text{ ml}$$

Direction-of-Change Method. From the statement of the problem we have learned that the temperature is going to increase, which means that the volume of the gas is also going to increase. In other words, we are going to have to multiply the initial volume by some number having a value greater than 1. Using the two temperatures, we can create two fractions: 273°K/310°K and 310°K/273°K. The second fraction has a value greater than 1 and so we multiply the initial volume by the second fraction:

$$\text{Final volume} = \frac{310°\text{K}}{273°\text{K}} (500 \text{ ml})$$

$$\text{Final volume} = 568 \text{ ml}$$

Sample problem 7.2 A sample of gas measured 100 ml when collected at 20°C. To what temperature must we raise the gas in order to change its volume to 150 ml?

Solution We wish to change the volume of the gas from 100 ml to 150 ml, which is an increase in volume. According to Charles' Law, this is brought about by raising the temperature. Using the proportion method, we get

$$\frac{100 \text{ ml}}{150 \text{ ml}} = \frac{293°\text{K}}{\text{Final temperature}}$$

$$\text{Final temperature} = \frac{(150 \text{ ml})(293°\text{K})}{(100 \text{ ml})}$$

$$\text{Final temperature} = 440°\text{K or } 167°\text{C}$$

The direction-of-change method applies to this problem as well.

It was mentioned earlier that the kinetic molecular theory describes an idealized gas and that real gases deviate slightly from the ideal. The behavior of gases at extremely low temperatures is an example of one kind of deviation. According to the theory, the particles of a gas are assumed to have little or no attraction for each other. This is true for particles having high kinetic energies; however, at extremely low temperatures the particles move so slowly that the van der Waals forces of attraction become significant. The increased attraction between particles reduces the number of collisions with the walls of the container, which has the effect of reducing the volume below that predicted by Charles' Law.

7.5 A FEW COMMENTS ON EXPERIMENTAL DESIGN

You probably have spent 12 or more years in close association with books. You know from your experience that it is possible to learn something of chemistry by reading this book because it channels your thinking about the subject and contains the answers to a number of questions. You also know, of course, that you learned many things before you were old enough to read books, simply by observing and asking questions. In this respect you were in much the same situation as that which faces a scientist who is engaged in research. Books were useless to you because you could not read, and they are of limited value to a scientist because they do not contain the answers he seeks. He learns by observing, just as you did, and by asking questions. But you had someone to direct your questions to, whereas the scientist has only his mute apparatus; and it is from this apparatus that many of his answers must come. When a scientist designs an experiment, he is putting his question into such a form that he can deduce the answer to his question from the result, as shown by the following example.

Suppose we have noticed that the volume of a gas changes as it is heated, and we would like to know what kind of relationship exists between volume and temperature. The simplest way to carry out the investigation is to observe the volume that the gas occupies at several different temperatures. The temperature and the volume are called variables because they are expected to change during the course of the experiment. Since the temperature is purposely changed, it is called the *independent variable,* and the volume, which we believe will respond to the temperature change in some regular fashion, is termed the *dependent variable.*

If we are to be certain that any changes in volume that we observe are actually due to temperature variations, we must take care that all other experimental conditions remain unchanged throughout the experiment. When we record the temperature and volume, for example, we must be sure that the pressure is the same as it was at the start of the experiment, and that none of the gas sample has leaked out. Unless there is a way of checking these factors, our experimental design is faulty and our results cannot be trusted. There should be only one independent and one dependent variable for each experiment of this type.

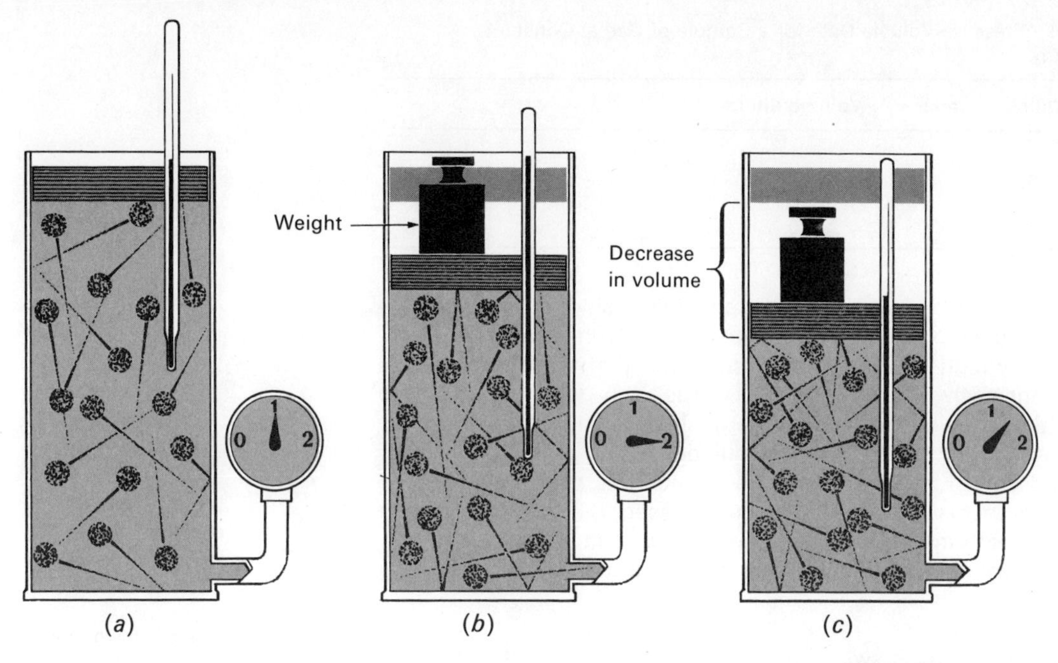

Figure 7.5
The effect of increased pressure on the volume of a gas. (a) The initial pressure is 1 atmosphere. (b) The addition of a weight increases the external pressure and the internal temperature. (c) After the temperature has returned to its initial value, the gas is found to occupy a smaller volume.

7.6 PRESSURE-VOLUME RELATIONSHIPS (BOYLE'S LAW)

The effect of increased pressure on the volume of a gas sample is apparent to anyone who has ever operated a bicycle pump or blown up a balloon. Depending upon the pressure, a large volume of gas can be squeezed into a rather small container. We can investigate the pressure-volume relationship by using the Charles' Law apparatus.

A gas sample under a pressure of 1 atmosphere is placed in the cylinder and the volume and temperature are recorded (Figure 7.5a). Weights are placed on the piston to increase the external pressure. As the piston moves downward, the particles are driven closer together. The total energy of the gas sample is now concentrated within a smaller volume, causing a rise in temperature (Figure 7.5b). The excess heat must be removed, otherwise the temperature will become a third variable. The piston stops moving when the external pressure and the internal pressure are equal. The external pressure is calculated from the weights, and the new volume of the gas sample is recorded (Figure 7.5c). Table 7.1 shows the results of a typical experiment.

These results show clearly that the volume is reduced by half each time the pressure is doubled. The relationship is expressed in *Boyle's Law,* which states that, if the temperature remains constant, the volume of a confined gas varies inversely with the pressure on it.

You will recall that the temperature-volume relationship is not uniform for gases at extremely low temperatures. In other words, gases at

TABLE 7.1 Pressure-Volume Data for a Sample of Gas at Constant Temperature

Pressure (atmospheres)	Volume (liters)
1	1.00
2	0.49
3	0.32
4	0.23

low temperatures do not behave in an ideal fashion. The pressure-volume relationship also deviates from the ideal when the gas is subjected to very high pressure. The assumption that the volume of the particles themselves is negligible is true only when the spaces between the particles are relatively large. If the particles already have been crowded tightly together, then subjecting the gas to still greater pressure does not reduce its volume as much as predicted because the particles themselves cannot be compressed. Since we will not be dealing with extremes of pressure, however, we can apply Boyle's Law to pressure-volume relationships as demonstrated by the following problem.

Sample problem 7.3 A swimmer at a depth of 30 feet exhales an air bubble that has a volume of 100 ml. The pressure at this depth is 1440 torr. What will be the volume of the bubble when it reaches the surface?

Solution The pressure at the surface will be one atmosphere, or 760 torr. According to Boyle's Law, the volume of the bubble will increase because the external pressure is being reduced. This inverse relationship is shown by the following proportion:

$$\frac{\text{Initial volume}}{\text{Final volume}} = \frac{\text{Final pressure}}{\text{Initial pressure}}$$

Substituting, we get

$$\frac{100 \text{ ml}}{\text{Final volume}} = \frac{760 \text{ torr}}{1440 \text{ torr}}$$

$$\text{Final volume} = \frac{(100 \text{ ml})(1440 \text{ torr})}{(760 \text{ torr})}$$

$$\text{Final volume} = 190 \text{ ml}$$

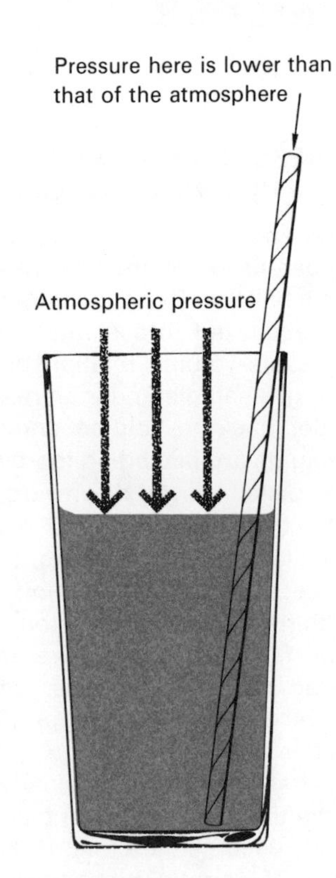
Pressure here is lower than that of the atmosphere
Atmospheric pressure

Figure 7.6
A drinking straw. Liquid rises in the straw because of pressure differences.

7.7 BREATHING—AND OTHER APPLICATIONS OF BOYLE'S LAW

Many familiar processes depend upon differences in pressure. Gas flows from the jets in the laboratory when the valves are opened because the pressure exerted on the gas at the pumping station is greater than that of the air in the room. Liquid moves upward in a straw when we suck on it, not because we lift it by sucking, but because it is forced upward by the difference in pressure (Figure 7.6). A difference in pressure also forces fresh air into our lungs and exhausts the depleted air.

The lungs resemble two spongy bellows that are in contact with the lining of the thoracic cavity. The thoracic cavity is separated from the abdominal cavity by a muscular membrane called the diaphragm. In its relaxed position the diaphragm curves slightly upward in the center, and the ribs incline slightly downward from the spine, as shown in Figure 7.7a. Contraction of the diaphragm muscle causes it to become flat rather than arched, and increases the volume of the thoracic cavity. The same effect is produced by upward movement of the rib cage (Figure 7.7b). According to Boyle's Law, increasing the volume of the chest cavity lowers the pressure in the space between the lining and the lungs. The result is that the slightly higher pressure of the atmosphere forces air into the lungs, and they inflate. Relaxing the diaphragm and lowering the rib cage increases the pressure in the cavity, causing the air in the lungs to be expelled.

A number of years ago when a person was unable to breathe for himself, possibly because of paralysis of the diaphragm, he was usually placed in an iron lung (Figure 7.8). The patient's body, from the neck down, was enclosed within an airtight chamber. The pressure within the chamber was alternately increased and decreased, causing compression and expansion of the chest cavity. As the chest was compressed, the lungs emptied; and as the pressure was released, the lungs inflated. Iron lungs have been largely replaced by less bulky devices that enclose only the upper torso.

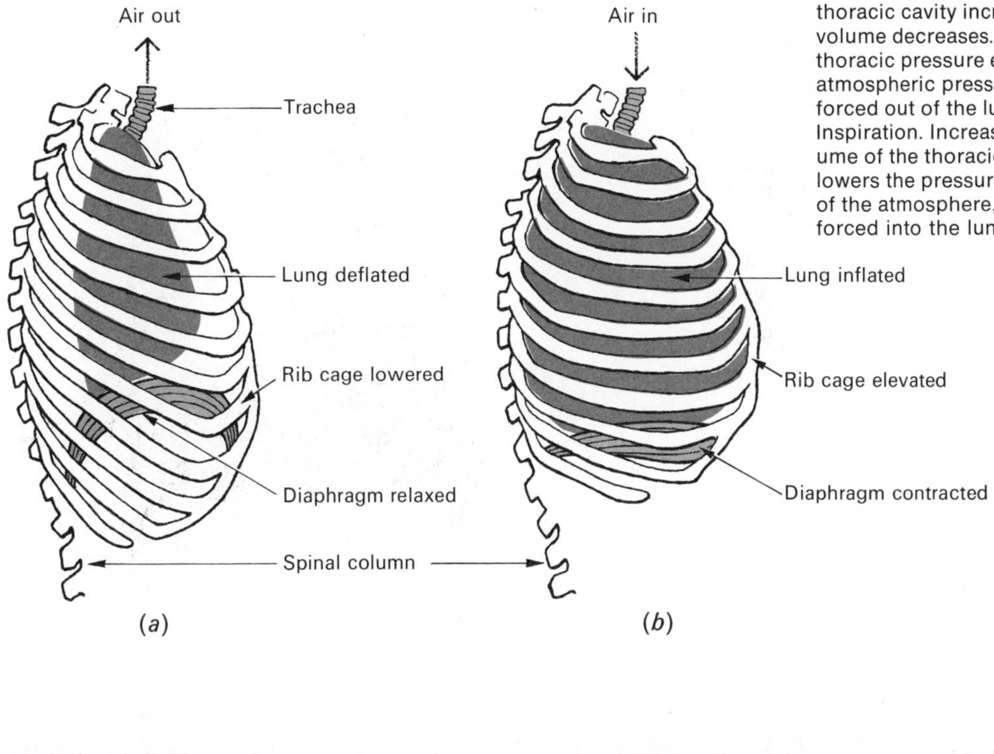

Figure 7.7
The role of pressure differences in breathing. (a) Expiration. The pressure in the thoracic cavity increases as its volume decreases. When the thoracic pressure exceeds atmospheric pressure, air is forced out of the lungs. (b) Inspiration. Increasing the volume of the thoracic cavity lowers the pressure below that of the atmosphere, and air is forced into the lungs.

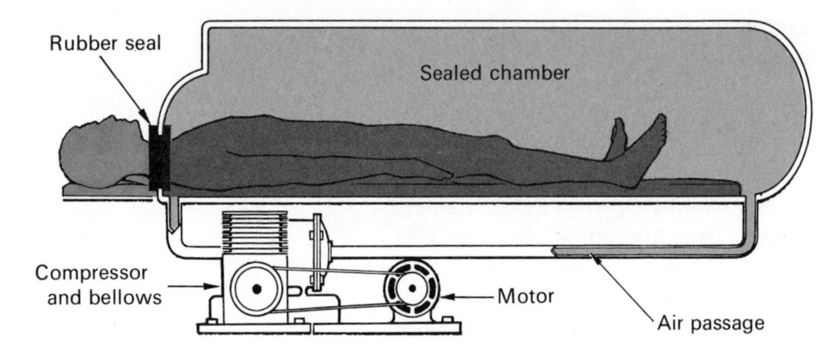

Figure 7.8
An iron lung. The pressure in the chamber is alternately increased and decreased, causing compression and expansion of the chest cavity. Air enters the lungs during expansion and leaves as the chest is compressed.

The same general principle is used in administering artificial respiration by the prone-pressure methods. It should be pointed out that techniques involving compression of the chest cavity are not as effective as mouth-to-mouth resuscitation, and may even be harmful if there has been an injury to the chest or internal organs.

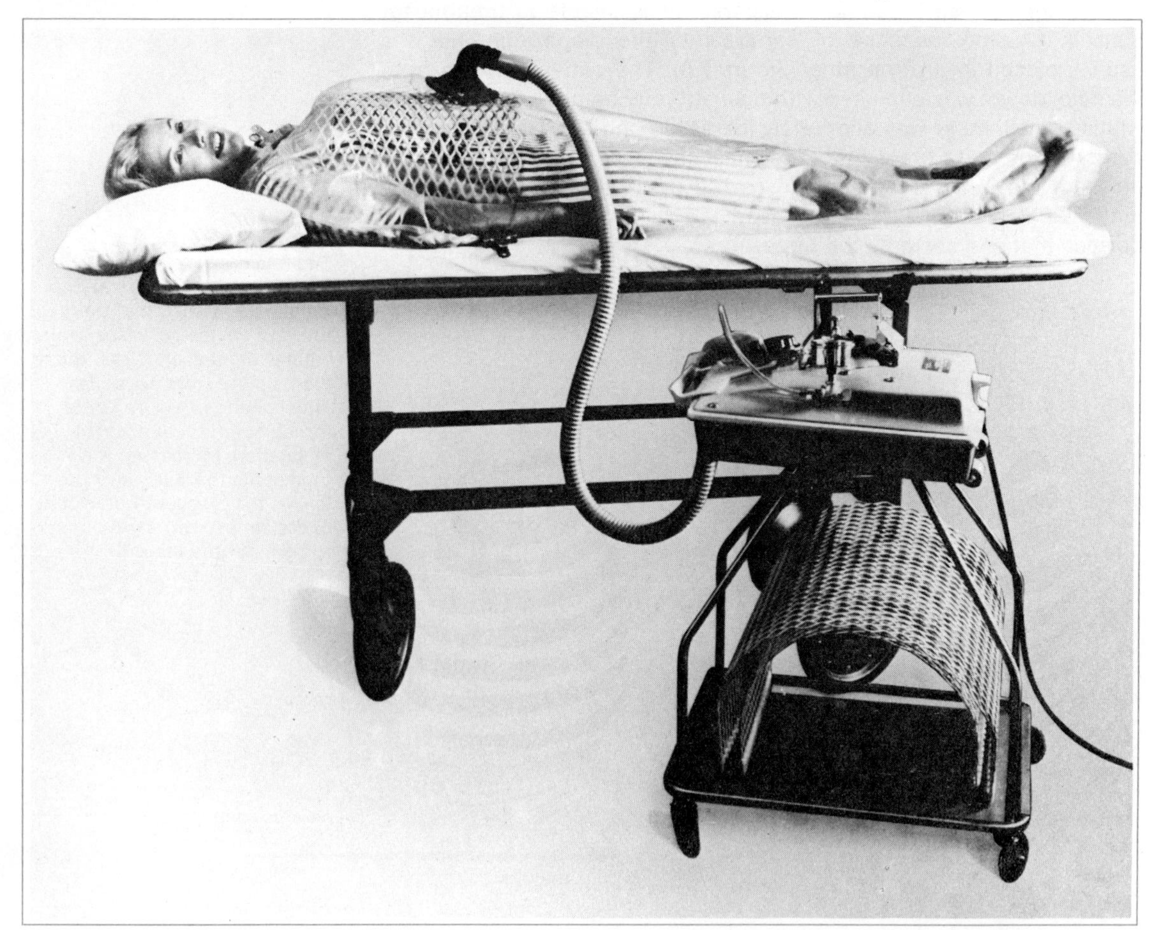

A chest respirator. This device aids a patient's breathing by producing a cyclic variation in pressure on the chest cavity. (Courtesy J. H. Emerson Company)

It is easy to see that an injury in which either the diaphragm or the chest wall is punctured may be extremely serious because the pressure within the cavity becomes the same as the atmospheric pressure, and the lungs may collapse. Such a condition is called *pneumothorax*. Sometimes the cavity is intentionally opened and a lung is collapsed in order to give it a rest. The function can be restored by simply closing the cavity.

7.8 STANDARD TEMPERATURE AND PRESSURE

Gas volumes are so greatly affected by pressure and temperature that a comparison of two samples is impossible unless the conditions are identical. A temperature of 0°C (273°K) and a pressure of 1 atmosphere (760 torr) have been designated as standard conditions of temperature and pressure (STP) for gases. The volume that a gas would occupy at STP can be calculated by using Boyle's and Charles' Laws. It is also common practice to state the densities of gases in grams per liter at standard temperature and pressure.

There is a rather intriguing relationship between the density of a gas at STP and the molecular weight of the gas. One mole of any gas contains Avogadro's number of atoms or molecules. The weight of a mole of gas is equal to its molecular weight in grams. Dividing the molecular weight in grams by the density of a gas at STP gives the volume that one mole of the gas will occupy under these conditions. For example, the molecular weights of methane and nitrogen are 16 and 28, respectively. The density of methane at STP is 0.717 g/liter and that of nitrogen is 1.25 g/liter. Dividing the molecular weight by the density gives

$$\text{Nitrogen} \quad \frac{28\ \text{g}}{1.25\ \text{g/liter}} = 22.4\ \text{liters}$$

$$\text{Methane} \quad \frac{16\ \text{g}}{0.717\ \text{g/liter}} = 22.3\ \text{liters}$$

For most gases the volume occupied by one mole at STP is very close to 22.4 liters. This is known as the *molar gas volume.* It is from calculations of this sort that we can check on the validity of *Avogadro's Hypothesis,* which states that equal volumes of gases at the same temperature and pressure contain equal numbers of molecules.

The relationship between density and molecular weight provides us with a convenient way to approximate the density of a gas: we simply divide the molecular weight by 22.4. The density of hydrogen calculated in this manner is 0.090 g/liter at STP, which shows excellent agreement with the observed density of 0.0899 g/liter.

7.9 DISSOLVED GASES (HENRY'S LAW)

You have probably noticed that a bottle of soda or beer makes a hissing sound as it is opened, and that for some time afterward tiny bubbles form in the liquid and rise to the surface. The gas in carbonated beverages is carbon dioxide, and its solubility, like that of all gases, depends upon the pressure. Numerous measurements have provided us with an understanding of the pressure-solubility relationship which is summarized in *Henry's Law*. According to Henry's Law, the solubility of a gas in a liquid at a given temperature is directly related to the pressure of the gas on the liquid surface.

The increased solubility of a gas under pressure is responsible for a condition known as caisson disease—also called the *bends,* or decompression sickness. Although nitrogen is taken into the lungs when we breathe, very little of it finds its way into the blood because it is not very soluble at body temperature. However, when a person, such as a diver, breathes air under pressure, the amount of dissolved nitrogen becomes appreciable. If the diver is brought to the surface too quickly, the pressure of the dissolved gas will be greater than atmospheric pressure. Nitrogen comes out of solution and tiny bubbles of it may form in the tissues. The bubbles may disrupt nerve pathways in the brain or spinal cord, and bubbles in the peripheral nerves can cause severe pain. The only treatment is to reapply the pressure and then very slowly remove it. In this way the escape of the gas is so gradual that bubbles do not form. Divers who must work at great pressures or remain below for extended periods may breathe a mixture of oxygen and helium instead of air. Helium may be used as a substitute for nitrogen because both nitrogen and helium are physiologically inert. Since helium is only about 40% as soluble as nitrogen at the same pressure, the problem of bubble formation is eliminated.

Using pressure to increase the solubility of a gas is an effective technique in the treatment of infections caused by anaerobic bacteria. Anaerobes are organisms whose growth is inhibited by molecular oxygen. They include the bacteria responsible for tetanus (*Clostridium tetani*) and gangrene (*Clostridium welchii*). Antibiotics are not very effective in the treatment of anaerobic infections. Instead, the patient is placed in a *hyperbaric chamber*—that is, a chamber containing oxygen at a pressure of 3-4 atmospheres. The tissues become saturated with oxygen, and the bacteria cannot grow. Hyperbaric chambers are also used for treatment of carbon monoxide poisoning and for acute myocardial infarctions.

7.10 MIXTURES OF GASES (DALTON'S LAW)

A basic premise of our model is that pressure is exerted by a gas as a result of collisions between its particles and the walls of its container. Raising the temperature of a gas increases its pressure by increasing the average kinetic energy of its particles, and thus increasing the frequency with which they strike the walls. It is logical to assume that the pressure can also be raised by increasing the concentration of particles; that is, by adding more particles to the ones that are already there.

Consider, for example, two 100-ml samples of oxygen at the same temperature. Gauges attached to the containers show a pressure of 2

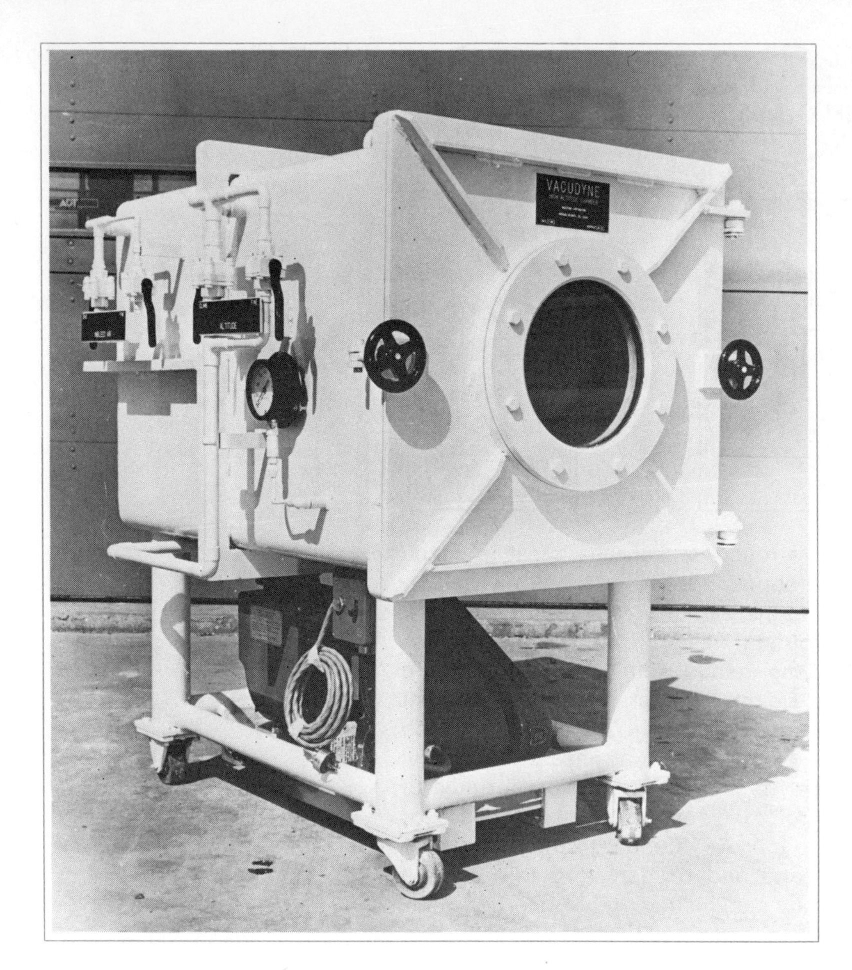

A hyperbaric chamber designed for small experimental animals. (Courtesy Vacudyne Altair, Chicago, Illinois)

atmospheres in one and 4 atmospheres in the other. The difference in pressure cannot be related to kinetic energy differences since both samples are at the same temperature. It must be that the sample having twice as much pressure also contains twice as many oxygen molecules. In other words, if a given number of molecules in a confined space produces a certain pressure, then twice that number of molecules in the same space should produce twice that pressure. According to this hypothesis, if all of the molecules are placed in one container, then their total pressure will be exerted against the walls of that container. This is indeed the case, and the pressure of the combined samples is 6 atmospheres.

It should be noted that the pressure exerted by a gas is independent of the size, shape, or chemical nature of the particles, and depends only upon their number, their average kinetic energy, and the volume of the container. Thus, had the containers held two different nonreactive gases, mixing them would still have produced a pressure of 6 atmospheres.

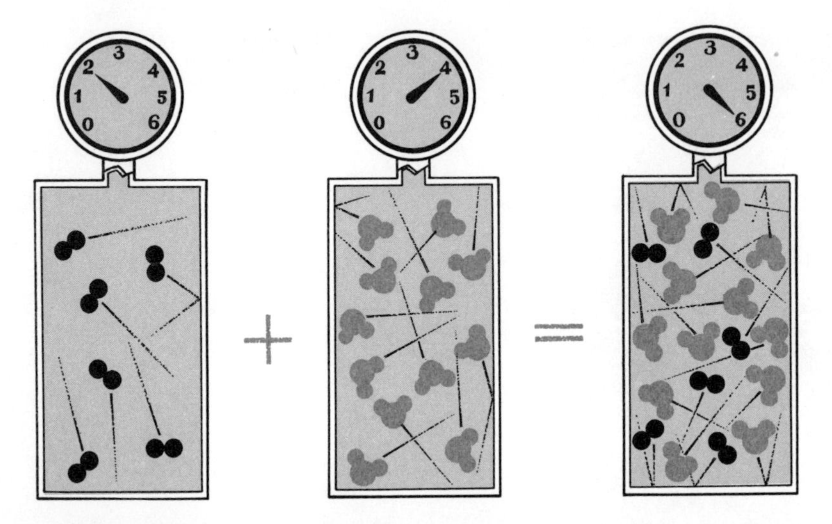

Figure 7.9
An illustration of Dalton's Law. The pressure of a mixture of gases is equal to the sum of their separate pressures.

A gas exerts the same pressure regardless of whether it is mixed with other gases or alone in the container. The pressure exerted by each gas in a mixture is known as its *partial pressure,* and the total pressure of the mixture can be found by adding the partial pressures of all the gases (Figure 7.9). This relationship is summarized in *Dalton's Law of Partial Pressures.* Dalton's Law states that, at constant temperature and volume, the pressure of a mixture of gases is equal to the sum of the partial pressures of the individual gases. The partial pressure of a gas is represented by a capital "P" with the formula of the gas as a subscript. In a mixture of oxygen and carbon dioxide, for example, the total pressure of the mixture is equal to the sum of the partial pressures of carbon dioxide and oxygen, and this can be expressed in the following symbolic statement:

$$P_{total} = P_{CO_2} + P_{O_2}$$

Nonpolar gases are only slightly soluble in water and are usually collected by water displacement (Figure 7.10). Water in the liquid state exerts an appreciable vapor pressure, and gases in contact with water

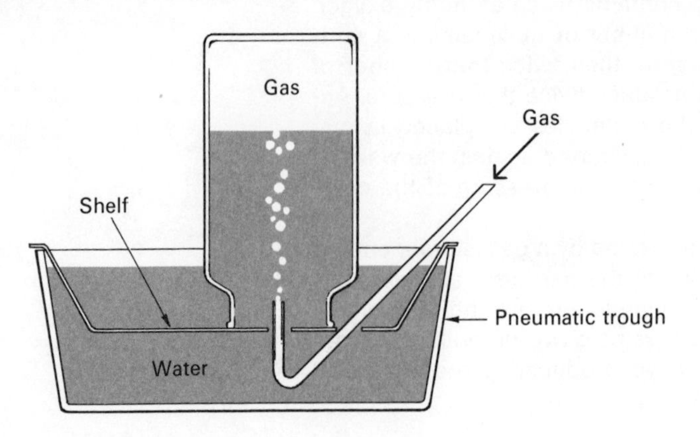

Figure 7.10
Collection of a gas by displacement of water.

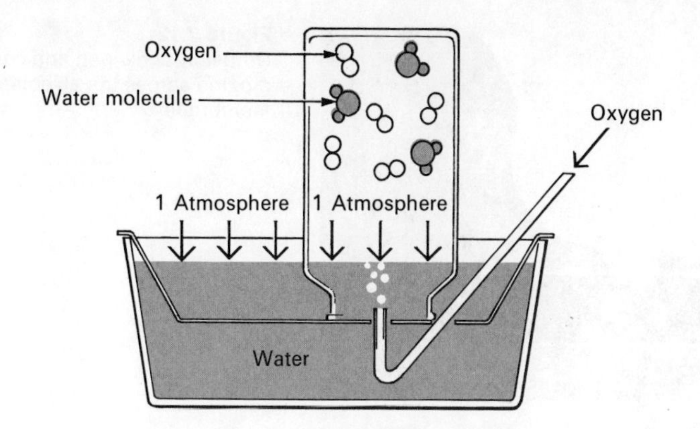

Figure 7.11
Oxygen being collected over water at a pressure of 1 atmosphere. The total pressure in the bottle is the sum of the partial pressures of oxygen and water.

invariably contain water molecules. The total pressure of the mixture is then equal to the partial pressure of the gas plus the vapor pressure of water at that particular temperature. Figure 7.11 shows oxygen being collected by water displacement at a pressure of 1 atmosphere and a temperature of 20°C. If the water in the bottle is at the same level as the water in the trough, the pressure of the oxygen-water vapor mixture will be equal to atmospheric pressure, or 760 torr:

$$P_{total} = P_{O_2} + P_{H_2O} = P_{atm} = 760 \text{ torr}$$

The vapor pressure of water at 20°C is 17.5 torr, and so the pressure exerted by the oxygen alone is 760 torr − 17.5 torr, or 742.5 torr.

7.11 PARTIAL PRESSURES AND THE DIFFUSION OF RESPIRATORY GASES

As air is inspired, it comes into contact with the moist membranes lining the respiratory passages and becomes humidified. At a temperature of 37°C the partial pressure of water in inspired air is 47 torr. Since the total pressure of the air entering the alveoli cannot exceed atmospheric pressure, the pressure of the other components amounts to 760 torr minus 47 torr, or 713 torr. The major components of air and their approximate partial pressures in inspired air are: nitrogen, 557 torr; oxygen, 149 torr; and argon, 6.7 torr. The concentration of carbon dioxide in air entering the alveoli is only about 0.03% and so its partial pressure is less than 1 torr.

Within the alveolar spaces the inspired air mixes with residual air containing a higher concentration of carbon dioxide and a lower concentration of oxygen. The P_{O_2} falls to 104 torr and the P_{CO_2} rises to about 40 torr. It will be shown that the movement of oxygen and carbon dioxide across the alveolar membrane is a direct consequence of their partial pressures in alveolar air.

Adjacent to the alveolar membrane are capillaries carrying blood on its return trip from the tissues. The metabolic activities of the tissues have reduced the P_{O_2} in this blood to about 40 torr and have increased the P_{CO_2} to about 45 torr. This means that alveolar air has a higher P_{O_2}

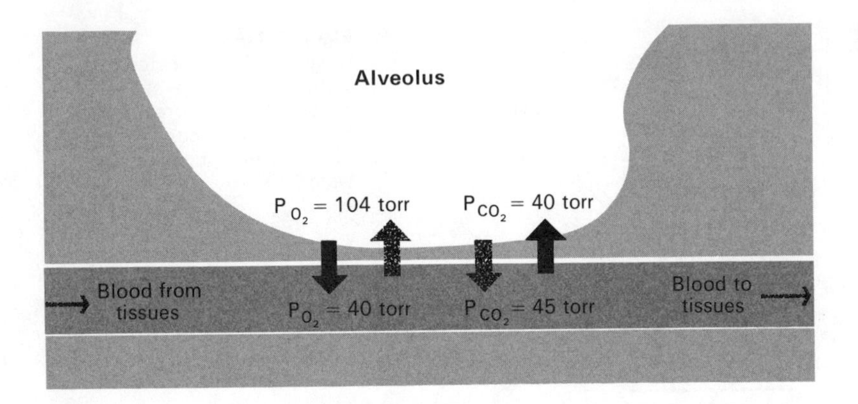

Figure 7.12
Diffusion of oxygen and carbon dioxide across the alveolar membrane.

and a lower P_{CO_2} than venous blood; therefore, oxygen diffuses from the alveolus into the blood, and CO_2 from the blood into the alveolus (Figure 7.12). Despite its smaller pressure differential, the diffusion of CO_2 is extremely rapid, possibly owing to its reversible combination with the moisture on the surface of the membrane.

By the time the blood leaves the lungs on its return trip to the tissues, its P_{O_2} has risen to 104 torr and the P_{CO_2} has dropped to 40 torr. Since oxygen is always being used by the cells, the P_{O_2} within a cell (about 6 torr) is always lower than that of the fluid surrounding the cell (30-40 torr). Oxygen thus diffuses from the arterioles into the extracellular fluid, and from this fluid into the cell (Figure 7.13). The consumption of oxygen is accompanied by the production of CO_2, which reaches a partial pressure of 47-50 torr within the cell, causing it to diffuse into the extracellular fluid (about 47 torr) and from there into the blood. The diffusion is reversed when the venous blood once again reaches the lungs.

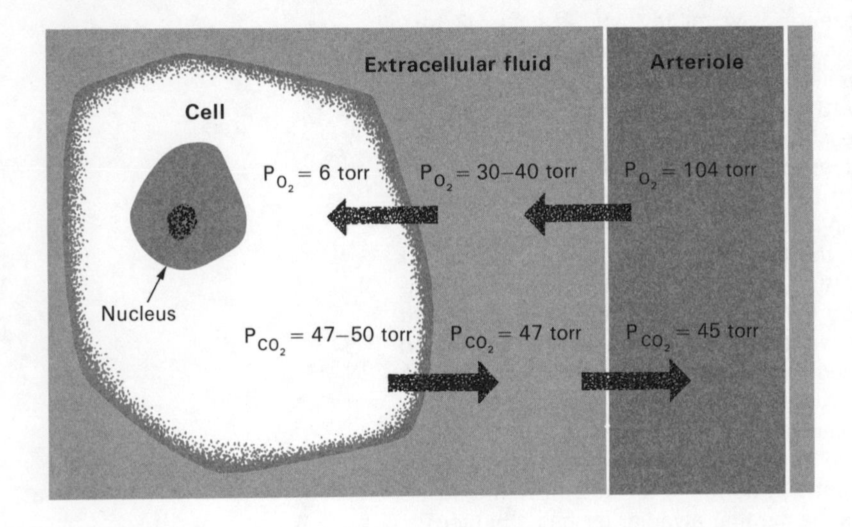

Figure 7.13
The influence of partial pressures on the diffusion of respiratory gases. The net diffusion of oxygen is from the arteriole into the cell, while carbon dioxide diffuses from the cell into the arteriole.

7.12 TRANSPORT OF RESPIRATORY GASES

The preceding discussion concerned the pressure differentials involved in the diffusion of respiratory gases across cell membranes. We shall now consider the mechanisms by which oxygen and carbon dioxide are transported in the blood.

The body of an average adult at rest requires about 350 ml of oxygen per minute. Because of the low solubility of oxygen in plasma, only about 2% of the required total is carried in solution. The remainder is transported in loose chemical combination with hemoglobin inside the red blood cells. The oxygen-hemoglobin combination is known as oxyhemoglobin. The blood of a normal individual contains approximately 15 g of hemoglobin per 100 ml. The condition wherein the blood contains significantly less hemoglobin than this is known as anemia.

The formation of oxyhemoglobin is affected by the partial pressure of oxygen, the temperature, and the concentration of hydronium ions. The association is favored by a high partial pressure of oxygen, a slight cooling of the blood, and a decrease in the concentration of hydronium ions—conditions which the blood encounters as it passes through the capillaries of the lungs.

Within the tissues the conditions are reversed. The metabolic processes of the cells have generated heat, reduced the partial pressure of oxygen, and produced carbon dioxide, which has reacted with water to increase the concentration of hydronium ions. (See Reactions with Oxides in Section 6.8.) The oxyhemoglobin dissociates and oxygen diffuses into the tissues.

Hemoglobin is also capable of forming a reversible association with carbon dioxide—an association that is favored by the conditions prevailing in the tissues. Only about 10% of the CO_2 is carried in this manner, however. Most of the remainder is transported in the form of bicarbonate ion, while 5-10% is carried in simple solution. When the blood once again reaches the capillaries of the lungs, the bound CO_2 is released, and a cellular catalyst reverses the reaction that formed the bicarbonate ion.

Carbon dioxide and oxygen are not the only gases able to combine with hemoglobin. Carbon monoxide binds so strongly that it effectively prevents the transport of oxygen, and may lead to brain damage or death by asphyxiation.

7.13 OXYGEN: FRIEND AND FOE

Oxygen is a colorless, odorless, tasteless gas having a density of 1.43 g per liter at STP. It is only sparingly soluble in water. It is prepared commercially by the fractional distillation of liquid air (Chapter 2). In the laboratory oxygen is usually collected by water displacement and is produced by the following methods:

1. Decomposition of mercury(II) oxide:

$$2\ H_gO_{(s)} \xrightarrow{\Delta} 2\ H_{g(s)} + O_{2\,(g)}$$

2. Decomposition of potassium chlorate:

$$2\ KClO_{3(s)} \xrightarrow{MnO_2} 2\ KCl_{(s)} + 3\ O_{2(g)}$$

3. Reaction of sodium peroxide with water:

$$2\ Na_2O_{2(s)} + 2\ H_2O_{(l)} \longrightarrow 4\ NaOH_{(aq)} + 3\ O_{2\,(g)}$$

4. Electrolysis of water:

$$2\ H_2O_{(l)} \longrightarrow 2\ H_{2\,(g)} + O_{2\,(g)}$$

Oxygen is one of the most reactive elements, combining with metals as well as nonmetals to form compounds called oxides. The formation of an oxide is generally accompanied by the release of energy. At low temperatures and low oxygen concentrations oxide formation generally proceeds slowly. At higher temperatures or in a higher oxygen concentration the reaction may become so rapid that the reacting substance is said to burn. However, the total amount of energy released in a reaction is the same regardless of its rate. The following are representative examples of reactions with oxygen:

1. Reaction with a metal:

$$2\ Mg_{(s)} + O_{2\,(g)} \xrightarrow{\Delta} 2\ MgO_{(s)}$$

2. Reaction with a nonmetal:

$$C_{(s)} + O_{2\,(g)} \xrightarrow{\Delta} CO_{2\,(g)}$$

3. Reaction with glucose:

$$C_6H_{12}O_{6(s)} + 6\ O_{2(g)} \xrightarrow{\text{catalysts}} 6\ H_2O_{(l)} + 6\ CO_{2(g)}$$

The last reaction is called cellular respiration, and occurs in most living cells, although a very small number of organisms (called anaerobes) release the energy stored in glucose without the use of molecular oxygen. Although the equation correctly summarizes the overall change in the glucose, it is somewhat misleading since it infers that the change involves a single step, which is not the case. This reaction is considered in detail in Chapter 12.

In certain abnormal conditions the amount of oxygen reaching the cells is insufficient for cellular respiration. In pneumonia, for example, fluid often accumulates in the alveolar spaces and reduces the rate at which oxygen diffuses into the blood. Damage to the heart or major blood vessels may decrease the rate of blood flow, resulting in a deficiency of oxygen in the tissues. Destruction or thickening of the alveolar membranes may also occur in certain diseases. In all of these cases it is necessary to increase the partial pressure of oxygen in the alveolar air in order to increase the diffusion rate. When the oxygen deficiency is the result of reduced blood flow, increasing the partial pressure of oxygen to 600 torr causes appreciable amounts of oxygen to dissolve in the plasma, which supplements the oxygen carried by the red blood cells. Oxygen must also be mixed with anaesthetic gases such as nitrous oxide and cyclopropane for use in the operating room.

Pure oxygen is such a reactive substance that elaborate precautions should be taken when it is used. Many materials that are normally regarded as inert become fiery torches when mixed with oxygen and heated. If the conditions are right, a spark from an electrical appliance or static electricity can ignite the mixture. Whenever oxygen is in use, the following precautions should be observed:

1. Be certain that all electrical devices (shavers, radios, television sets, etc.) are turned off.
2. The use of flammable substances such as alcohol and oil should be avoided.
3. Nylon or silk clothing should not be worn because they accumulate static electricity. Shoes should have leather soles because rubber is an insulator.

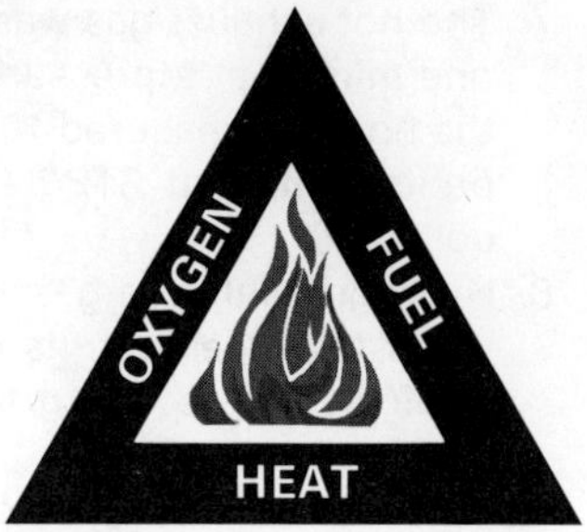

Figure 7.14
The fire triangle. Remove any one of the three components and the fire goes out.

In order for a substance to burn, three things are necessary: oxygen, a fuel, and heat. They are often referred to as the *fire triangle* (Figure 7.14). If any one of them is removed, the fire goes out. Spraying water on a fire helps to extinguish it by reducing the heat. Smothering a fire cuts off the oxygen supply. However, the best practice is to make sure that the three components of the fire triangle never get together.

APPLICATION OF PRINCIPLES

1. The air near the ceiling of a room is always slightly warmer than the air close to the floor. Explain this observation.
2. A sample of gas at a pressure of 1 atmosphere is composed of oxygen and carbon dioxide. If the oxygen molecules are four times as numerous as carbon dioxide molecules, what is the partial pressure of the CO_2?
3. A tank of medical oxygen contains 2.5 cubic feet of gas at a pressure of 2500 psi. Assuming that there is no change in temperature, what will be the volume of the gas at a pressure of 1 atmosphere?
4. A sample of expired air is collected from a patient. After the water vapor has been removed, will the volume of the air sample be larger or smaller than it was initially?
5. Calculate the volume of 10 g of oxygen gas (O_2) at STP.
6. One liter of dry oxygen gas at a pressure of 1 atmosphere was collected from the decomposition of $KClO_3$. If the temperature of the gas was 45°C when it was collected, what would be its volume at 20°C? Assume that the pressure remains unchanged.

7. The hot exhaust gas was collected from an automobile engine for one minute at a pressure of 50 psi and a temperature of 400°C. If the hot gas measured 100 liters when it was collected, what would be its volume at STP? Assume that the gas was dry when it was collected.
8. Hydrogen chloride gas is too soluble in water to be collected by water displacement. Its density is 1.64 g/liter at STP. The density of air at STP is 1.29 g/liter. Suggest a method for collecting the gas.
9. When water is sprayed on burning oil, the oil floats on top of the water and continues to burn. Suggest a method, based on the fire triangle, for extinguishing an oil fire.
10. How would you expect the rate of diffusion of ammonia gas to compare with that of methane under the same conditions? Explain your answer.
11. Calculate the approximate density of carbon dioxide gas at STP.
12. Cite evidence to show that oxygen gas dissolves in water.
13. Before starting out on a trip, a man checks the pressure in his automobile tires and notes that it is 32 psi. After an hour of driving, he again checks the pressure and it has increased. Explain what causes this.

A student performs a titration to determine the concentration of acid in a solution. (Photograph by J. Asdrubal Rivera)

8

ACIDS, BASES, AND BUFFERS

LEARNING OBJECTIVES

1. Compare the Arrhenius, Bronsted-Lowry, and Lewis models of acids and bases.
2. Distinguish between weak acids and strong acids.
3. Describe the relationship between pH and the concentration of hydronium ions.
4. Describe a procedure for measuring the total acidity of a solution.
5. Explain how the pH of a solution determines the color of an acid-base indicator.
6. Describe the relationship between the two members of a conjugate acid-base pair.
7. Predict the acid-base reaction of a hydrolyzed salt.
8. Describe the formation of a zwitterion and explain how zwitterions act as buffers.
9. Outline the chemical interactions that take place during the transport of carbon dioxide in the blood.
10. Explain how a buffer stabilizes the pH of a solution.
11. Apply the explanation in Objective 10 to the buffering mechanism of the body.
12. Distinguish between respiratory and metabolic acidosis, and describe the adjustments that the body makes as it attempts to counteract each of these conditions.

KEY TERMS AND CONCEPTS

acidosis	dissociation
alkalosis	hydronium ion
amide	indicator
amine	ketosis
amino acid	Lewis acid or base
amphoteric	neutralization
Arrhenius acid or base	pH
Brönsted-Lowry acid or base	polymerization
buffer	salt
carboxylic acid	strong acid
chloride shift	titration
conjugate acid or base	total acidity
dehydration synthesis	weak acid
	zwitterion

Acids and bases may be regarded as chemical opposites: that is, the distinctive chemical properties of acids disappear upon reaction with bases, and vice versa. Until the first theoretical model was proposed in 1884 by Svante Arrhenius, acids and bases were defined operationally; that is, in terms of their properties. According to its operational definition, an acid is a substance whose aqueous solution has the following properties:

1. Tastes sour.
2. Reacts with most metals to produce hydrogen gas.
3. Turns an organic dye called litmus from blue to red.
4. Conducts electricity.
5. Produces carbon dioxide upon reaction with carbonates.
6. Reacts with a base to produce a salt and water.

Operationally, a base is a substance whose aqueous solution

1. Tastes bitter.
2. Feels slippery when rubbed on the skin.
3. Turns litmus from red to blue.
4. Conducts electricity.
5. Reacts with active metals to produce hydrogen gas.
6. Reacts with an acid to produce a salt and water.

A definition stated in operational terms is of limited value because it is nothing more than a summary of observations. In contrast, Arrhenius' model attempted to account for these observations by showing that the properties of acids and bases resulted from specific molecular structures. His model represented the first stage in the evolution of the modern acid-base concept.

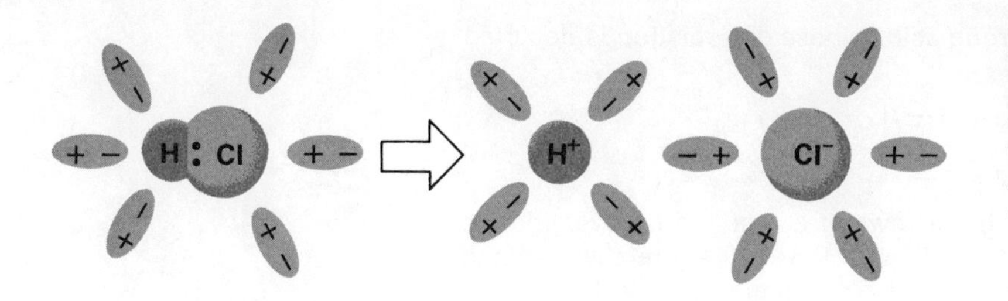

Figure 8.1
Dissociation of hydrogen chloride. Polar solvent molecules weaken the bond between the hydrogen and its neighboring atom, resulting in the formation of hydrated ions.

8.1 THE ARRHENIUS MODEL OF ACIDS AND BASES

Arrhenius Acids

Arrhenius observed that all of the substances which could be operationally defined as acids contained the element hydrogen. He proposed that any substance which dissociated in water to give H^+ ions—that is, protons—was an acid. *Dissociation* involves the interaction of a substance, in this case the acid, with molecules of polar solvent (water). This weakens the bond between hydrogen and its neighboring atom and causes the acid to split into a hydrogen ion and an anionic species (Figure 8.1). The presence of these ions accounts for the observation that acid solutions conduct electricity.

Although Arrhenius' model is successful in accounting for many of the properties of acids, we now know that a free proton is a highly reactive species, and is unlikely to exist for very long in aqueous solution. Because a proton is a very small ion carrying a relatively large positive charge, it should be attracted by any species having unshared valence electrons. Since the oxygen atom in each water molecule has two such pairs of electrons, water is able to form a coordinate covalent bond with a proton. The resulting ion is called a *hydronium ion* (H_3O^+). This interaction may be shown as

$$H^+ + :\ddot{\underset{\cdot\cdot}{O}}:H \rightleftharpoons \left[H:\ddot{\underset{\cdot\cdot}{O}}:H \right]^+$$
$$\qquad\quad H \qquad\qquad\quad H$$

The existence of the hydronium ion has been confirmed experimentally. Thus an Arrhenius acid is more properly defined as a substance which dissociates in water to produce hydronium ions.

The dissociation of acid molecules is a reversible process; an equilibrium exists between the molecular and the ionized forms. This means that at no one time are *all* of the molecules of acid dissociated. Ionization of the hydrogen in an acid molecule is due to the polarity of its bond and also depends upon the polarity of the solvent: the greater the polarity, the greater the degree of ionization. Acids which ionize almost completely in dilute solutions (approximately 0.1 *M*) are classified as *strong acids.* Strong acids can cause severe burns and are destructive to most natural and man-made materials. Rayon, silk, leather, and wool deteriorate rapidly in an acid environment. Hydrochloric and nitric acid are strong acids whose dissociation is depicted in the following equations:

$$HCl_{(g)} + H_2O_{(l)} \rightleftharpoons H_3O^+_{(aq)} + Cl^-_{(aq)}$$

$$HNO_{3(l)} + H_2O_{(l)} \rightleftharpoons H_3O^+_{(aq)} + NO_{3(aq)}$$

Note the relative lengths of the arrows. If a reaction is reversible as this one is, two arrows must be shown. However, the relative lengths of the arrows tells you which reaction predominates, the forward or the reverse. In this case the forward dissociation predominates.

Hydrochloric acid is produced by specialized cells in the lining of the stomach and is essential for the digestion of proteins. Excessive production of HCl can cause digestive upset (hyperacidity) and is often a factor in the development of gastric ulcers.

The dissociation of hydrogen sulfate is observed to produce twice as many hydronium ions per molecule as either HCl or HNO_3. It is concluded from this that both of its hydrogen atoms are capable of ionization. A two-stage dissociation has been proposed to account for this behavior:

1. $H_2SO_{4(l)} + H_2O_{(l)} \rightleftharpoons H_3O^+_{(aq)} + HSO_4^-{}_{(aq)}$

2. $HSO_4^-{}_{(aq)} + H_2O_{(l)} \rightleftharpoons H_3O^+_{(aq)} + SO_4^{2-}{}_{(aq)}$

Carbonic acid (H_2CO_3) also contains two hydrogen atoms per molecule, but in contrast to sulfuric acid, fewer than 0.17% of its molecules (about 17 of each 10,000) are ionized in a 0.1 *M* solution. Note the relative lengths of the arrows in the equilibrium expression for carbonic acid:

1. $H_2CO_{3(\,)} + H_2O_{(l)} \rightleftharpoons H_3O^+_{(aq)} + HCO_3^-{}_{(aq)}$

2. $HCO_3^-{}_{(aq)} + H_2O_{(l)} \rightleftharpoons H_3O^+_{(aq)} + CO_3^{2-}{}_{(aq)}$

Because it is only slightly ionized in dilute solution, carbonic acid is classified as a *weak* acid. It is found in carbonated beverages, where it is produced by the reaction of CO_2 with water, and in the blood, where it plays a major role in the elimination of metabolic waste (see Section 8.12).

Acetic acid ($HC_2H_3O_2$) is another weak acid. It is formed in wines by the action of yeasts and/or bacteria on ethyl alcohol. Vinegar is a 4-5% solution of acetic acid in water. It should be noted that, although it is classified as a weak acid on the basis of its limited ionization, a concentrated solution of acetic acid can cause severe burns.

Lactic acid production accompanies muscle contraction in the body, and pyruvic acid is one of the most important compounds produced by the body during the breakdown of sugars. The structures of these two acids, and the roles they play in metabolism, are described in Chapter 12. Table 8.1 shows the degree of ionization and classification of several important acids. Although hydrochloric, nitric, and sulfuric acids are shown in the table as being completely ionized, it is doubtful that any material dissociates to the last molecule.

TABLE 8.1 Ionization and Acid Strength

Acid	Degree of Ionization[1]	Classification
Hydrochloric	Complete	Strong
Nitric	Complete	Strong
Sulfuric	Complete	Strong
Phosphoric	Moderate	Strong
Acetic	Slight	Weak
Carbonic	Slight	Weak
Lactic	Very slight	Weak
Pyruvic	Very slight	Weak

[1] In 0.1 *M* solution.

Arrhenius Bases

An Arrhenius base is a substance which dissociates in water to produce hydroxide ions. The difference in the taste of acids and bases can be attributed to the fact that H_3O^+ ions stimulate the taste buds on the sides of the tongue that transmit a sensation of sourness to the brain, while OH^- ions affect taste buds on the back of the tongue where the sensation is one of bitterness.

Arrhenius bases are soluble metallic hydroxides such as NaOH and KOH. Sodium hydroxide (also called lye) is a strong base that is used in oven and drain cleaners and in soap-making. Strong bases are corrosive to tissues and are damaging to rubber, plastic, fabrics, and paint. Strong bases, like strong acids, are almost completely dissociated in water as shown by the following equations:

$$NaOH \rightleftharpoons Na^+ + OH^-$$

$$KOH \rightleftharpoons K^+ + OH^-$$

Calcium hydroxide, $Ca(OH)_2$ is a strong base that is only slightly soluble in water. A saturated solution of calcium hydroxide is called limewater, and is used in testing for CO_2. When CO_2 is bubbled through limewater, a white precipitate of calcium carbonate is formed. Carbon dioxide is the only gas to give this result. The equation for the reaction is

$$CO_{2\,(g)} + Ca(OH)_{2(aq)} \longrightarrow CaCO_{3(s)} + H_2O_{(l)}$$

Another important weak base is $Mg(OH)_2$, also known as milk of magnesia. In small amounts milk of magnesia acts as an antacid, and larger doses have a laxative effect.

Reactions of Acids and Bases

Reactions between Arrhenius acids and bases produce water and ionic compounds called salts, and are called *neutralization* reactions. A *salt* may be thought of as a substance formed by combining the anion from an acid with the cation from a base. A representative neutralization reaction takes place when milk of magnesia is used as an antacid. The ionic equation for the reaction is

$$Mg^{2+}_{(aq)} + 2\ OH^-_{(aq)} + 2\ H_3O^+_{(aq)} + 2\ Cl^-_{(aq)} \longrightarrow Mg^{2+}_{(aq)} + 2\ Cl^-_{(aq)} + 4\ H_2O_{(l)}$$

The net ionic equation for a neutralization reaction is

$$OH^-_{(aq)} + H_3O^+_{(aq)} \longrightarrow 2\ H_2O_{(l)}$$

Antacid Ingredients and their Reactions with Stomach Acid

Aluminum hydroxide $Al(OH)_3$

$Al(OH)_3 + 3\ HCl \rightarrow AlCl_3 + 3\ H_2O$

Dihydroxyaluminum sodium carbonate $NaAl(OH)_2CO_3$

$NaAl(OH)_2CO_3 + 4\ HCl \rightarrow NaCl + AlCl_3 + 3\ H_2O + CO_2$

Magnesium hydroxide $Mg(OH)_2$

$Mg(OH)_2 + 2\ HCl \rightarrow MgCl_2 + 2\ H_2O$

Sodium citrate $Na_3C_6H_5O_7$

$Na_3C_6H_5O_7 + 3\ HCl \rightarrow 3\ NaCl + 2\ H_2O + H_3C_6H_5O_7$

8.2 MEASUREMENT OF HYDRONIUM ION CONCENTRATION (pH)

It was pointed out in Chapter 6 that water molecules react with each other to form equal numbers of OH^- and H_3O^+ ions. The reversible reaction is

$$H_2O_{(l)} + H_2O_{(l)} \rightleftharpoons H_3O^+_{(aq)} + OH^-_{(aq}$$

The concentration of H_3O^+ ions in pure water is approximately one ten-millionth of a mole (10^{-7} mole) per liter. Because acids dissociate in water to produce H_3O^+ ions, the hydronium ion concentration of acid solutions is greater than 10^{-7} mole per liter. And since strong acids dissociate to a greater degree than weak acids, their solutions will contain correspondingly higher concentrations of hydronium ion. For example, HCl is a strong acid and is completely dissociated in dilute solutions; therefore, a 0.1 *M* solution of HCl contains 0.1 mole of H_3O^+ ions per liter. By comparison, a liter of 0.1 *M* acetic acid contains only about 0.00134 mole of H_3O^+, which means that its hydronium ion concentration is roughly 1/100 that of the stronger acid. Therefore, there may be a difference between the hydronium ion concentration of a given solution and its *total acidity.* The former refers to the number of ions actually present in the solution, while the latter is the number which would be present if the acid were 100% dissociated. Thus the number of H_3O^+ ions in 0.1 *M* HCl is roughly 100 times as great as that in 0.1 *M* $HC_2H_3O_2$, but the total acidity of the two solutions is the same.

The hydronium ion concentration of a solution is often so low that it becomes awkward to express it as a decimal fraction. For example, in milk of magnesia it is only about 0.00000000003 mole per liter. Such numbers may also be written in exponential form (3×10^{-11}), or they may be shown on a logarithmic scale devised by Soren Sorensen in 1909. The units of his scale, which runs from 0 through 14, are called *pH* units (for power of hydrogen). Sorensen defined pH as the negative logarithm of the hydronium ion concentration. Stated symbolically,

$$pH = -\log_{10}[H_3O^+]$$

The molar concentration of a particular chemical species is indicated by enclosing its symbol or formula in brackets: thus $[H_3O^+]$ means "the molar concentration of hydronium ion." The logarithm of a number expressed in exponential form is approximately equal to its exponent. For example, pure water has a hydronium ion concentration of 0.0000001 (1×10^{-7}) mole per liter. The logarithm of this number is -7.0, which is the same as its exponent. Since pH is defined as the negative logarithm, the sign of the exponent must be multiplied by -1. Thus the pH of pure water is 7.0.

As the hydronium ion concentration increases, the size of the exponent decreases, and so does the pH. A solution having 1×10^{-6} mole of H_3O^+ ions per liter would have a pH of 6.0 and would be 10 times as acidic as water, while a solution 100 times as acidic as water would have a concentration of 1×10^{-5} mole per liter and a pH of 5.0. A solution whose pH is greater than 7.0 is less acidic than water, and is considered to be basic. The relationship between pH and hydronium ion concentration is shown in Table 8.2.

TABLE 8.2 The Relationship between pH and $[H_3O^+]$

$[H_3O^+]$	pH	Classification
1×10^{-1}	1.0	Strongly acidic
1×10^{-2}	2.0	↑
1×10^{-3}	3.0	
1×10^{-4}	4.0	Increasing acidity
1×10^{-5}	5.0	
1×10^{-6}	6.0	
1×10^{-7}	7.0	Neutral
1×10^{-8}	8.0	
1×10^{-9}	9.0	
1×10^{-10}	10.0	Increasing basicity
1×10^{-11}	11.0	
1×10^{-12}	12.0	
1×10^{-13}	13.0	↓
1×10^{-14}	14.0	Strongly basic

Many chemical processes, especially those of a biological nature, are pH-dependent. For example, azaleas and camelias thrive in an acid soil, while bottle brush and baby's breath prefer soil that is alkaline. The digestive enzymes in gastric juice require extremely acidic conditions, but enzymes in the intestine have an optimum pH of about 7.5. The growth of yeasts and molds is favored by low or acidic pH, and meat can be tenderized by soaking in acidic solutions such as wine or vinegar. The multiple reactions that take place within the cells of the body are extremely sensitive to variations in pH. The pH of the cells is normally maintained within a very narrow range (7.35-7.45) by acid-base systems containing buffers, which are described later in this chapter. The approximate pH of some familiar substances is shown in Table 8.3. We have already indicated the relationship between electrical conductivity and hydronium ion concentration. This relationship can be used to measure the pH of a solution using an instrument called a pH meter. A pH meter utilizes a special electrode that is selectively sensitive to hydronium ions. The pH of a solution may also be determined, though less precisely, from the reaction of the solution with indicators, as explained in the following section.

TABLE 8.3 The Approximate pH of Some Common Liquids

Gastric juice	1.8	Increasingly acidic ↑
Lemon juice	2.3	
Vinegar	3.0	
Soft drinks	3.0	
Grapefruit juice	3.2	
Black coffee	5.0	
Urine	5.0–7.0	
Rain water	6.2	
Pure water	7.0	Neutral
Saliva	6.2–7.4	
Intestinal juice	7.0–8.0	
Blood	7.35–7.45	
Bile	7.8–8.6	
Sea water	8.5	
Milk of magnesia	10.5	↓ Increasingly basic

8.3 ACID-BASE INDICATORS

In addition to explaining neutralization reactions and the electrical conductivity of solutions containing acids or bases, the Arrhenius theory suggests a mechanism for the behavior of acid-base indicators. *Indicators* are unique dyes that undergo slight changes in molecular structure with changes in pH. Since the color of a dye is related to its molecular structure, changing the pH also changes the color. The juices from beets, grapes, red cabbage, and berries act as indicators; litmus, one of the most frequently used indicators, is extracted from a primitive plant. Litmus is a weak acid whose molecular form is red and whose ionic form is blue. The equilibrium reaction between the molecular (Ind-H) and ionic (Ind^-) forms may be represented in the following way:

$$\underset{}{H_2O} + \underset{\text{(red)}}{\text{Ind-H}} \underset{(2)}{\overset{(1)}{\rightleftharpoons}} \underset{\text{(blue)}}{\text{Ind}^-} + H_3O^+$$

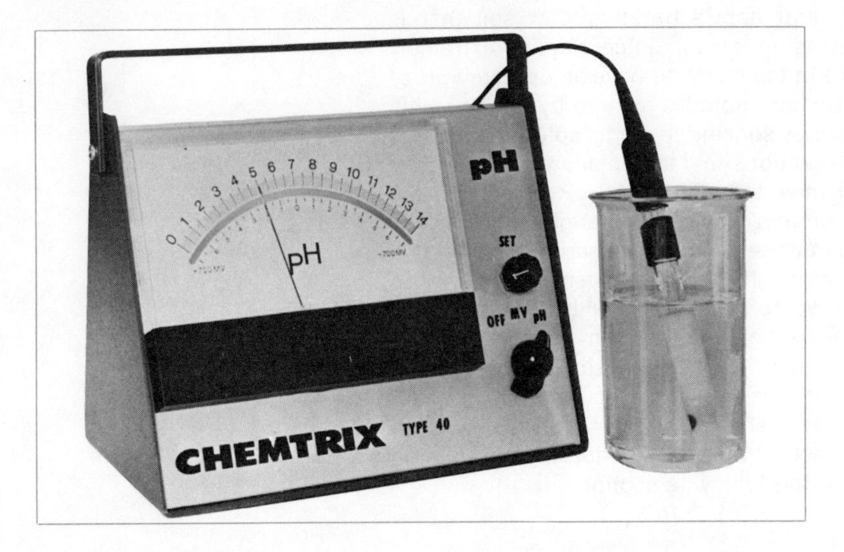

A pH meter. The glass electrode is selectively permeable to hydronium ions. (Courtesy Chemtrix, Inc., Beaverton, Oregon)

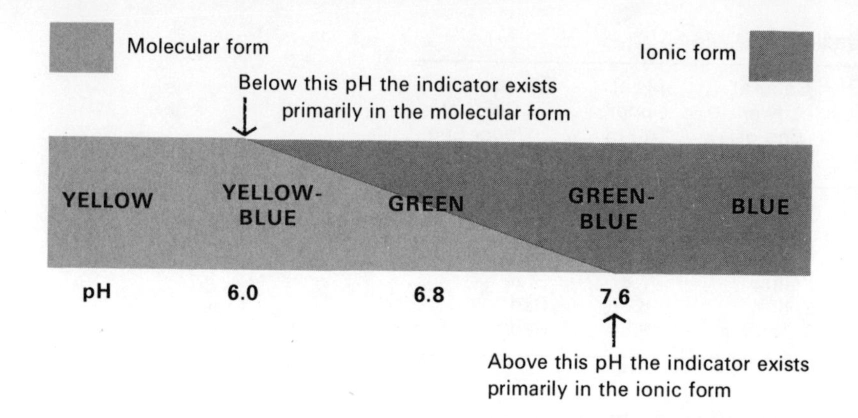
Molecular form
Ionic form
Below this pH the indicator exists primarily in the molecular form
YELLOW
YELLOW-BLUE
GREEN
GREEN-BLUE
BLUE
pH
6.0
6.8
7.6
Above this pH the indicator exists primarily in the ionic form

Figure 8.2
Color changes of bromothymol blue indicator with changes in pH. The indicator exists in two different colored forms, and the changing ratio between these forms accounts for the variation in color.

In high concentrations of hydronium ions reaction 2 is favored, and the indicator exists mostly in its molecular form. When a base is added, the OH^- ions from the base react with and decrease the concentration of H_3O^+ ions, shifting the equilibrium to the right and forming more of the blue indicator ion:

$$OH^- + \text{Ind-H} \longrightarrow \text{Ind}^- + H_2O$$

The color of an indicator solution at a particular pH depends upon the ratio between its molecular and ionic forms. For example, bromothymol blue is yellow at a pH of 6.0 or below, and blue at a pH of 7.6 or above. If the pH is gradually changed from 6.0 to 7.6, the varying ratio between the ionic and molecular forms causes successive color changes from yellow to yellow-green, from yellow-green to green, from green to greenish-blue, and finally to blue (Figure 8.2). At the midpoint of its effective range (about pH 6.8), bromothymol blue is actually green.

Indicators are used extensively in microbiology because they may be added directly to the media in which bacteria or fungi are grown, where they provide visual evidence of changing growth conditions. Indicators are also used in physiology and biochemistry. One test of kidney function involves the use of an indicator called phenolsulfonphthalein (PSP). A solution of the dye is injected into the patient's bloodstream. If the kidneys are functioning normally, 50-60% of the dye should be eliminated in the urine within two hours.

The pigment in red cabbage is a natural indicator. It is green in strongly basic and blue in slightly basic solutions, purple in a neutral solution, and red when acidic. Similar indicators can be extracted from beets and purple grapes.

The colors and effective pH ranges for several common indicators are shown in Table 8.4. Since the useful range of an individual indicator generally covers less than 2 pH units, mixtures of indicators are sometimes used to extend the pH range. One commercial mixture, covering a pH range from 2.0 to 10.0, is called Universal Indicator Solution because of the wide range it covers.

The approximate pH of a solution can often be determined by the color which it imparts to an indicator. Suppose, for example, that samples of a given solution caused methyl orange to show an orange-yellow color, and bromocresol purple to appear yellow. Bromocresol purple is yellow at or below pH 5.2, while methyl orange is orange-

TABLE 8.4 Selected Acid–Base Indicators

Common Name	pH at Lower End of Range	Color at Lower End of Range	pH at Upper End of Range	Color at Upper End of Range
Methyl orange	3.1	Red	4.4	Orange-yellow
Alizarin red	3.7	Yellow	4.2	Pink
Bromocresol purple	5.2	Yellow	6.8	Purple
Bromothymol blue	6.0	Yellow	7.6	Blue
Cresol red	7.2	Yellow	8.0	Red
Phenolphthalein	8.9	Colorless	9.8	Red

yellow at the upper end of its effective range—that is, at pH 4.4 or higher. Thus the pH of the solution must lie between 4.4 and 5.2. Indicators are also used in the measurement of total acidity, as described in the following section.

8.4 THE MEASUREMENT OF TOTAL ACIDITY: TITRATION

The pH of a solution is a measure of its hydronium ion concentration, while its total acidity includes undissociated acid as well. The total acidity of a sample may be determined by adding a standard solution of base until chemically equivalent amounts of acid and base are present. Visual evidence that this point has been reached is supplied by the change in color of an indicator that has been added to the reaction vessel. The indicator should be one whose color change takes place at a pH close to that of the mixture when equivalent amounts of acid and base are present. Since mixtures containing equivalent amounts of acid and base are usually neutral, an appropriate indicator would be one which changes color at pH 7.0. Phenolphthalein comes closest to this requirement, although its color changes sharply from colorless to red at a pH slightly on the basic side.

The procedure in which a standard solution is used to determine the concentration of another solution is called *titration*. The solutions are dispensed from calibrated tubes called burettes, shown in Figure 8.3. In practice a sample of acid is measured into a beaker or a flask along with a few drops of indicator, and then standard base solution is added a little at a time with stirring. The endpoint of the titration is reached when one drop of base causes a distinct change in the color of the indicator. The concentration of the acid is calculated from the volumes of acid and base and the concentration of the base. The following sample problem will illustrate the procedure.

Sample problem 8.1 If 20 ml of an acid is neutralized by 30 ml of 0.1 *M* NaOH, what is the concentration of the acid?

Solution A 0.1 *M* solution of NaOH contains 0.1 mole of OH^- ion per

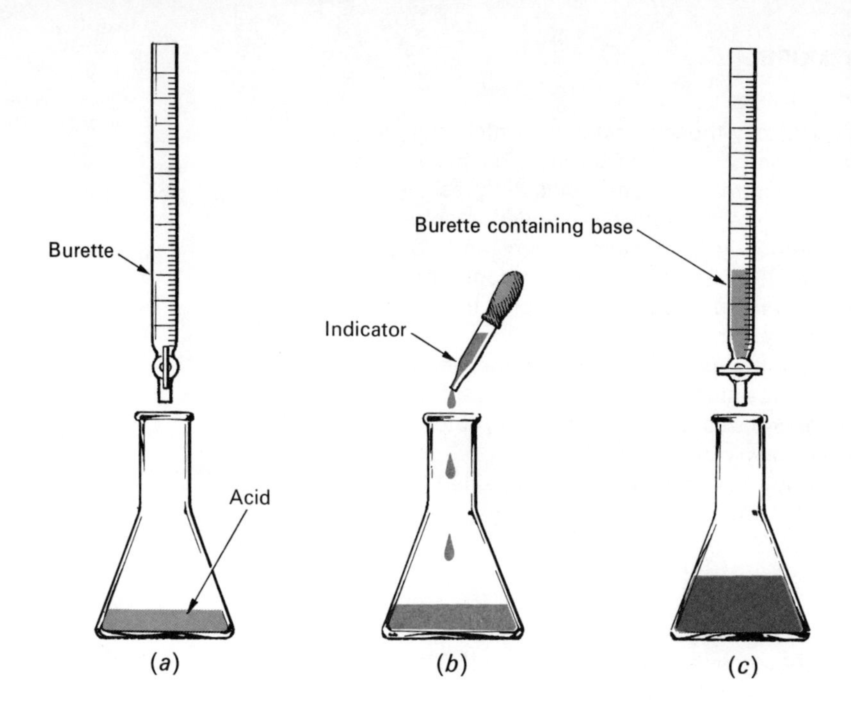

Figure 8.3
Titration. (a) Sample of acid is dispensed from burette. (b) Indicator is added. (c) Base is added until the indicator changes color.

liter, so in 30 ml there would be 30/1000 of 0.1 mole or 0.003 mole. This is chemically equivalent to the ionizable hydrogen in 20 ml of acid. In other words, 20 ml of acid furnishes 0.003 mole of H_3O^+ ion. By simple proportion if 20 ml furnishes 0.003 mole, then 1000 ml would furnish 0.15 mole. Thus the concentration of the acid solution is 0.15 *M*.

Sample problem 8.2 The acetic acid in 50.0 ml of vinegar is neutralized by 20.0 ml of 2.0 *M* KOH solution. What is the concentration of the acetic acid?

Solution Since 1 liter of 2.0 *M* KOH contains 2.0 moles of OH^- ion, 20.0 ml of the solution would contain $20.0/1000 \times 2.0$ moles, or 0.040 mole. This is equal to the amount of H_3O^+ ion in 50.0 ml of acid. By proportion, if 50.0 ml of acid contains 0.040 mole, then 1000 ml would contain 1000/50.0, or 20, times as much. Thus there are 20×0.040, or 0.80 mole or H_3O^+ ion in a liter of the vinegar, and so its concentration is 0.80 *M*.

Titration is not restricted to acids and bases, but can be used to determine the concentration of any chemical species which undergoes a distinct change at the point of chemical equivalency. This technique is used to measure such diverse characteristics as the amount of the digestive enzyme in saliva, the quantity of chloride in sea water, and the percent of silver in an alloy.

8.5 THE BRÖNSTED-LOWRY MODEL OF ACIDS AND BASES

According to the Arrhenius model, only those substances which contain the OH^- ion and dissolve in water are considered to be bases. Since there is no OH group in ammonia, it does not qualify as an Arrhenius base, although OH^- ions are present in its aqueous solutions. A model proposed independently by Thomas Lowry and Johannes Brönsted not only accounts for the behavior of ammonia, but it extends the acid-base concept to include chemical species that do not qualify as Arrhenius acids or bases.

According to the *Brönsted-Lowry* theory, an acid is a proton donor and a base is a proton acceptor. Since all Arrhenius acids are capable of donating a proton, and Arrhenius bases can combine with protons, they are included in the Brönsted-Lowry categories.

The presence of OH^- ions in solutions of ammonia is explained by assuming that NH_3 acquires a proton from water in the following reaction:

$$NH_{3\,(g)} + H_2O_{(l)} \rightleftharpoons NH_4{}^+{}_{(aq)} + OH^-{}_{(aq)}$$

It is important to note that in this reaction water acts as an acid by donating a proton, and ammonia becomes a base by accepting it. This does not mean, however, that water should always be regarded as an acid. The classification of a species as a Brönsted-Lowry acid or base depends upon the specific conditions and the other chemical species which are present. Thus water acts as an acid in the reaction above, but it is behaving as a base when it accepts a proton to form the hydronium ion. A substance that is capable of acting as either an acid or a base is said to be *amphoteric.* Other amphoteric species are $HCO_3{}^-$, $HSO_4{}^-$, and the amino acids (discussed later in this chapter).

Reactions between Arrhenius acids and bases always produce water and a salt, while Bronsted-Lowry acids and bases react to form a different acid and a different base. For example, the reaction between ammonia (a base) and water (an acid) produces $NH_4{}^+$ ion (a different acid) and OH^- ion (a different base). Ammonium ion is a base because it is able to donate a proton to the hydroxide ion. The $NH_4{}^+$ ion is said to be the *conjugate acid* of NH_3, and the OH^- ion is the *conjugate base* of water. Each acid-base reaction thus produces the conjugate base of the reacting acid and the conjugate acid of the reacting base. In other words, the members of a conjugate acid-base pair can be formed from each other by the transfer of a proton, as shown by the following example:

Conjugate acid of H_2O — Conjugate base of $HC_2H_3O_2$

$$\underset{\text{Acid}_1}{HC_2H_3O_{2(aq)}} + \underset{\text{Base}_1}{H_2O_{(l)}} \rightleftharpoons \underset{\text{Acid}_2}{H_3O^+{}_{(aq)}} + \underset{\text{Base}_2}{C_2H_3O_2{}^-{}_{(aq)}}$$

Some common acids and their conjugate bases are shown in Table 8.5.

TABLE 8.5 Some Common Acids and Their Conjugate Bases

Acid	Conjugate Base	Acid	Conjugate Base
HCl	Cl^-	H_2O	OH^-
H_3O^+	H_2O	H_2SO_4	HSO_4^-
HNO_3	NO_3^-	HCO_3^-	CO_3^{2-}
NH_4^+	NH_3	$HC_2H_3O_2$	$C_2H_3O_2^-$

There is an inverse relationship between the strength of an acid and that of its conjugate base. For example, HCl is a strong acid while Cl^- ion (its conjugate base) is an extremely weak base. Water, on the other hand, is a weak acid, but its conjugate base (OH^- ion) is one of the strongest. A comparison of several conjugate acid-base pairs is shown in Table 8.6.

TABLE 8.6 Relative Strengths of Some Conjugate Acid-Base Pairs

	Acid	Conjugate Base	
Strong acid ↑	HCl	Cl^-	Weak base ↓
	H_2SO_4	HSO_4^-	
	H_3O^+	H_2O	
	$HC_2H_3O_2$	$C_2H_3O_2^-$	
	H_2CO_3	HCO_3^-	
	NH_4^+	NH_3	
Weak acid	H_2O	OH^-	Strong base

8.6 THE LEWIS MODEL OF ACIDS AND BASES

The most recent step in the evolution of acid-base theory is the model proposed by G. N. Lewis. Whereas ionizable hydrogen is an essential component of all Arrhenius and Brönsted-Lowry acids, Lewis acids do not have to contain hydrogen. A *Lewis acid* is any species having a valence orbital that is empty, while a *Lewis base* is any species capable of donating a pair of electrons for formation of a coordinate covalent bond. Stated simply, a Lewis acid is an electron pair acceptor, and a Lewis base is an electron pair donor. Although this model includes most Brönsted-Lowry acids, it is interesting to note that water cannot be a Lewis acid because it does not have an available empty orbital. Since the acid-base concepts discussed in this book are fairly limited, we need not discuss the Lewis model in detail.

8.7 ACID-BASE PROPERTIES OF HYDROLYZED SALTS

Hydrolysis is a chemical reaction between water and another compound wherein bonds within the compound are broken. The hydrolysis of acids and bases produces hydronium and hydroxide ions, respectively. Many hydrolytic reactions involving salts also produce acidic or basic solutions. For example, a solution of sodium acetate has a pH greater than 7.0, indicating an excess of OH^- ions. Since the salt itself contains no hydroxide ions, they must have been provided by the molecules of water. Recall that the dissociation of water produces hydronium and hydroxide ions in equal numbers. Any reaction that preferentially removes H_3O^+ ions would result in a basic solution, while the removal of OH^- ions would cause the solution to become acidic. The excess of hydroxide ions in a sodium acetate solution suggests that one of the components of the salt forms an association with hydronium ions and removes them from solution. The ionic equation for the reaction may indicate how this is done.

Sodium acetate, like most salts, dissociates completely in aqueous solution, and is shown in the equation as aqueous sodium and acetate ions:

$$Na^+_{(aq)} + C_2H_3O_2^-{}_{(aq)} + H_3O^+_{(aq)} + OH^-_{(aq)} \longrightarrow$$

The products of the reaction are H_2O, NaOH, and $HC_2H_3O_2$. Since NaOH is a strong base, it will dissociate into Na^+ and OH^- ions. However, $HC_2H_3O_2$ is a weak acid, meaning that it exists primarily in the molecular form. The completed equation clearly shows why the solution is basic:

$$Na^+_{(aq)} + C_2H_3O_2^-{}_{(aq)} + H_3O^+_{(aq)} + OH^-_{(aq)} \rightleftharpoons H_2O_{(l)} + HC_2H_3O_{2(aq)} + Na^+_{(aq)} + OH^-_{(aq)}$$

Additional examples of hydrolytic reactions of salts are shown in Table 8.7.

TABLE 8.7 Hydrolysis of Salts

Acidic Salts	$CuSO_{4(s)} + 2H_2O_{(l)} \rightleftharpoons Cu(OH)_{2(aq)} + H_3O^+_{(aq)} + SO_4^{2-}{}_{(aq)}$
	$NH_4Cl_{(s)} + 2H_2O_{(l)} \rightleftharpoons NH_4OH_{(aq)} + H_3O^+_{(aq)} + Cl^-_{(aq)}$
Neutral Salts	$KCl_{(s)} + 2H_2O_{(l)} \rightleftharpoons K^+_{(aq)} + OH^-_{(aq)} + H_3O^+_{(aq)} + Cl^-_{(aq)}$
	$NH_4C_2H_3O_{2(s)} + 2H_2O_{(l)} \rightleftharpoons NH_4OH_{(aq)} + HC_2H_3O_{2(aq)}$
Basic Salts	$NaHCO_{3(s)} + 2H_2O_{(l)} \rightleftharpoons H_2CO_{3(aq)} + Na^+_{(aq)} + OH^-_{(aq)}$
	$NaC_2H_3O_{2(s)} + 2H_2O_{(l)} \rightleftharpoons HC_2H_3O_{2(aq)} + Na^+_{(aq)} + OH^-_{(aq)}$

From a careful study of the examples given in Table 8.7, we can formulate a set of guidelines for predicting the outcome of similar reactions.

1. The solution of a salt will be acidic if the salt was formed from a strong acid and a weaker base.

2. The solution will be basic if the salt was formed from a strong base and a weaker acid.
3. The solution of a salt will be neutral if the strengths of the acid and base were approximately equal.

It will be shown in a later section that salts play important roles in maintaining the pH of body fluids within a narrow range.

8.8 CARBOXYLIC (ORGANIC) ACIDS

You may have noticed that the structure of acetic acid differs slightly from that of the other acids that have been discussed thus far. Acetic acid contains hybridized carbon atoms and therefore is an organic compound. Its ionizable hydrogen atom is part of a cluster of atoms called a *carboxyl group.*

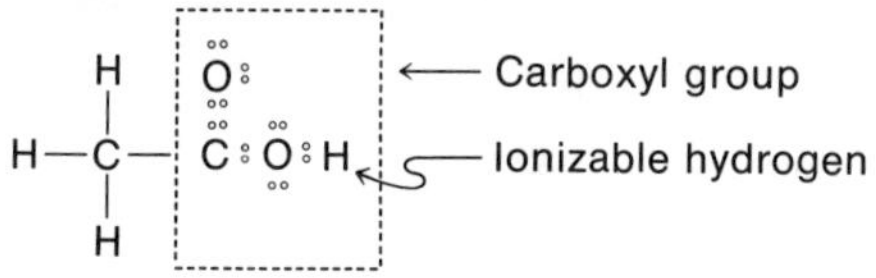

$$\mathrm{H{-}\overset{H}{\underset{H}{C}}{-}\boxed{\overset{\ddot{O}:}{C}:\ddot{\underset{..}{O}}:H}}$$

← Carboxyl group

← Ionizable hydrogen

Note the doubly bonded oxygen atom adjacent to the hydrogen in the carboxyl group. The strong electron-attracting ability of this atom weakens the covalent bond between hydrogen and oxygen, making it possible for polar water molecules to remove the hydrogen:

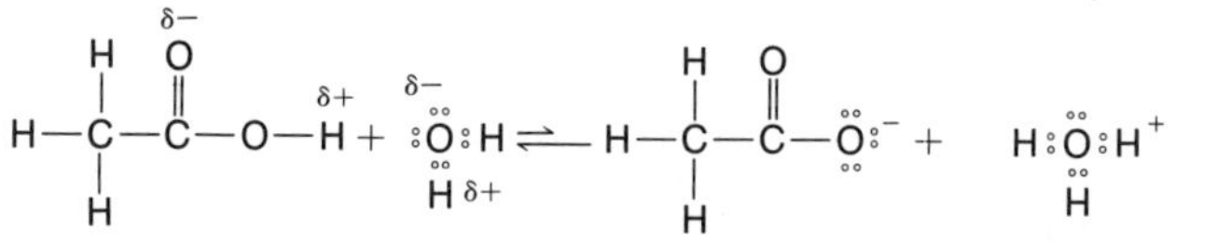

$$\underset{\text{Acetic acid}}{\mathrm{H{-}\overset{H}{\underset{H}{C}}{-}\overset{\overset{\delta-}{O}}{\overset{\|}{C}}{-}O{-}\overset{\delta+}{H}}} + \underset{\text{Water}}{\mathrm{\overset{\delta-}{:\ddot{O}:}H}_{\,\underset{H\,\delta+}{}}} \rightleftharpoons \underset{\text{Acetate ion}}{\mathrm{H{-}\overset{H}{\underset{H}{C}}{-}\overset{O}{\overset{\|}{C}}{-}\ddot{\underset{..}{O}}:^{-}}} + \underset{\text{Hydronium ion}}{\mathrm{H:\underset{H}{\ddot{O}}:H^{+}}}$$

Acids containing the carboxyl group are called *carboxylic acids.* Carboxylic acids are generally weak acids because their conjugate bases are strong. Many of them are solids under ordinary conditions, and only those having fewer than six carbon atoms are soluble to any extent in water. The procedure for naming carboxylic acids was discussed in Chapter 5. Their unique properties are described in Chapter 11.

8.9 AMINES-ORGANIC BASES

Amines are nitrogen-containing compounds whose properties are somewhat similar to those of ammonia. Like ammonia, they are quite water soluble, and their aqueous solutions are basic. Also like ammonia, they react with acids to produce salts. Amines are Lewis and Brönsted-Lowry bases because they can donate a pair of electrons or accept a proton. Structurally, the amines can be thought of as derivatives of ammonia in which one or more of the hydrogens has been replaced by a carbon chain, as shown below:

Ammonia

Methyl amine (a primary amine)

Ethyl amine (a primary amine)

Dimethyl amine (a secondary amine)

Trimethyl amine (a tertiary amine)

Note that in primary amines only one of the hydrogens has been replaced, while secondary amines have two carbon chains and tertiary amines have three. Amines react with organic acids to form *amides* as shown by the following equation:

$$CH_3\overset{O}{\overset{\|}{C}}-OH + H-\underset{H}{\underset{|}{N}}-CH_3 \longrightarrow CH_3\overset{O}{\overset{\|}{C}}-\underset{H}{\underset{|}{N}}-CH_3 + H_2O$$

Acetic acid (ethanoic acid) + Methyl amine → Acetamide + Water

The formation of an amide is an example of *dehydration synthesis,* a reaction in which the joining of two molecules is accompanied by the removal of a water molecule. Dehydration synthesis is the mechanism by which glucose molecules are combined to form cellulose and starch, and amino acids are joined to form proteins. It is also involved in the formation of nucleic acids, fats, and many other molecules of biological importance. Additional discussions of dehydration synthesis are found in Chapters 11 through 14.

DEHYDRATION SYNTHESIS IS ONE OF THE MOST IMPORTANT REACTIONS THAT TAKES PLACE IN THE BODY. THE REVERSE OF THIS REACTION IS KNOWN AS HYDROLYSIS. DIGESTION IS A HYDROLYTIC REACTION.

Hydrogen bonding occurs extensively in amides, as shown in the following example:

$$CH_3-\overset{O}{\overset{\|}{C}}-\underset{H}{\underset{|}{N}}-CH_3$$

$$\vdots \quad \text{Hydrogen bonding}$$

$$CH_3-\overset{O}{\overset{\|}{C}}-\underset{H}{\underset{|}{N}}-CH_3$$

The influence of these secondary bonds on the shapes of protein molecules is quite important and is discussed in Chapter 14.

8.10 AMINO ACIDS

Amino acids are amphoteric molecules containing both a carboxyl group and an amino (—NH_2) group. The structures of two common amino acids, glycine and alanine, are shown below:

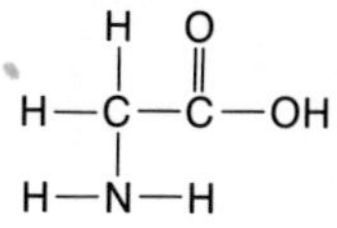

```
    H  O                     H   H   O
    |  ||                    |   |   ||
 H—C—C—OH                 H—C—C—C—OH
    |                        |   |
 H—N—H                       H   N—H
                                 |
                                 H
 Glycine                  Alanine
 (aminoacetic acid)       (α-amino propanoic acid)
```

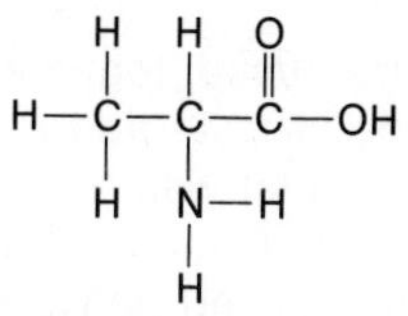

The carboxyl group is capable of donating a proton, while the lone pair of electrons in the amino group acts as a proton acceptor. An amino acid responds to the addition of base by forming water and an aqueous cation:

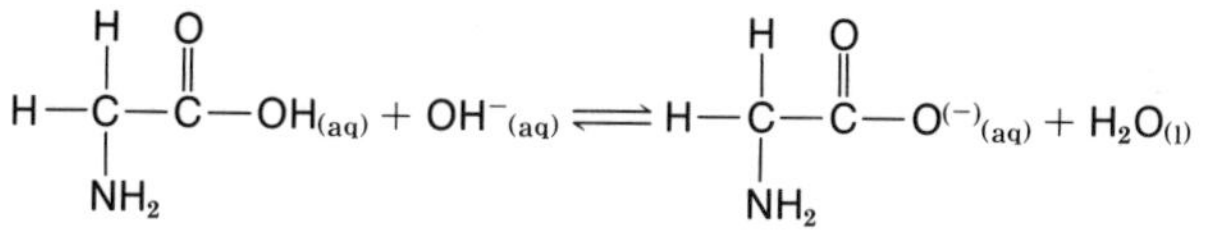

```
    H  O                             H  O
    |  ||                            |  ||
 H—C—C—OH(aq) + OH⁻(aq) ⇌ H—C—C—O(−)(aq) + H2O(l)
    |                                |
    NH2                              NH2
```

Its response to the addition of acid is to form water and an aqueous anion:

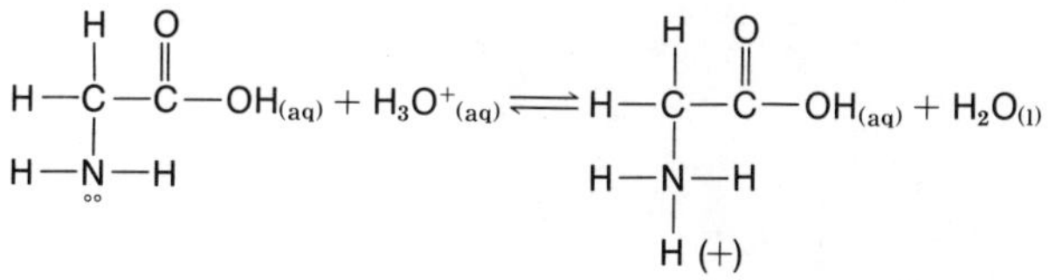

```
    H  O                                H  O
    |  ||                               |  ||
 H—C—C—OH(aq) + H3O+(aq) ⇌ H—C—C—OH(aq) + H2O(l)
    |                                   |
 H—N—H                               H—N—H
    ..                                  |
                                        H (+)
```

Amino acids tend to form dipolar ions when dissolved in water. The amino group removes protons from hydronium ions because it is a stronger base than water. The decreased concentration of H_3O^+ ion in the solution increases the dissociation of the carboxyl group. The result is a dipolar ion called a *zwitterion.* The formation of a zwitterion is shown below:

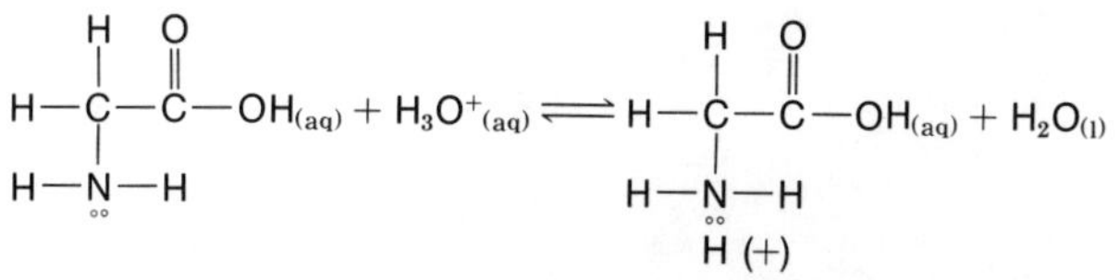

```
    H  O                                H  O
    |  ||                               |  ||
 H—C—C—OH(aq) + H3O+(aq) ⇌ H—C—C—OH(aq) + H2O(l)
    |                                   |
 H—N—H                               H—N—H
    ..                                  ..
                                        H (+)
```

1. Amino groups remove protons from hydronium ions.

$$\begin{array}{ccccccc}
 & H & O & & & H & O \\
 & | & \| & & & | & \| \\
H- & C- & C-OH_{(aq)} + H_2O_{(l)} & \rightleftharpoons & H- & C- & C-O^{(-)}_{(aq)} + H_3O^+_{(aq)} \\
 & | & & & & | & \\
H- & N-H & & & H- & N-H & \\
 & H\ (+) & & & & H\ (+) & \\
 & & & & & \text{Zwitterion} &
\end{array}$$

2. The decrease in concentration of H_3O^+ ion increases the dissociation of the carboxyl group.

As mentioned earlier, two amino acids can be fused together by dehydration synthesis. The same reaction can be used to add a third molecule to the first two, and so on, until the molecular weight is quite large. The process in which a number of individual molecules are linked together to form a single large molecule is called *polymerization*. Polymerization of amino acids may produce molecules whose molecular weights range from several hundred to several million. Amino acid polymers having molecular weights of more than 6000 are called proteins. Like the amino acids from which they are constructed, proteins have the ability to combine with acids or bases and to form dipolar ions. The contribution of proteins to the regulation of pH will be discussed shortly. The effect of their dipolar nature on their overall structure is described in Chapter 14.

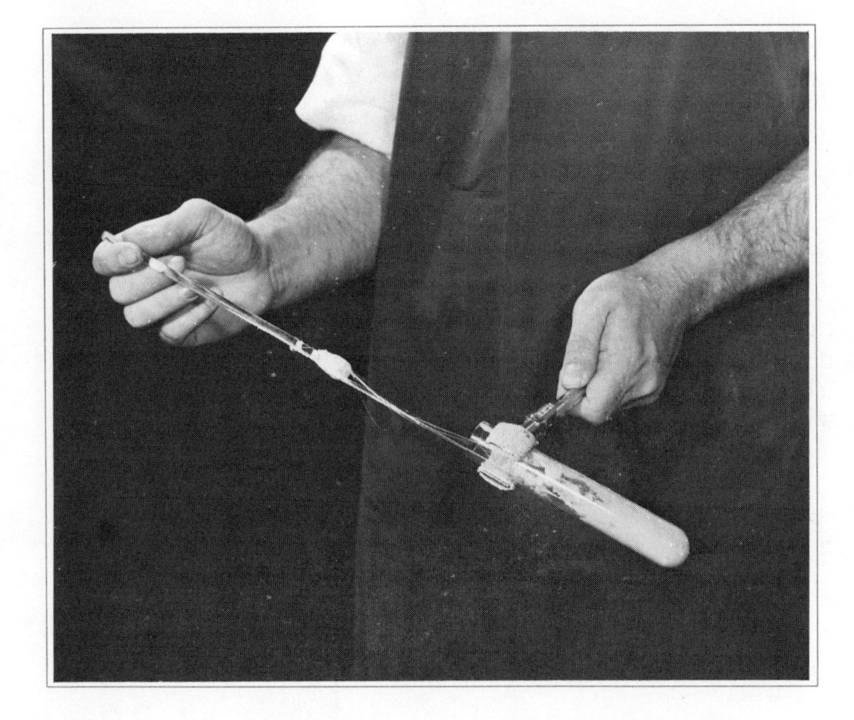

A chemist demonstrates the properties of a nylon-like polymer that he has produced in a test tube. (Courtesy E. I. du Pont de Nemours & Co., Wilmington, Delaware)

8.11 METABOLIC PRODUCTION OF ACID

The tissues of a living organism generate a constant stream of metabolic wastes, of which CO_2 is the major component. Carbon dioxide is a normal product of the oxidation of fats, carbohydrates, and proteins, which takes place within the cells. In higher organisms the CO_2 diffuses from the tissues into the blood, which carries it to the lungs, where it is excreted. Some of the CO_2 dissolves in the blood and is carried in simple solution, but the majority is transported in the form of HCO_3^- ion. The reactions involved in the elimination of CO_2 are important examples of acid-base chemistry and will be described in detail.

Because of concentration differences, some of the CO_2 diffuses into the red blood cells, where the enzyme carbonic anhydrase catalyzes its conversion into carbonic acid:

$$CO_{2(g)} + H_2O_{(l)} \xrightleftharpoons{\text{carbonic anhydrase}} H_2CO_{3(aq)}$$

Carbonic acid dissociates to form hydronium and bicarbonate ions:

$$H_2CO_{3(aq)} + H_2O_{(l)} \rightleftharpoons H_3O^+_{(aq)} + HCO_3^-{}_{(aq)}$$

The bicarbonate ion associates with K^+ ions, and the hydronium ions are neutralized by anionic protein components of the red blood cell—primarily those proteins that are part of the hemoglobin molecule. As the concentration of HCO_3^- ion increases within the red blood cells, there is a tendency for these ions to diffuse through the cell membranes into the plasma. However, the movement of negatively charged particles in one direction across a membrane would create a charge imbalance. Such an imbalance could be prevented either by the simultaneous movement of positively charged particles in the same direction, or by the passage of negatively charged particles in the opposite direction. Since the membrane of the red blood cell is relatively impermeable to cations, the passage of HCO_3^- ions is compensated by the diffusion of Cl^- ions from the plasma into the cell (Figure 8.4). This reversal in the positions of HCO_3^- and Cl^- ions is called the *chloride shift.* When this diffusion process reaches an equi-

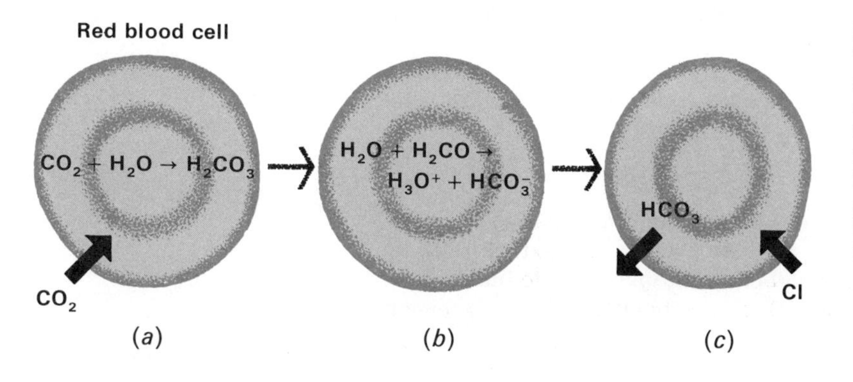

Figure 8.4
The chloride shift. (a) Carbon dioxide diffuses into red blood cells where it is converted into carbonic acid. (b) Carbonic acid dissociates into bicarbonate and hydronium ions. (c) As bicarbonate ions diffuse out, they are replaced by chloride ions.

librium, about 60% of the bicarbonate ions are in the plasma and the other 40% are in the red blood cells.

Because most of the H_3O^+ ions from carbonic acid are neutralized by proteins, the ratio between undissociated acid molecules (H_2CO_3) and bicarbonate ions is approximately 1:20. In other words, there are about 20 HCO_3^- ions for every H_2CO_3 molecule.

When the blood reaches the lungs, the entire set of reactions is reversed. Bicarbonate ion reenters the cells and is converted to carbonic acid, which is catalyzed by carbonic anhydrase into CO_2 and H_2O, and expelled from the lungs.

8.12 BUFFER SYSTEMS AND THE CONTROL OF pH

The mixture of carbonic acid and bicarbonate ion is more than just a convenient form for the transfer of a waste product; it is also part of a remarkable system which keeps the pH of the body within a very narrow range. Among the other components of the system are hemoglobin, the proteins of the blood, and salts of phosphoric acid. These substances are called *buffers* because they prevent changes in pH by neutralizing added acids or bases. A simple buffer consists of a weak acid and its salt. Reaction of the buffer with hydronium ions produces more of the acidic component, while the addition of base results in the formation of more of the acid salt. Thus the buffering action simply produces greater amounts of the buffer. In the discussion that follows, it should be remembered that the various components of the body's buffering system work together as a whole. However, because of the complexity of the system, it is easier to consider each component as if it were the only one.

The Carbonic Acid-Bicarbonate Buffer

In the event that hydronium ions are formed in metabolism and find their way into the blood, bicarbonate ions unite with them to form molecules of carbonic acid and water.

$$H_3O^+{}_{(aq)} + HCO_3^-{}_{(aq)} \longrightarrow H_2CO_{3(aq)} + H_2O_{(l)}$$

Excess carbonic acid is decomposed into carbon dioxide and water, which are excreted. Thus hydronium ions are removed from the blood and a change in pH is prevented.

Conversely, if hydronium ions are being lost from the blood, they are replaced by the ionization of molecular carbonic acid, and a rise in pH is prevented.

$$H_2CO_{3(aq)} + H_2O_{(l)} \longrightarrow HCO_3^-{}_{(aq)} + H_3O^+{}_{(aq)}$$

Note that hydroxide ions have not been mentioned: that is because the production of OH^- ion does not take place to any extent in living organisms. Notice also that the carbonic acid-bicarbonate buffer consists of a weak acid and its salt. Such combinations make excellent buffers.

A typical example of the buffering action of bicarbonate ion is its reaction with lactic acid. Lactic acid is produced from glucose during muscle contraction. The dissociation of lactic acid tends to lower the pH rather drastically. Since lactic acid is stronger than carbonic acid, its conjugate base (lactate ion) is weaker. The stronger base (bicarbonate ion) combines with the hydrogen from lactic acid, forming undissociated carbonic acid.

$$\underset{\substack{\text{Lactic acid}\\ \text{(acid)}}}{CH_3COH\overset{\overset{\displaystyle O}{\|}}{C}-O-H_{(l)}} + \underset{\substack{\text{Bicarbonate}\\ \text{(base)}}}{HCO_3^-{}_{(aq)}} \rightleftharpoons \underset{\substack{\text{Carbonic acid}\\ \text{(conj. acid)}}}{H_2CO_{3(aq)}} + \underset{\substack{\text{Lactate ion}\\ \text{(conj. base)}}}{CH_3COH\overset{\overset{\displaystyle O}{\|}}{C}-O^-{}_{(aq)}}$$

Phosphate Buffer

The phosphate buffer consists of HPO_4^- and $H_2PO_4^-$ ions. The buffering action of this ion pair is shown by the following equation:

$$H_3O^+_{(aq)} + HPO_4^{2-}{}_{(aq)} \rightleftharpoons H_2O_{(l)} + H_2PO_4^-{}_{(aq)}$$

The hydronium ion concentration can be reduced by combination with HPO_4^{2-} (the forward reaction) or increased by the hydrolysis of $H_2PO_4^-$ (the reverse reaction). Phosphate buffer is found primarily in the plasma, where the ions are associated with sodium.

Protein Buffers

The protein molecules which function as buffers include hemoglobin, serum albumin, fibrinogen, and serum globulins. At the normal pH of the body they exist as negatively charged proteinate ions. Proteinate ions combine readily with hydronium ions, as described earlier. Using the symbol Hb^- for the anionic form of hemoglobin, the reactions of hemoglobin are

$$Hb^- + H_3O^+_{(aq)} \rightleftharpoons HHb + H_2O_{(l)}$$

Hemoglobin accounts for about half the total buffering capacity of the blood.

8.13 ACIDOSIS AND ALKALOSIS

When the body is subjected to severe stress, as in prolonged muscle contraction or acute starvation, the buffer system may be unable to keep the pH within the normal range. If the pH rises above 7.45, the condition is called *alkalosis;* if it falls below 7.35, it is called *acidosis.* Acidosis does not refer to the excess acidity of gastric fluids, which is correctly described by the term hyperacidity. Acidosis is an abnormal condition resulting either from the overproduction of acidic substances or from failure of the excretory system. The consequences of acidosis are serious indeed. The nerve cells are extremely sensitive to changes in the pH of the surrounding fluid. They become less active as the pH decreases, and at a pH of 7.0 are so inactive that a comatose state is produced. Conversely, nerve cells become more irritable as the pH rises, and at a pH of about 7.5 cerebral convulsions may occur.

Physiologists often distinguish between respiratory acidosis and metabolic acidosis, although the distinction is an artificial one. *Respiratory acidosis* is a condition (often caused by slow or shallow breathing) in which hydronium ions accumulate because CO_2 is not being excreted fast enough. (Remember that hydronium ions are removed from the blood by conversion into molecular carbonic acid, which subsequently decomposes into CO_2 and water.) Breathing that is too rapid (hyperventilation) has the effect of removing too many hydronium ions from the blood, and causes respiratory alkalosis.

Metabolic acidosis refers to an uncompensated increase in hydronium ions from causes other than failure to excrete CO_2. For example, diarrhea can produce metabolic acidosis because the intestinal contents are alkaline, and the body replaces the alkaline components with bicarbonate ion drawn from the blood. This reduces the buffering ability of the blood and results in an increase in acidity. Conversely, vomiting may produce metabolic alkalosis because of the loss of acidic gastric juices. Prolonged acidosis causes nausea, which is followed by depression of the central nervous system, severe dehydration, deep coma, and finally death.

Prolonged muscle contraction can cause metabolic acidosis because it produces large amounts of lactic acid. Recall that lactic acid is normally converted to lactate ion by reaction with bicarbonate ion. If too much lactic acid is produced, the demand for bicarbonate ion is greater than the supply, and lactic acid ionizes to produce hydronium ions.

Starvation often leads to acidosis because the reserves of glycogen are quickly used up and the body then derives its energy from the oxidation of fats, and the products of this oxidation are acidic. For this reason persons who attempt to lose weight by crash diets in which the intake of food is severely restricted may bring about a condition of mild to moderate acidosis. Fad diets, such as the one in which only grapefruit and black coffee are taken, are especially dangerous.

One of the most common causes of metabolic acidosis is uncontrolled diabetes mellitus. The condition resembles starvation since the individual cannot utilize carbohydrates and metabolizes abnormal amounts of fats with the consequent production of acidic products—chief among them being acetoacetic acid (β-ketobutyric acid) and β-hydroxybutyric acid, whose formulas are shown below:

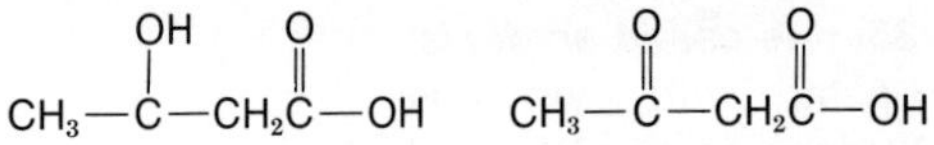

$$CH_3-\overset{\overset{\displaystyle OH}{|}}{C}-CH_2\overset{\overset{\displaystyle O}{\|}}{C}-OH \qquad CH_3-\overset{\overset{\displaystyle O}{\|}}{C}-CH_2\overset{\overset{\displaystyle O}{\|}}{C}-OH$$

β-hydroxybutyric acid β-ketobutyric acid

The condition that results from the production of excessive amounts of these products is known as *ketosis.*

The kidneys attempt to compensate for acidosis or alkalosis by varying the type of urine that is excreted. The urine normally contains phosphate ions which are derived from the metabolism of proteins. If the blood is too acidic, the kidneys excrete NaH_2PO_4; and if it is too alkaline, Na_2HPO_4 is found in the urine. The ability of the kidneys to compensate in this way is, of course, limited by the availability of sodium and phosphate ions.

Acidosis is prevented in diabetics by injection of insulin, which permits a more normal utilization of carbohydrates, or by ingestion of tolbutamide, which mimics the action of insulin in some individuals. Acidosis is treated by administration of a solution of sodium bicarbonate. Its chemical action aids in the removal of hydronium ions by forming un-ionized carbonic acid. The carbonic acid thus formed is eliminated via the lungs.

APPLICATION OF PRINCIPLES

1. What is the pH of 0.01 *M* solution of hydrochloric acid?
2. The acidity in a 20-ml sample of gastric fluid is neutralized by 15 ml of 0.15 *M* NaOH. Assuming that all of the acid in gastric fluid is HCl, what is its molar concentration?
3. Write equations to show the complete dissociation of phosphoric acid (H_3PO_4) in water.
4. Write a balanced equation for the hydrolysis of ammonium sulfate and predict whether the resulting solution will be acidic, neutral, or basic.
5. Solution A has a pH of 4.5, while the pH of solution B is 3.5. Which solution is more acidic? What is the difference in their hydronium ion concentration?
6. Which of the following chemical species are Lewis bases?
   (a) CH_4 (b) NH_4^+ (c) S^{2-} (d) HCl
7. Why should an excessive use of indicator be avoided in a titration?
8. Write the formulas for the conjugate bases of the following acids:
   (a) HSO_3^- (b) HCOOH (c) H_3PO_4 (d) $HClO_4$
9. Show by an equation the buffering action of the following zwitterion:

$$\underset{\underset{+NH_3}{|}}{CH_3CH_2}\overset{\overset{O}{\|}}{C}O\!:^-$$

10. Since NaOH is a stronger base than $Mg(OH)_2$, why isn't it used to neutralize excess stomach acid in cases of hyperacidity? Give at least two reasons.

11. What is the approximate pH of a solution in which bromocresol purple is purple and cresol red is yellow?
12. Would the following ion be most likely to exist in an acidic or a basic solution? Justify your answer.

$$CH_3CH(\overset{+}{N}H_3)\overset{O}{\overset{\|}{C}}OH$$

13. Would you expect the breathing of a person in a diabetic coma to be slower than normal or faster than normal? Why?

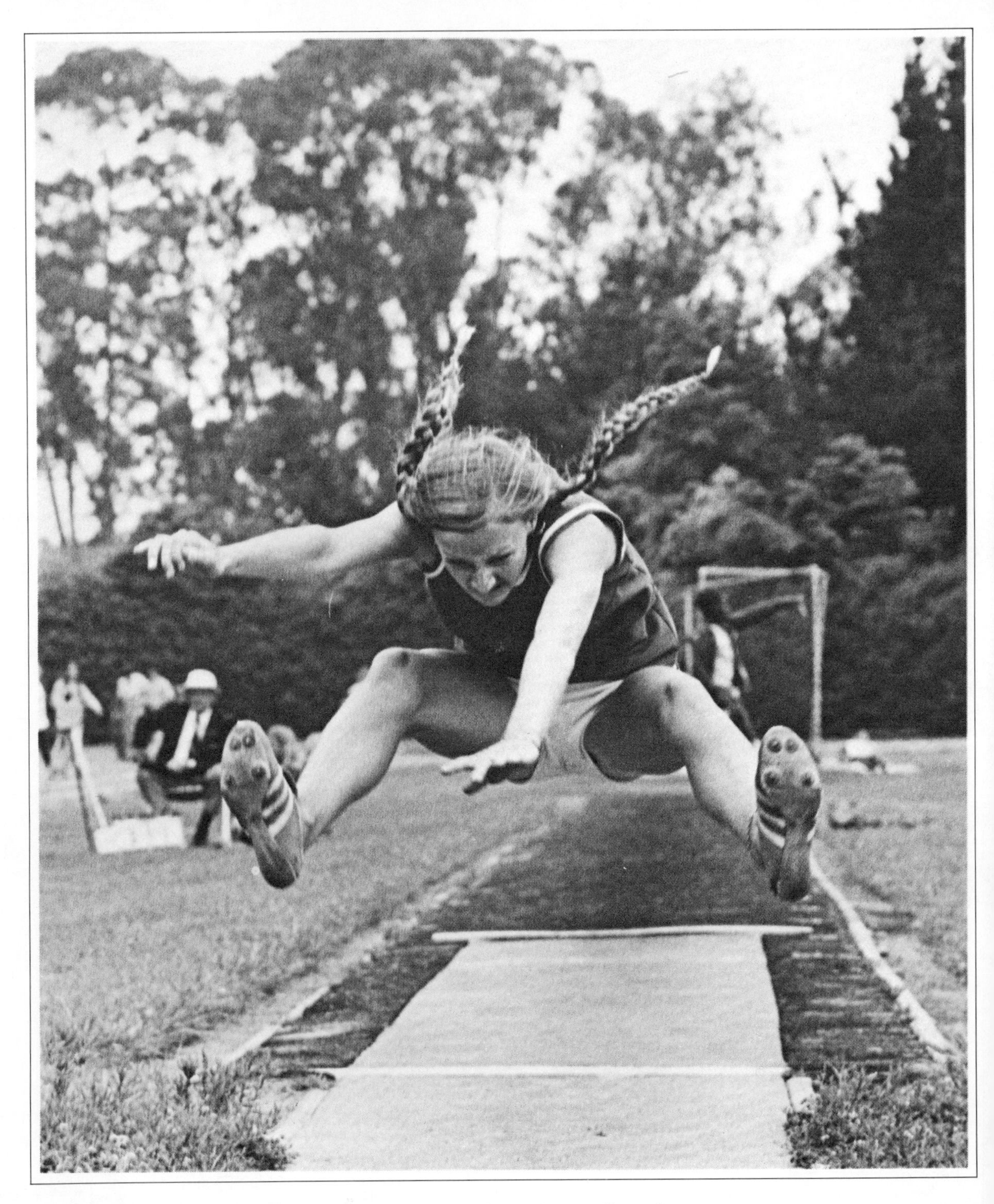

A transformation of chemical energy into kinetic energy. (Santa Barbara News-Press)

9

ENERGY RELATIONSHIPS IN CHEMICAL PROCESSES

LEARNING OBJECTIVES

1. Summarize the characteristics of energy conversions.
2. Relate the concept of entropy to a living organism.
3. Describe the conditions that will cause a chemical reaction to be spontaneous.
4. Compare exergonic and endergonic reactions in terms of energy entering and leaving the system.
5. Use a potential-energy diagram to describe a reaction.
6. Use graphs to explain how the rate of a reaction is affected by such variables as temperature, catalysts, concentration, surface area, and the nature of the reactants.
7. Describe the role of phosphate compounds in biological energy transformations.

KEY TERMS AND CONCEPTS

activated complex
activation energy
ATP
coenzyme
electron transport system
endergonic reaction
entropy
exergonic reaction
exothermic
First Law of Thermodynamics (Law of Conservation of Energy)
high-energy phosphate bond
Law of Mass Action
potential-energy diagram
reaction rate
Second Law of Thermodynamics
spontaneous reaction

One of the most interesting concepts of the universe is that of an incredibly complex device which was set in motion countless billions of years ago and, like the mechanism of a windup toy, is gradually running down. Although the hypothesis is difficult to prove, it is strongly supported by observations concerning energy transformations.

9.1 CHARACTERISTICS OF ENERGY CONVERSIONS

First of all, it is almost a certainty that the amount of energy in the universe does not change. Scientists have failed to discover a single physical, chemical, or biological process in which energy is either created or destroyed, although energy is readily converted from one form into another. Such conversions may be as simple as the burning of a candle or as complex as the explosion of an atomic bomb, yet in every case it has been possible to account for the energy both before and after the conversion. This fact is the basis for the *First Law of Thermodynamics* (also called the *Law of Conservation of Energy*), which states that the quantity of energy is not affected by its conversion from one form into another.

The energy obtained by the complete conversion of 1 pound of matter into energy would:

- equal 11 billion kilowatt hours of electricity or 10 billion calories
- run an electric iron for 1 million years
- operate a home furnace continuously for 25–50,000 years
- drive a car 180,000 times around the earth.

The discovery that there is a constant amount of energy in the universe would appear to be an argument against the hypothesis that the universe is running down. It would seem that, regardless of the form in which it appeared, the energy would still be available, and could be used to keep the various dynamic processes going. This logical argument fails because of a second general characteristic of energy conversions: every known method of energy conversion is somewhat inefficient. This means that, although all of the energy can be accounted for, part of it is no longer available for useful purposes. For example, consider the process that converts the chemical energy in gasoline into the kinetic energy of a moving automobile. A mixture of fuel and air is ignited in a chamber fitted with a movable piston. The burning mixture produces heat and light, as well as a number of gaseous substances. Some of the heat is absorbed by the piston and by the walls of the cylinder, which raises their temperature, and the rest of it is absorbed by the gases in the chamber, which increases the pressure that they exert. (Recall the temperature-pressure relationship described in Chapter 7.) As a result of this pressure increase, the piston is set in motion, and its motion is transmitted through a system of gears to the wheels of the car (Figure 9.1).

Only about 20% of the energy released by burning gasoline is actually used to propel an automobile. The heat absorbed by the cylinders and pistons does no work, and must be removed by the cooling system or the engine will be damaged; and part of the useful energy must be expended to overcome the friction between moving parts. Finally, energy is lost through the exhaust system as the hot gases are expelled to make room for more of the fuel-air mixture. In other words, this energy is not available to do the work of propelling the vehicle.

You mav argue that there are more efficient energy converters than the internal combustion engine—which is true. Steam engines may approach 40% efficiency (a strong point in favor of steam-driven vehicles), and fuel cells, which produce electricity and have no moving parts, are even more efficient (Figure 9.2). Nevertheless, in *every* energy transformation some of the energy is converted into a form that cannot be used to perform work. For example, the energy that is used in overcoming friction does not disappear, but for all practical purposes it is no longer available. This situation is reflected in the *Second Law of Thermodynamics,* which states that every energy transformation results in a reduction of the usable energy of the system.

In a sense energy is the currency of the universe. In every system, regardless of its nature, order is created and maintained by the expenditure of energy. And when the useful energy within a system decreases, the disorder within that system increases. Consider, for example, an aerosol spray can containing paint. Such a system is highly ordered (or organized) because the gaseous component is confined within a small space contrary to its natural tendency to spread out. It is also a system capable of doing work. When the valve is opened, the gas drives the paint out of the can and breaks it up into tiny droplets. Dispersing the gas molecules into the air decreases the amount of useful work that they can be called upon to perform; and the arrangement of the molecules is less orderly when they are scattered. Physicists use the word *entropy* to mean disorder, and describe the dispersion of the gas by saying that the entropy of the system has increased (Figure 9.3).

It can also be said that an organized arrangement of gas molecules is less probable than one in which the molecules are spread out. It is

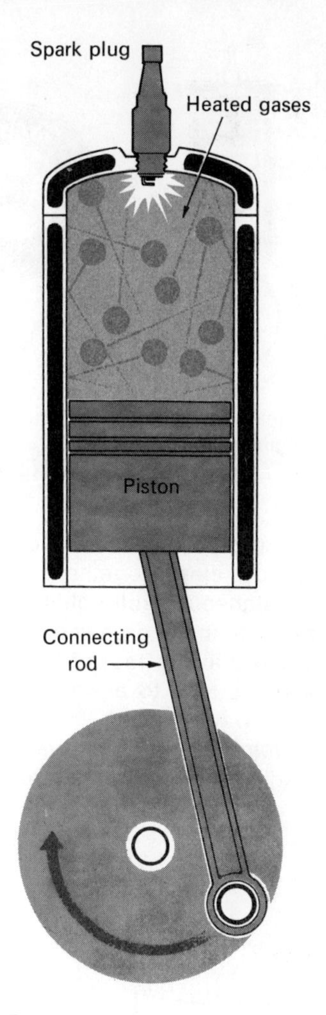

Figure 9.1
Conversion of chemical energy into kinetic energy in an automobile engine. Pressure exerted by the heated gases causes the piston to move. The motion of the piston is transmitted through the drive train to the wheels.

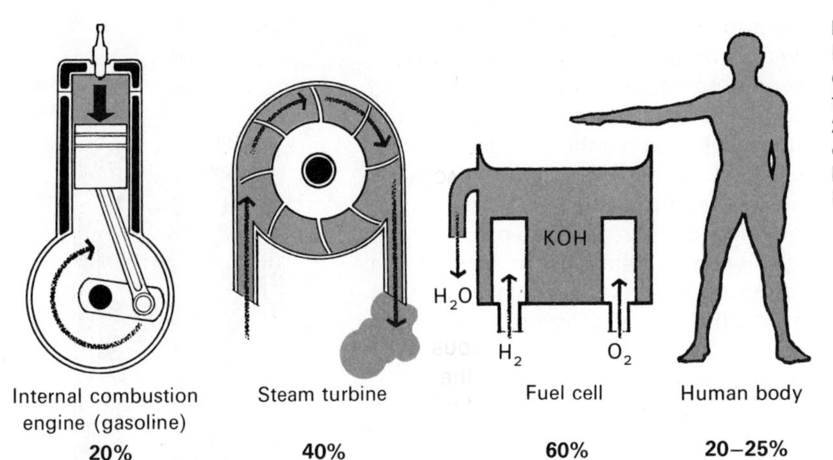

Figure 9.2
Relative efficiencies of some energy converters. Note that the body as an engine is only 20–25% efficient, since 75–80% of the energy released into the body is given off in the form of heat.

less probable because molecules of a gas have never been observed to move closer together spontaneously. Likewise, since objects do not spontaneously roll uphill, a boulder resting on a mountain top represents a more organized (therefore less probable) situation than a boulder at the foot of the mountain. It is a general observation that systems tend to assume more probable arrangements; which is another way of saying that the entropy of a system tends to increase. *All systems have a natural tendency to become disordered,* and changes leading to this condition are those that are most likely to occur.

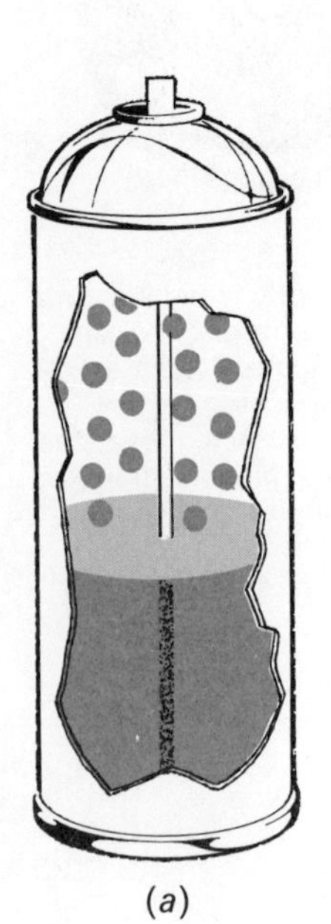

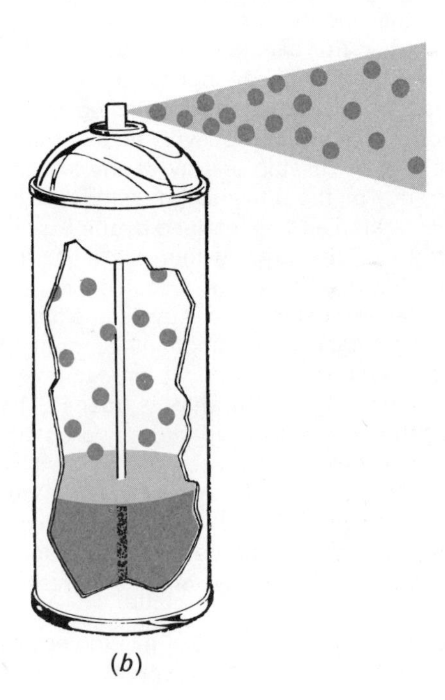

Figure 9.3
A spray can containing paint. (a) Arrangement is orderly: entropy is low. (b) Arrangement less orderly: entropy has increased.

A house of cards. The entropy of this system is very low since it represents a highly ordered (and improbable) arrangement of the cards. (Photograph by J. Asdrubal Rivera).

9.2 ENERGY REQUIREMENTS OF LIVING SYSTEMS

A living organism is a highly complex, and therefore a most improbable, arrangement of chemicals. Unless there is a constant expenditure of energy, the chemicals will revert to a random, more probable, arrangement. In other words, an organism must constantly fight against entropy. Energy is also required for the processes of growth, repair, and reproduction; and in organisms such as man, it is needed for muscle contraction, for pumping blood, and for the maintenance of a constant body temperature. The energy for these processes is obtained from chemical reactions. The energy relationships in these reactions will be described in the sections that follow.

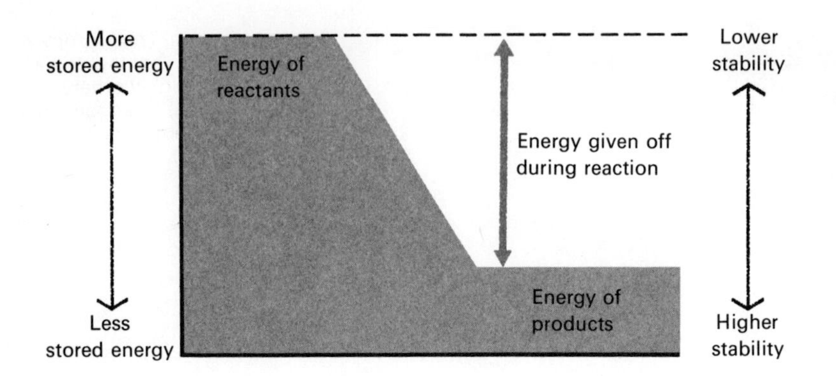

Figure 9.4
A spontaneous reaction. The products are more stable than the reactants because they contain less stored energy. Some of the energy of the reactants is changed into a different form during the reaction.

9.3 THE CAUSE OF CHEMICAL CHANGES

A chemical reaction is an exchange or a rearrangement of electrons in which bonds are broken and new bonds are formed. Under a specific set of conditions a given reaction is either spontaneous (probable) or nonspontaneous (improbable). Reactions are generally spontaneous when the products are more stable (i.e., have a smaller amount of stored energy) than the reactants (Figure 9.4). Since there is a decrease in the amount of stored (or *potential*) energy during the reaction, it would seem that some of the energy has disappeared—which would be contrary to the First Law of Thermodynamics. The energy has not been destroyed, however: it has simply been converted into a different form (generally electromagnetic radiation) and released from the system.

Reactions which release more energy than they absorb are called *exergonic.* (When the energy released in an exergonic reaction is mainly in the form of heat, the reaction is said to be *exothermic.*) Most reactions that occur spontaneously at body temperature are exergonic. However, spontaneous reactions are not always exergonic. In some instances a spontaneous reaction will be *endergonic;* that is, it will absorb more energy than it releases. Endergonic reactions occur because they result in an increase in entropy. As a general rule, an endergonic reaction will be spontaneous if the entropy of the products is greater than that of the reactants.

Using the word *spontaneous* to describe reactions that are probable under a given set of conditions may lead to the false impression that all such reactions are rapid, or that they occur as soon as the reactants are brought together. The truth is that spontaneous reactions, especially those involving covalently bonded substances, are often slow, and every reaction requires an initial input of energy to get it started. The energy used to start a reaction is called *activation energy.* It is convenient to think of activation energy as energy used to break the bonds in the reactants, although activation energy may also be used to join reactants in an arrangement (called an *activated complex*) which, because it is unstable, decomposes to form the products, which are stable.

9.4 ENERGY PROFILES OF CHEMICAL REACTIONS

The general form of an exergonic reaction is

$$\text{Reactants} = \text{Products} + \text{Energy}$$

while that of an endergonic reaction is

$$\text{Reactants} + \text{Energy} = \text{Products}$$

Examples of exergonic reactions are

$$C_{(s)} + O_{2\,(g)} \longrightarrow CO_{2\,(g)} + 94 \text{ kcal/mole}$$

$$CaO_{(s)} + H_2O_{(l)} \longrightarrow Ca(OH)_{2(aq)} + 15.6 \text{ kcal/mole}$$

$$2\ H_{2\,(g)} + O_{2\,(g)} \longrightarrow 2\ H_2O_{(l)} + 136.6 \text{ kcal/mole}$$

$$\text{Glucose} + \text{Oxygen} \longrightarrow \text{Pyruvic acid} + \text{Water} + 141 \text{ kcal/mole}$$

The following are examples of endergonic reactions:

$$2\ HgO_{(s)} + 43.3 \text{ kcal/mole} \longrightarrow 2\ Hg_{(l)} + O_{2\,(g)}$$

$$H_{2\,(g)} + I_{2(s)} + 12.4 \text{ kcal/mole} \longrightarrow 2\ HI_{(g)}$$

$$6\ CO_{2\,(g)} + 12\ H_2O_{(l)} + 690 \text{ kcal/mole} \longrightarrow C_6H_{12}O_{6(s)} + 6\ O_{2\,(g)} + 6\ H_2O_{(l)}$$

$$\text{Phosphoric acid} + \text{Adenosine diphosphate} + 7.4 \text{ kcal/mole} \longrightarrow \text{Adenosine triphosphate}$$

Equations written in this fashion are useful because they provide an energy profile of the reaction: that is, they show the relative potential energies of reactants and products. For instance, in the reaction between carbon and oxygen (the first of the exergonic examples above), the combined potential energy in one mole of carbon and one mole of oxygen is 94 kcal greater than the potential energy in one mole of carbon dioxide. The relative energies of the reactants and products are perhaps more clearly shown by a potential-energy diagram (Figure 9.5). Time is shown on the horizontal axis and potential energy on the vertical axis. Thus the diagram represents the change in potential energy as the reaction progresses. The diagram clearly depicts an exergonic reaction because the potential energy of the products is lower than that of the reactants. It also shows the energy barrier, described earlier, which must be overcome before the reaction can take place. We know that this barrier exists since carbon must be heated before it will combine with oxygen. Once a portion of the carbon starts to burn, it is not necessary to continue heating it, since the remainder of the reactants obtain their activation energy from the reaction itself.

Figure 9.6 depicts the endergonic decomposition of solid mercury(II) oxide into liquid mercury and gaseous oxygen. Despite the fact that the products are less stable than the reactant, the reaction is still spontaneous because it leads to an increase in entropy. The entropy of the products is greater than that of the reactant because liquids and gases represent greater amounts of disorder than solids.

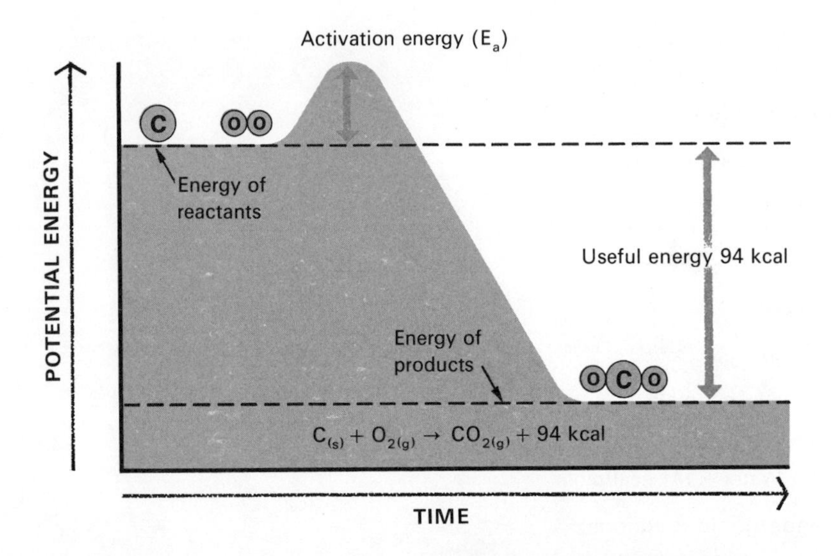

Figure 9.5
A potential-energy diagram of the exergonic reaction:

$C_{(s)} + O_{2(g)} \longrightarrow CO_{2(g)} + \text{Energy}$

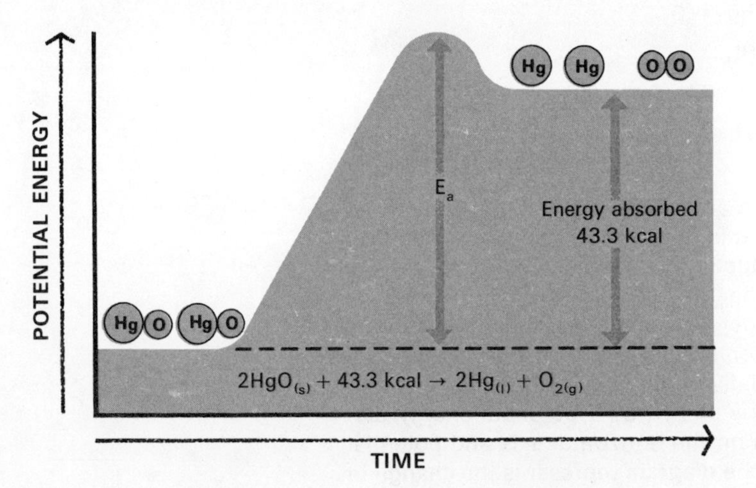

Figure 9.6
A potential-energy diagram for the endergonic reaction:

$2\ HgO_{(s)} + \text{Energy} \longrightarrow 2\ Hg_{(l)} + O_{2\ (g)}$

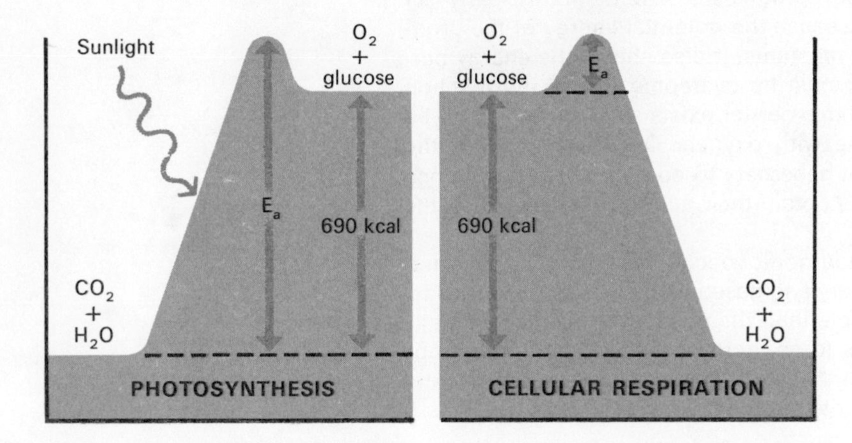

Figure 9.7
Potential-energy diagrams for photosynthesis and cellular respiration.

From a biological standpoint there are two energy transformations of particular importance—photosynthesis and cellular respiration. Their potential-energy diagrams are shown in Figure 9.7. *Photosynthesis* is the series of reactions by which energy obtained from sunlight is converted into chemical (potential) energy. The end products of photosynthesis are oxygen and glucose. *Cellular respiration* makes the chemical energy of glucose available to an organism through reactions that gradually oxidize the molecule to carbon dioxide and water. These reactions are described in more detail in Chapter 12.

9.5 REACTION RATES

The *rate of a chemical reaction* is the quantity of reactant that is converted into product during a specified period of time. Among the factors affecting reaction rates are temperature, catalysts, concentration, and the nature of the reactants. In general, ionic substances react faster than covalent ones; and reaction rates tend to increase with increases in concentration or temperature. The effect of each of these variables is possibly best described by a model in which a chemical change is the result of fruitful collisions between competent chemical species. In other words, the reactants must contact each other in a specific way before any exchange or rearrangement of electrons can take place; therefore, the rate of a reaction is controlled by variables that influence the number of contacts and the energy and relative positions of the reactants during contact. The effect of each variable is described next.

Concentration

As mentioned previously, the reacting species must come together before a reaction can take place. Remember that molecular motions are random, and so these contacts occur by chance. The likelihood of a collision is increased if the concentration of either reactant increases, as shown by the following hypothetical situation.

Let us assume that we are dealing with a reaction of the type $A + B \rightarrow C$, and that we have a box containing one particle of each reactant. A reaction can occur only if A and B collide. Since the motion of the particles is random, the probability that this will happen is rather low (Figure 9.8a). However, if the concentration of one of the reactants is doubled (Figure 9.8b), the probability that a collision will occur becomes twice as great, and the rate of the reaction should increase proportionately. If the concentration of both reactants is doubled (Figure 9.8c), the number of potential collisions will be four times as great as it was when the box contained only one particle each of A and B.

The relationship between reaction rate and the concentrations of the reactants is summarized in the *Law of Mass Action,* which states that the rate of a reaction is proportional to the product of the concentrations of the reactants. Thus if concentrated sulfuric acid is spilled

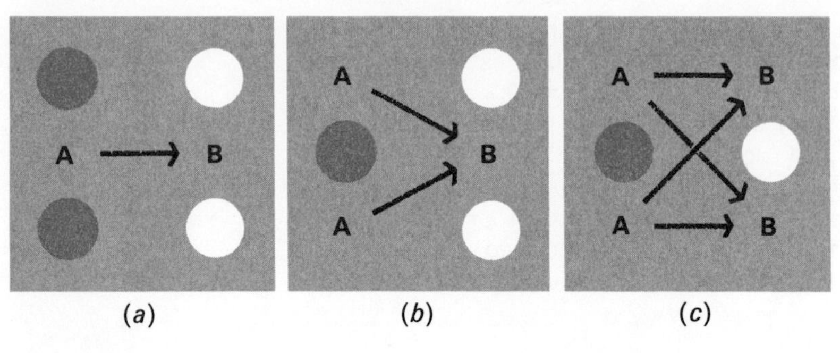

Figure 9.8
The relationship between concentration and reaction rate. As the concentration of either reactant increases, the number of possible fruitful combinations also increases.

on the skin or clothing, the reaction will be much faster and more damaging than that of an equal volume of dilute acid. The effect of concentration on reaction rates may be demonstrated by heating steel wool in air and in pure oxygen. The metal merely glows in air, but it burns in a brilliant shower of sparks in oxygen. You may recall that three astronauts died in an intense fire that literally incinerated their oxygen-filled Apollo space capsule in 1967. The chemical reactions in the body are also speeded up by high concentrations of oxygen.

Interior of Saturn spacecraft following a fire which killed astronauts Virgil Grissom, Edward White, and Roger Chaffee. The intensity of the fire was due to the high concentration of oxygen inside the capsule. (Courtesy NASA)

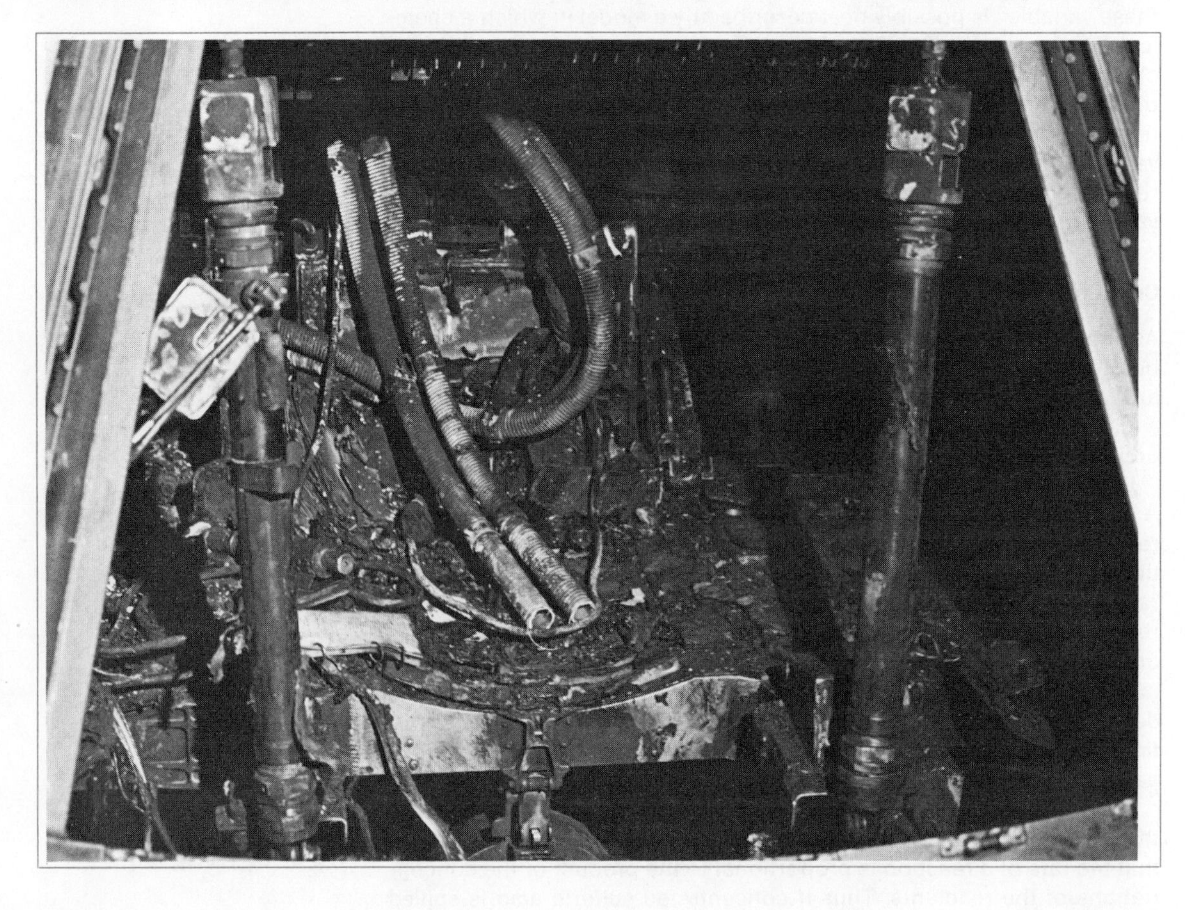

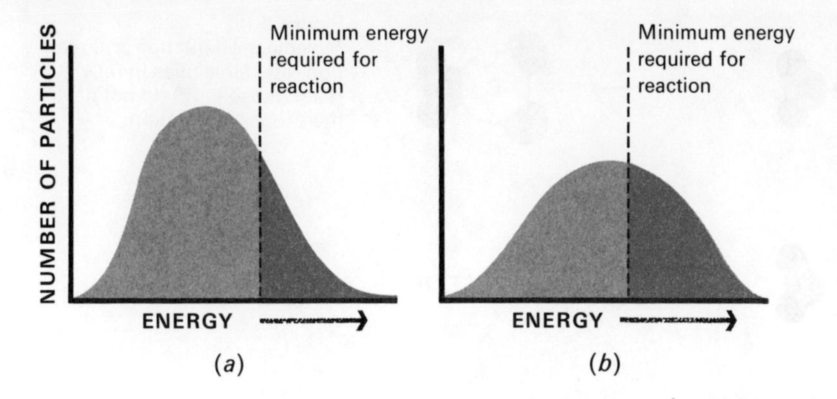

Figure 9.9
The effect of temperature on a reaction rate. (a) Particles to the left of the energy barrier fail to react because they do not possess the minimum required energy. (b) Increasing the temperature raises the average kinetic energy and increases the number of particles that possess the minimum energy for reaction.

Temperature

As described in Section 7.2, there is a wide variation in velocity among the individual particles in a sample of gas. This is also true of particles in the liquid and solid states. Since the kinetic energy of a particle is directly proportional to its velocity, we can say that in any sample of a substance the particles will not all have the same kinetic energy.

If we assume that the reactants must not only collide, but that they must possess a certain minimum amount of energy before the collision will result in an exchange or a rearrangement of electrons, then some of the reactants may collide without forming products (Figure 9.9a). Raising the temperature generally increases the rate of a reaction because it raises the average kinetic energy of the reactants, thus increasing the number of particles that possess the minimum reaction energy (Figure 9.9b).

On the practical side, the relationship between reaction rate and temperature is used to advantage in the preservation of food, the reduction of pain, and certain surgical procedures. In each of these examples the desired effect is produced by lowering the temperature: thus refrigeration retards the spoilage of food by slowing the reactions that cause spoilage; temperatures in the neighborhood of 0°C reduce the activity of the nerves and can be used to produce anaesthesia in the skin; and heart surgery can be carried out more easily after the action of the heart has been slowed by cooling.

The Nature of the Reactants

Reactions between molecules are usually much slower than ionic reactions. Individual ions exert an attractive force in all directions; therefore, it is not necessary for the ions to collide in a specific way before reaction can take place. However, the orbitals in a molecule have precise locations, and an exchange or rearrangement of electrons can occur only when the proper orbitals in the reacting molecules come together. This means that molecular species, especially

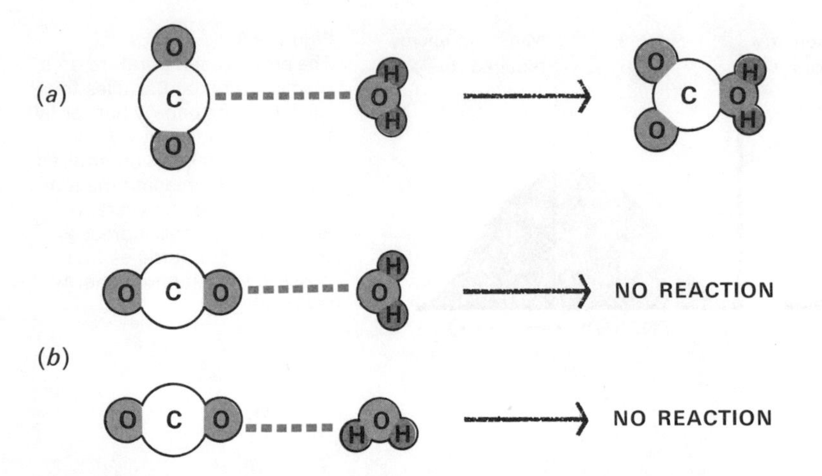

Figure 9.10
Molecular orientation and reaction rate. Molecules in (a) will react; those in (b) do not have the proper orientation.

those that are large or complex, must have a certain orientation when they collide, or else they will not react.

The influence of spatial orientation on reaction rate is shown in Figure 9.10. A molecule of carbon dioxide will react with a water molecule only when its carbon atom collides with the oxygen atom of water. Collisions involving other parts of the two molecules are ineffective.

The reactions that take place during the digestion of food involve large complex molecules, and require extended periods of time for completion. This fact should be kept in mind the next time you are tempted to engage in strenuous activity immediately after eating.

Catalysts

A catalyst is a substance that alters the rate of a chemical change but is neither a reactant nor a product of the reaction. It is believed that a catalyst affects the rate of a reaction by raising or lowering its activation energy. A catalyst does not initiate a reaction—it merely changes the rate of a reaction that is already taking place. Catalysts, and catalytic effects, are so diverse that it is difficult, if not impossible, to further describe them in general terms. The following statements will indicate the complexity of the subject. A catalyst may be an element, an ion, or a molecule. A catalyst that changes the rate of a particular reaction may have no effect on a similar reaction. A large percentage of reactions appear to be unaffected by the presence of chemicals other than the reactants or products.

By studying a few selected reactions, chemists have been able to increase their understanding of the catalytic mechanism. One such reaction is the conversion of hydrogen and oxygen into water in the presence of platinum metal. The reaction appears to take place in two stages. In the first stage, hydrogen molecules form an activated complex with platinum:

$$2\,Pt + H_2 \longrightarrow 2\,PtH$$

During this process, the hydrogen molecules are split and the individual atoms are adsorbed or held on the surface of the platinum. The spacing between the hydrogen atoms is such that each atom in the oxygen molecule can make contact with a pair of them. As bonds form between hydrogen and oxygen, the platinum is freed to repeat the process:

$$4\ PtH + O_2 \longrightarrow 2\ H_2O + 4\ Pt$$

The adsorption of hydrogen by the platinum facilitates its reaction with oxygen and lowers the activation energy, as shown in Figure 9.11.

Catalysts in certain reactions have an inhibitory effect. For example, acetamide is added to solutions of hydrogen peroxide to slow their decomposition into water and oxygen. Inhibitors such as butylated hydroxytoluene (BHT) and butylated hydroxyanisole (BHA) are incorporated into the packaging material of cereals and potato chips to retard the reactions that produce changes in flavor or appearance. Other inhibitors are used in rubber products to prevent gradual loss of flexibility. The widespread use of such compounds has given rise to a new consciousness of the chemicals in food, leading to the growing popularity of so-called natural foods.

Surface Area

Potential reactions often involve reactants that are in different physical states. For example, solid carbon combines with gaseous oxygen, liquid water combines with gaseous carbon dioxide, and aqueous hydronium ions react with solid aluminum. In such cases the reaction rate increases when more opportunities are provided for the reactants to contact each other. Crushing or grinding a solid increases its area of contact and usually increases its reaction rate as well. One often

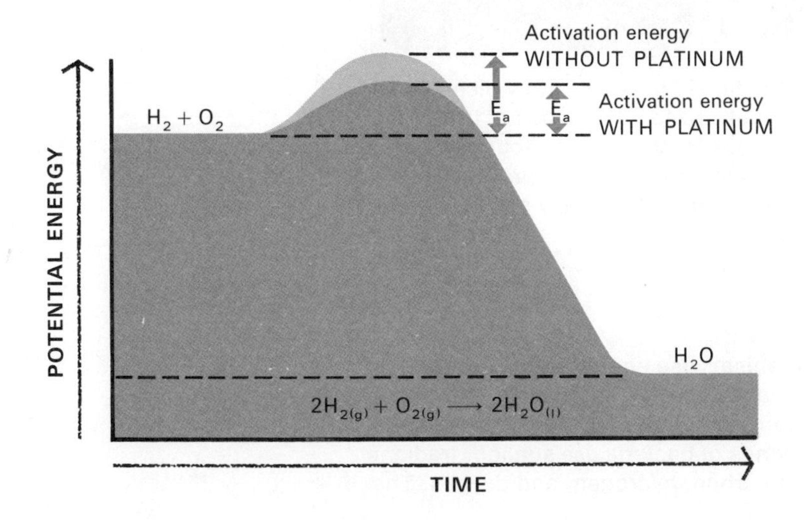

Figure 9.11
The effect of a platinum catalyst on the formation of water from hydrogen and oxygen. The catalyst lowers the activation energy.

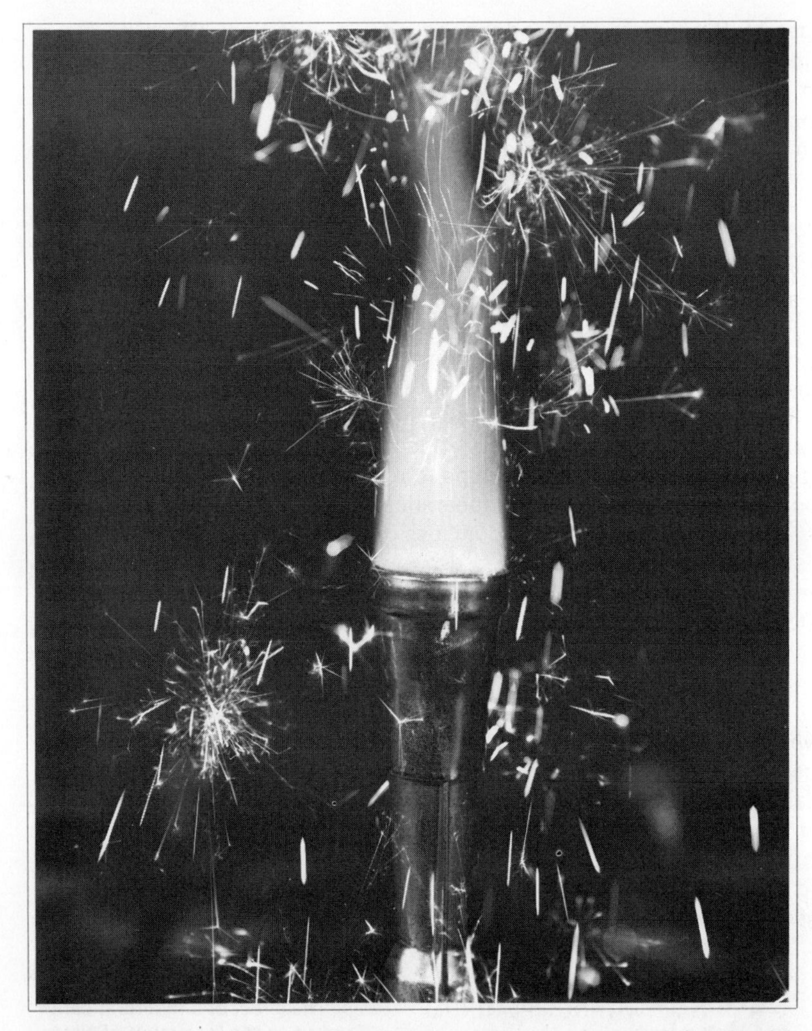

The effect of surface area on reaction rate. The brilliant shower of sparks was produced by sprinkling powdered iron in the flame. (Photograph by J. Asdrubal Rivera)

tragic example of this principle in action is the explosive violence with which coal burns after it has been pulverized. Thousands of lives have been lost as a result of dust explosions in coal mines, flour mills, and woodworking shops. Extreme care should be taken to prevent sparks or flames in dust-laden atmospheres, since almost any solid will burn vigorously if it is in powdered form.

9.6 PHOSPHATES AND ENERGY TRANSFORMATIONS

The activities for which living organisms use energy were described in Section 9.2. It is a general observation that all organisms, regardless of type, ultimately obtain the energy for their activities from the sun. Green plants, algae, and certain kinds of bacteria use sunlight (radiant energy) to link together atoms of carbon, hydrogen, and oxygen. The energy becomes available for use when the bonds between these atoms are broken.

Molecules containing phosphate groups have key roles in both the energy-storing and energy-releasing reactions. The structure of one of the most important of these molecules, adenosine triphosphate (ATP), is shown below:

$$\underbrace{\text{Adenosine}}\text{—}CH_2\text{—O—}\overset{\overset{\displaystyle O}{\|}}{\underset{\underset{\displaystyle OH}{|}}{P}}\text{—O}\sim\overset{\overset{\displaystyle O}{\|}}{\underset{\underset{\displaystyle OH}{|}}{P}}\text{—O}\sim\overset{\overset{\displaystyle O}{\|}}{\underset{\underset{\displaystyle OH}{|}}{P}}\text{—OH}$$

Adenosine

ATP has been found in the cells of every living organism. It is composed of a molecule of adenosine coupled with three molecules of phosphoric acid—hence the name adenosine *tri*phosphate. The bonds indicated by the symbol $\sim$ are of special importance. They are called *high-energy phosphate bonds* to distinguish them from other bonds involving phosphoric acid. When one of these bonds is broken, approximately 7 kcal of energy are made available to the cell. Hydrolysis of the terminal phosphate group converts ATP into adenosine disphosphate (ADP). The reaction is

$$ATP + H_2O \xrightleftharpoons{\text{catalyst}} ADP + H_3PO_4 + 7.4 \text{ kcal}$$

Hydrolytic removal of a second phosphate group changes ADP into adenosine monophosphate (AMP):

$$ADP + H_2O \xrightleftharpoons{\text{catalyst}} AMP + H_3PO_4 + 6.8 \text{ kcal}$$

The phosphate bond in AMP is not a high-energy bond. Removal of the third phosphate group leaves adenosine:

$$AMP + H_2O \xrightleftharpoons{\text{catalyst}} \text{Adenosine} + H_3PO_4 + 2.2 \text{ kcal}$$

Equivalent amounts of potential energy are stored by the reverse reactions.

The general pattern of energy transfer in a cell is as follows: (1) a molecule containing a large amount of chemical energy (such as glucose) undergoes stepwise oxidation, releasing energy with each step; (2) some of the energy takes the form of heat, while the remainder is used in the formation of high-energy phosphate bonds; (3) the cell obtains useful energy by hydrolyzing the phosphate bonds. The overall process is depicted in Figure 9.12.

It is important to note that the amount of potential energy in glucose is so much greater than that required for normal cellular activities

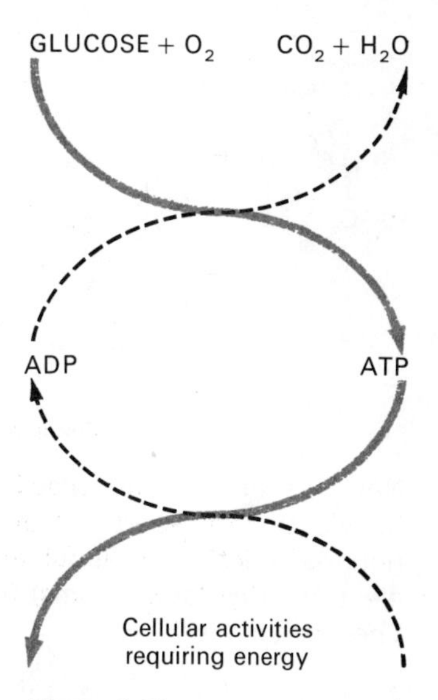

Figure 9.12
The general pattern of energy transfer in a living cell. The solid lines indicate the flow of energy.

that, were it released all at once, most of it would be wasted. By coupling the release of energy to the formation of phosphate bonds, this waste of energy is prevented.

The *pyrophosphates,* such as ADP and ATP, are not the only phosphate compounds which have active roles in energy-releasing processes. Many of the enzymes that bring about the stepwise oxidation of energy-rich compounds are inactive unless a specific molecule, called a *coenzyme,* is also present. A number of these coenzymes are phosphates, and several of them closely resemble the pyrophosphates. Two of the most important coenzymes are nicotinamide adenine dinucleotide (NAD), and flavin adenine dinucleotide (FAD), whose structures are

Nicotinamide adenine dinucleotide (NAD)

Flavin adenine dinucleotide (FAD)

NAD was formerly known as DPN. Both of these coenzymes are part of the electron transport system. This system consists of a series of compounds which bring about the oxidation of a molecule by removing electrons (hydrogen atoms) from it and transferring them to oxygen. The first step is

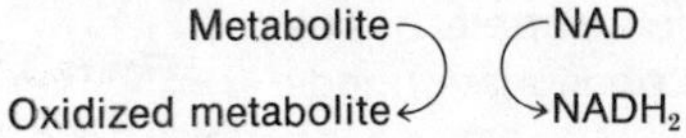

$$\text{Metabolite} \rightarrow \text{Oxidized metabolite} \qquad \text{NAD} \rightarrow \text{NADH}_2$$

The transfer of the electrons from $NADH_2$ to FAD is the second step and is accompanied by the conversion of ADP to ATP:

The Role of Vitamins in Metabolism

It has been established that a number of vitamins, especially those of the B complex, are structural components of coenzymes. For example, thiamine (vitamin B_1) is a principal part of the coenzyme that catalyzes the oxidation of pyruvic acid. Pantothenic acid is a component of coenzyme A, the carrier of the acetyl group in the Krebs cycle. Nicotinamide is found in both NAD and NADP, which have important roles as hydrogen acceptors during biological oxidations. Vitamin B_6 (pyridoxine) is a part of the coenzyme that is involved in transaminations, while riboflavin (vitamin B_2) is a hydrogen carrier in the cytochrome system.

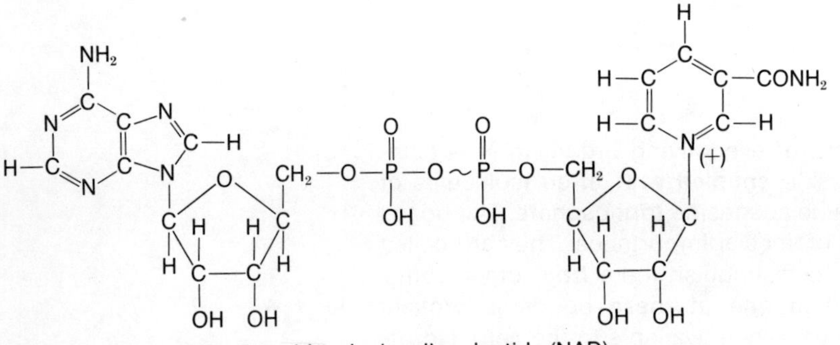

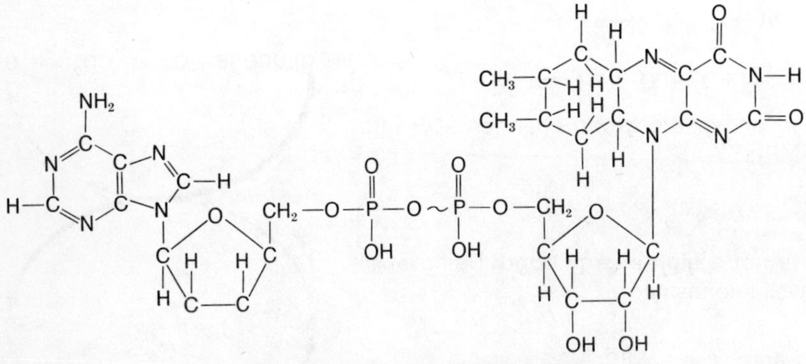

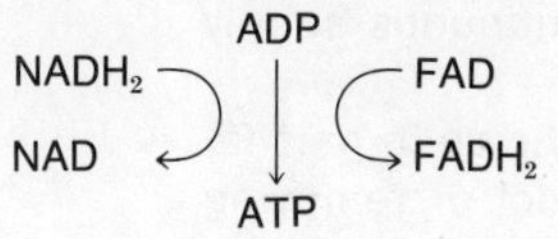

ADP
$NADH_2$ → NAD
FAD → $FADH_2$
ATP

From $FADH_2$ the electrons are transferred to Coenzyme Q, and thence successively to cytochromes b, c_1, a, and a_3. The cytochromes are similar in structure to the nonprotein portion of hemoglobin. In the final step, the electrons are passed from cytochrome a_3 to oxygen, and the end product is water.

Another important coenzyme is Coenzyme A, whose structure is

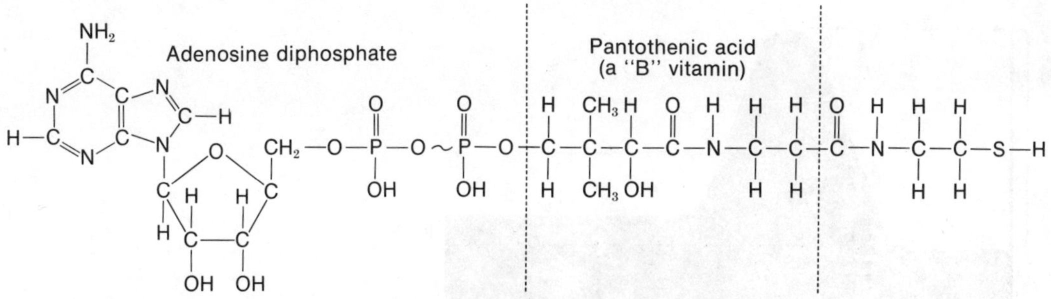

Adenosine diphosphate

Pantothenic acid (a "B" vitamin)

Coenzyme A (H—S—CoA)

The functional group of Coenzyme A is the sulfhydryl (—SH) group; thus Coenzyme A is usually abbreviated H—S—CoA. Coenzyme A is involved in the synthesis and the oxidation of fats, the oxidation of glucose, and several other important metabolic processes which are described in Chapters 12 and 13.

APPLICATION OF PRINCIPLES

1. Draw a potential-energy diagram for the reaction

$$\text{glucose} \longrightarrow 2\ CH_3CHOHCOOH + 52\ \text{kcal}$$

2. What effect will tripling the concentrations of both NO_2 and CO have on the rate of the following reaction?

$$NO_{2\,(g)} + CO_{(g)} \longrightarrow CO_{2\,(g)} + NO_{(g)}$$

3. Classify each of the following reactions as either endergonic or exergonic:

(a) $2\ SO_{2\,(g)} + O_{2\,(g)} \xrightarrow{Pt} 2\ SO_{3(g)} + 47\ \text{kcal}$

(b) $3\ H_{2\,(g)} + N_{2\,(g)} \rightleftharpoons 2\ NH_{3(g)} + 21.8\ \text{kcal}$

(c) $H_2O_{(g)} + C_{(s)} + 31.4\ \text{kcal} \longrightarrow CO_{(g)} + H_{2\,(g)}$

4. Explain why an endergonic reaction requires a continuous supply of energy.
5. Draw a potential-energy diagram for the reaction shown in part (a) of Question 3. On the same diagram show the effect of removing the platinum catalyst.
6. The potential-energy diagram for the reaction $2\ HI \rightarrow I_2 + H_2$ is shown below. (a) Is the reaction exergonic or endergonic? (b) Calculate the activation energy of the reaction. (c) How much useful energy could be derived from the conversion of two moles of HI into hydrogen and iodine?

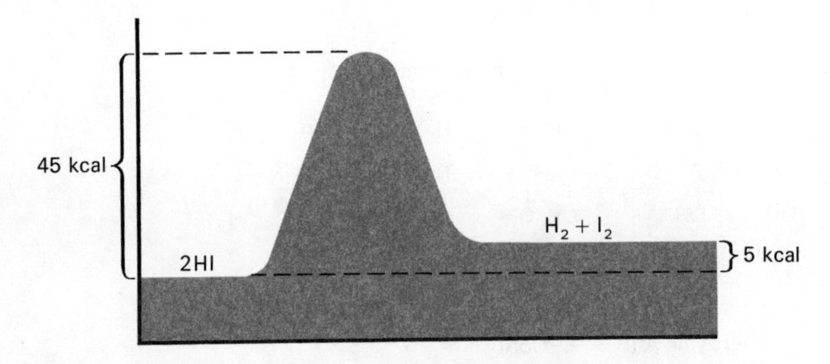

7. Does reaction (c) in Question 3 take place because it leads to an increase in entropy or because the products are more stable than the reactants? Justify your answer.

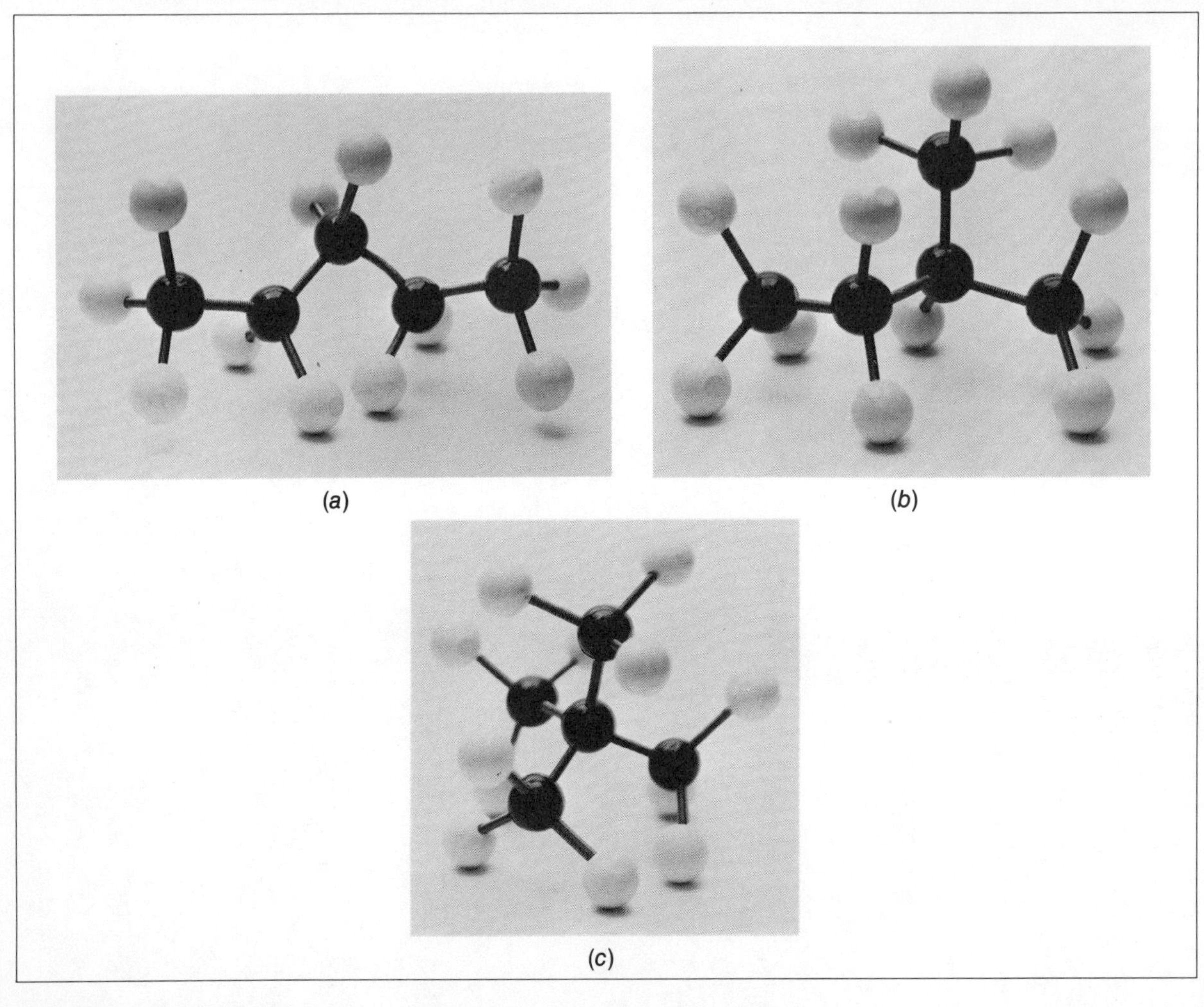

Isomers of C_5H_{12}: (a) n-pentane, (b) methylbutane (isopentane), and (c) 2,2-dimethylpropane. (Photographs by J. Asdrubal Rivera)

10

CARBON CHAINS, RINGS, AND HETEROCYCLIC COMPOUNDS

LEARNING OBJECTIVES

1. Compare the structures, general formulas, and chemical properties of saturated and unsaturated hydrocarbons.
2. Draw structural formulas for isomers of a given hydrocarbon.
3. Write the IUPAC names for hydrocarbons from their structural formulas.
4. Draw structural formulas for simple hydrocarbons from their IUPAC names.
5. Distinguish between aromatic and heterocyclic compounds.
6. Distinguish between aromatic and aliphatic hydrocarbons.

KEY TERMS AND CONCEPTS

- addition reaction
- alkane
- alkene
- alkyne
- aliphatic structure
- aromatic structure
- carboxyl group
- heterocyclic compound
- hydrocarbon
- hydrogenation
- isomer
- IUPAC system of nomenclature
- normal form
- saturation
- substitution reaction
- unsaturation

When the procedures for naming compounds were described in Chapter 5, a distinction was made between molecules that contain hybridized carbon atoms and molecules that do not. This is a logical subdivision because hybridized carbon atoms are capable of bonding together in chain-like arrangements, resulting in molecules that have a basic structural similarity. Many of these molecules belong to a class of compounds called hydrocarbons.

As the name implies, *hydrocarbons* are compounds composed of hydrogen and carbon. From a biological point of view they are relatively unimportant since, with rare exceptions, they are not found in living organisms. However, as shown by the example of pentane and pentanoic acid, many biologically important compounds may be considered as derivatives of hydrocarbons. The discussion that follows is centered around this idea.

10.1 HYDROCARBONS—THE PARENT MOLECULES

Recall the examples of pentane and pentanoic acid described in Chapter 5. The major structural feature of pentane is a 5-carbon chain, to which are attached 12 atoms of hydrogen. A similar carbon chain is found in pentanoic acid. A comparison of the two structures shows that they are identical in all respects except one—there are oxygen atoms attached to one of the terminal carbons in pentanoic acid:

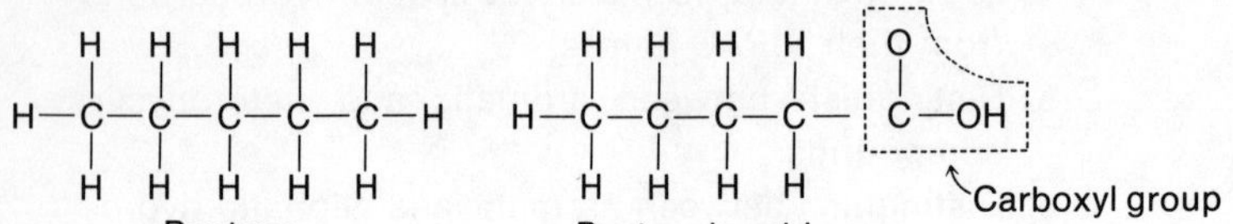

$$\begin{array}{ccccccccccc} & & H & & H & & H & & H & & H & & \\ & & | & & | & & | & & | & & | & & \\ H & - & C & - & C & - & C & - & C & - & C & - & H \\ & & | & & | & & | & & | & & | & & \\ & & H & & H & & H & & H & & H & & \end{array}$$

Pentane

$$\begin{array}{cccccccccccc} & & H & & H & & H & & H & & O & & \\ & & | & & | & & | & & | & & \| & & \\ H & - & C & - & C & - & C & - & C & - & C & - & OH \\ & & | & & | & & | & & | & & & & \\ & & H & & H & & H & & H & & & & \end{array}$$

Pentanoic acid

Carboxyl group ($-COOH$)

As you have already learned, carboxylic acids have properties which differ from those of the hydrocarbons. Since the only structural difference between an organic acid and its corresponding alkane is the carboxyl group, the differences between alkanes and organic acids must result from modification of the hydrocarbon structure. However, not all of the properties of the alkanes and the organic acids are different: characteristics due to the carbon chain remain the same. The nonpolar hydrocarbon chain that is found in both of these types of molecules makes them virtually insoluble in water and contributes to rather low intermolecular attractions. This being the case, it is logical to examine the hydrocarbons themselves before considering the properties that are associated with various modifications of the hydrocarbon structure.

The chain of carbon atoms in a hydrocarbon may be either *open* or *cyclic.* Cyclic means that the chain is closed so that a loop or ring is formed. The open-chain structures are also referred to as *aliphatic hydrocarbons.*

10.2 ALIPHATIC HYDROCARBONS

Aliphatic hydrocarbons are subdivided according to the kinds of carbon-carbon bonds they contain. There are three major groups: alkanes, alkenes, and alkynes.

Saturated Hydrocarbons (the Alkane Series)

The members of the alkane series are characterized by single covalent bonds between carbon atoms. This means that every carbon atom is bonded to four other atoms. Since the bonding orbitals of all of the carbon atoms are fully engaged, and there is no possibility of additional bonding, the alkanes are said to be *saturated.* The members of this series are also called the *paraffin hydrocarbons* because of their low chemical reactivity. The material called paraffin is a mixture of high-molecular-weight alkanes.

The names and molecular formulas of alkanes having 2 to 10 carbons were given in Table 5.4. Note that the formula of each member corresponds to the general formula C_nH_{2n+2}, where n is the number of carbon atoms in the molecule. For example, the number of hydrogen atoms in pentane is twice as great as the number of carbons, plus two more: $C_5H_{(2\times5)+2}$, or C_5H_{12}. There is a general formula for each of the three types of aliphatic hydrocarbons.

Hydrocarbons are nonpolar molecules; thus their intermolecular attractions are limited to the weak van der Waals forces. Since van der Waals forces depend upon the surface area of the molecules, intermolecular attraction increases as the molecules become larger. As a result, the boiling and melting points of the alkanes increase in a regular fashion. The first five members are gases at room temperature, the next 12 are liquids, and alkanes containing more than 17 carbons are solids.

Saturated hydrocarbons are relatively low in chemical activity, although the gaseous and liquid alkanes burn explosively when mixed with air, releasing large amounts of energy.

$$CH_{4\,(g)} + 2\ O_{2\,(g)} \longrightarrow CO_{2\,(g)} + 2\ H_2O_{(l)} + 210\ \text{kcal}$$

$$C_7H_{16(l)} + 11\ O_{2\,(g)} \longrightarrow 7\ CO_{2\,(g)} + 8\ H_2O_{(l)} + 1416\ \text{kcal}$$

The low-molecular-weight alkanes are important fuels. Natural gas is a mixture consisting mostly of methane, with small amounts of ethane and traces of other hydrocarbons. Since methane and ethane are essentially odorless, a sulfur compound called a mercaptan with a distinct, unpleasant smell is added to natural gas sold commercially so that leaks can be detected. Propane and butane are sold in pressurized metal containers as "bottled gas," which is used for cooking and heating in campers, trailers, boats, and in homes in some rural areas. Gasoline is a mixture of alkanes ranging in size from 5 to 12 carbons, and diesel fuel contains 10- to 16-carbon alkanes.

When combustion of fuel is complete, the products are water and carbon dioxide, both of which are used in photosynthesis by green plants. If the oxygen that is available is not sufficient for complete oxidation of the carbon, carbon monoxide, a poisonous gas, is produced:

$$2\ CH_{4\,(g)} + 3\ O_{2\,(g)} \longrightarrow 2\ CO_{(g)} + 4\ H_2O_{(l)}$$

$$2\ C_7H_{16(l)} + 15\ O_{2\,(g)} \longrightarrow 14\ CO_{(g)} + 16\ H_2O_{(l)}$$

One of the shortcomings of an automobile engine is that it produces large amounts of carbon monoxide. Thus in large cities, such as New York, where the air stagnates in pockets between the buildings, an overabundance of cars produces a real health hazard. Working in a closed garage with an engine that is running is a dangerous and foolish practice. Many deaths are also caused each year by improperly vented gas or oil heaters. It is a good idea to have all fuel-burning appliances checked and adjusted at least once a year, and it is absolutely essential that gas or oil heaters be vented to an outside environment where the air is circulating.

Another chemical characteristic of the alkanes is their reaction with halogens at room temperature in the absence of catalysts. This is a substitution (or displacement) reaction in which a halogen atom replaces one of the atoms of hydrogen in the alkane. The displaced hydrogen then combines with an unreacted halogen atom, as shown in the following example:

$$\mathrm{H{-}\underset{\underset{H}{|}}{\overset{\overset{H}{|}}{C}}{-}H + Cl{-}Cl \xrightarrow{\text{light}} H{-}\underset{\underset{H}{|}}{\overset{\overset{H}{|}}{C}}{-}Cl + H{-}Cl}$$

Monochloromethane

In the presence of excess halogen the reaction continues and additional chlorine is introduced into the molecule:

$$CH_3Cl + Cl_2 \xrightarrow{\text{light}} CH_2Cl_2 + HCl$$
Dichloromethane

$$CH_2Cl_2 + Cl_2 \xrightarrow{\text{light}} CHCl_3 + HCl$$
Trichloromethane
(chloroform)

$$CHCl_3 + Cl_2 \xrightarrow{\text{light}} CCl_4 + HCl$$
Tetrachloromethane
(carbon tetrachloride)

With each successive step the flammability of the molecule is lessened, but its toxicity is increased. Carbon tetrachloride was at one time widely used in fire extinguishers and as a solvent for grease. Because of its extreme toxicity, it is no longer used for these purposes. Chloroform (trichloromethane) has an advantage over most general anaesthetics because it is not flammable; however, evidence suggests that chloroform may have a toxic effect on the liver, and for this reason the use of chloroform is declining.

Unsaturated Hydrocarbons

A hydrocarbon is said to be *unsaturated* when it contains one or more multiple bonds between carbon atoms. The unsaturated hydrocarbons include the *alkenes* (double bonds) and the *alkynes* (triple bonds).

The Alkene Series

The members of the alkene series are also called *olefins*.[1] The names of the individual alkenes have the same ending as the name of the series. The first member of the alkene series has two carbons, and is called ethene (C_2H_4). There cannot be an alkene that corresponds to methane because there must be at least two carbon atoms in a molecule before a carbon-to-carbon double bond is possible. Ethene (also called ethylene) is a general anaesthetic, but forms an explosive mixture with air. It is also used to change the color of lemons and oranges which are ripe but have green skins when they are picked. The second member of the series is propene (C_3H_6), the third is butene (C_4H_8), and so on, with the stem of each name corresponding to the number of carbon atoms in the chain. The general formula of the alkene series is C_nH_{2n}.

Ball and stick model representing a molecule of ethene (Photograph by J. Asdrubal Rivera)

The alkenes are more reactive than the alkanes because two of the electrons forming the double bond are available under the right conditions for bonding with other atoms. The following example shows the addition of bromine to the double bond in ethene:

[1] The word *olefin* means oil-forming. The name was originally given to ethene, because it forms an oily liquid through an addition reaction with chlorine. The name now applies to all aliphatic hydrocarbons in which there are carbon-carbon double bonds.

$$Br{:}Br + H_2C{::}CH_2 \longrightarrow H_2C(Br){:}C(Br)H_2$$

Notice that the carbon atoms continue to share one pair of electrons, which means that the addition reaction has converted an unsaturated molecule into a saturated one. Notice also that, unlike the substitution reactions of alkanes, addition reactions produce only a single product. Since bromine is reddish-brown, and the addition product is colorless, the ability of a hydrocarbon to decolorize a bromine solution while forming a single product is taken as evidence (but not proof) that the hydrocarbon is unsaturated. Unsaturated hydrocarbons also reduce dilute solutions of permanganate ion, which is purple, to manganous ion, which is brown. Permanganate ion is also reduced by aldehydes and by both primary and secondary alcohols.

Hydrogen can be added to a double bond, but the reaction (called *hydrogenation*) requires high pressure and a catalyst such as platinum, nickel, or palladium. Complete hydrogenation converts an unsaturated molecule into the corresponding saturated structure (alkane):

$$H_3C{-}CH{=}CH_2 + H_2 \xrightarrow[\text{pressure}]{\text{Pt}} H_3C{-}CH_2{-}CH_3$$

Propene — Propane

The addition of hydrogen to unsaturated organic acids has both commercial and dietary significance, and will be discussed in Chapter 13.

Like alkanes, the alkenes may be oxidized to carbon dioxide and water, and could be used for fuels were it not for their scarcity.

The Alkyne Series

The members of the alkyne series are distinguished by a triple bond between two of their carbon atoms, and have names that end in *-yne*. The first member of the series is ethyne (C_2H_2), which is also known by its common name, acetylene. Molecules with triple bonds are extremely reactive and, with rare exceptions, are not found in nature. Acetylene, for example, readily adds hydrogen, and mixtures containing equal parts of air and acetylene may explode spontaneously. The general formula for the alkyne series is C_nH_{2n-2}.

10.3 AROMATIC HYDROCARBONS

Aromatic hydrocarbons have two features that distinguish them from their aliphatic counterparts. First, the members of this group all contain a 6-membered carbon ring similar to the structure of benzene. Second, the bonds in this ring are unusual in that they appear to be *hybrids,* as will be explained shortly. The name of the group stems from the fact that the first members to be studied have distinct, not unpleasant, odors.

For many years benzene was a chemical enigma. Its empirical formula is C_6H_6, which suggests a carbon ring having alternating single and double bonds:

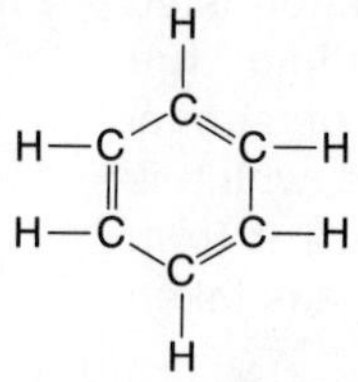

H
H—C C—H
H—C C—H
H

Because of the double bonds, such a structure would be expected to decolorize bromine and reduce permanganate ion; yet the reaction of benzene with these reagents resembles that of the saturated hydrocarbons. A German chemist named Kekule suggested that benzene behaves in this fashion because the double bonds are constantly changing positions, as shown here:

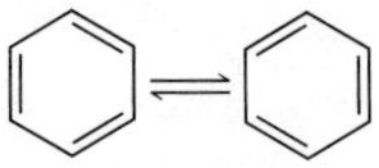

This abbreviated structure may be used if we remember there is a carbon atom with an attached hydrogen at each corner. Evidence to support Kekule's theory has never been produced, and the current view is that the bonds between the carbons are neither single nor double, but somewhere in between. This view is supported by comparisons of carbon-to-carbon bond lengths. The bonds in benzene are 1.40 Å in length, which is intermediate between single bonds (1.54 Å) and double bonds (1.34 Å). Apparently the electrons form an orbital which encompasses the entire ring. The concept of delocalized electrons may more accurately be represented by the symbol

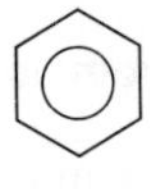

Aromatic hydrocarbons usually react in such a way that the ring remains intact: in other words, by substitution. For example, when benzene is reacted with monochloromethane in the presence of aluminum chloride, one of the ring hydrogens is replaced by a methyl ($—CH_3$) group.

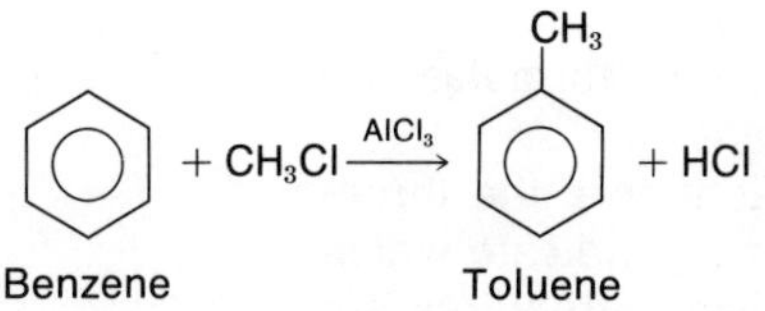

$$C_6H_6 + CH_3Cl \xrightarrow{AlCl_3} C_6H_5CH_3 + HCl$$

Benzene Toluene

The product of the reaction is another aromatic hydrocarbon called

toluene (or methylbenzene). Remember that the bonds to all of the hydrogens in benzene are equivalent; therefore, replacement of any one of the hydrogens by a methyl group will produce toluene. Toluene is a colorless liquid whose vapors act as a central nervous system depressant. A mixture of toluene and ethyl acetate is used as the solvent in some brands of glue. Inhalation of this mixture produces an initial effect similar to alcohol intoxication, but the combination is so toxic that persons who indulge in the practice of "glue sniffing" run the risk of nervous system damage, or even death. Medical personnel sometimes preserve urine samples by the addition of toluene, which is less dense than urine and floats on the surface and prevents it from contacting the air. Later, when the urine is analyzed, a sample is taken from under the layer of toluene.

A number of hydrocarbons contain ring structures that appear to be formed by the fusion of several molecules of benzene. Two of these are napthalene and phenanthrene:

Naphthalene Phenanthrene

Structures similar to napthalene appear in vitamin K and quinine, while sex hormones, cholesterol, bile salts, vitamin D, and the hormones of the adrenal cortex are structurally related to phenanthrene.

10.4 STRUCTURAL ISOMERISM

Compounds containing hybridized carbon outnumber all other compounds by a wide margin. One reason for their abundance is the fact that the same atoms can be assembled in a variety of ways. Consider propane, for example. The four carbon atoms in this molecule may bond together in two different ways. In one arrangement they form an unbranched chain $\begin{array}{c} | \quad | \quad | \quad | \\ -C-C-C-C- \\ | \quad | \quad | \quad | \end{array}$, in which every carbon atom is bonded to at least two hydrogen atoms. The four carbon atoms may also be joined in such a way that the chain is branched and the carbon atom at the junction is bonded to a single hydrogen, as shown by the structural formula $\begin{array}{c} | \quad | \quad | \\ -C-C-C- \\ | \quad | \quad | \\ -C- \\ | \end{array}$. Compounds such as these, which have the same molecular formula but different structural formulas, are called *isomers*. Thus there are two isomers of butane.

It is important to realize that structural formulas ignore the three-dimensional nature of molecules, and that they merely indicate which atoms are connected by a bond. All of the following structures are equivalent, and any one of them could be used to represent the unbranched chain of butane.

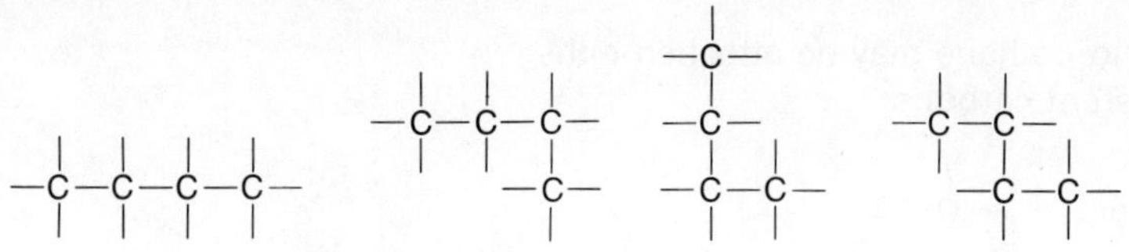

(1) (2) (3) (4)

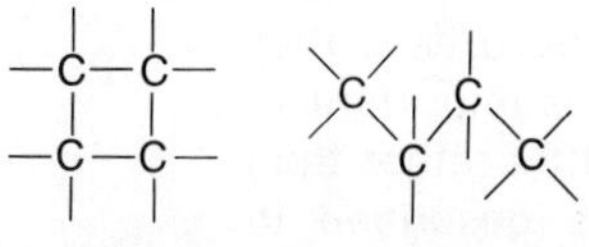

(5) (6)

Placing the carbon atoms in a straight line simply makes it easier to count them, or to identify side chains.

There is a geometric increase in the number of isomers as the carbon chain is lengthened. It has been calculated that there are 75 possible isomers of decane ($C_{10}H_{22}$), and 366,319 of eicosane ($C_{20}H_{42}$)! Needless to say, no one has even drawn the structural formulas for all of these isomers, but the numbers do give some indication of the large number and diversity of organic compounds.

In addition to their structural differences, isomers have slightly different physical properties. Melting and boiling points are generally lower for branched structures because branching reduces the surface area, and nonpolar molecules with smaller surfaces have lower intermolecular forces. This generalization is borne out by the example of pentane. There are three isomers of pentane, and their boiling points decrease as the amount of branching increases:

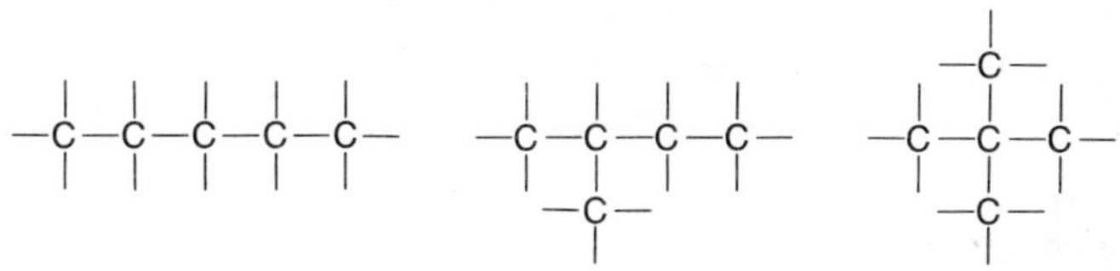

b.p. +36°C b.p. +28°C b.p. +9.5°C

The melting points do not always fit a regular pattern because they depend upon the ease with which the molecules fit into a crystalline lattice, which is a function of the molecular shape as well as its surface area.

Isomerism may also be due to differences in the position of a multiple bond, or of a substituted group such as a halogen atom. In butene, for example, the double bond may be located in the center of the chain, or it may connect the first and second carbons in the chain:

```
  |   |   |   |           |   |   |   |
—C—C=C—C—    or    —C=C—C—C—
  |       |                   |   |
```

And the two chlorine atoms in dichloroethane may be attached either to the same carbon atom or to different carbons:

```
     |   |                |   |
Cl—C—C—    or    —C—C—
     |   |                |   |
     Cl                  Cl  Cl
```

Structural differences among isomers are often responsible for differences in reactivity, and in some cases isomers behave as if they were entirely different compounds. For this reason it is important to have a system of nomenclature that is based on structure rather than composition. The framework of such a system was described in Chapter 5.

10.5 NAMING COMPOUNDS CONTAINING HYBRIDIZED CARBON ATOMS

As is the custom in a new field of study, the pioneers in organic chemistry exercised the right to name their discoveries. More often than not the names they chose offered no clue as to the nature of the compounds. This situation was corrected when, in 1961, the International Union of Pure and Applied Chemistry (IUPAC) met in Geneva and approved the rules which presently govern organic nomenclature. The provisions that apply to saturated hydrocarbons will be listed first, and they will then be used in naming the various isomers of pentane and several chlorinated hydrocarbons.

1. Pick out the longest continuous carbon chain in the molecule.
2. Number the carbon atoms in the chain so that the lowest numbers are assigned those atoms to which side chains or substituted groups are attached.
3. Name the compound as a derivative of the longest carbon chain and indicate the type, location, and number of the side chains or substituted groups by prefixes.

The structures of the three isomers of pentane are

```
                                                          |
                                                        —C—
  |   |   |   |   |           |   |   |   |          |   |   |
—C—C—C—C—C—      —C—C—C—C—      —C—C—C—
  |   |   |   |   |           |   |   |   |          |   |   |
                                    —C—                 —C—
                                      |                     |
       (1)                         (2)                  (3)
```

It is customary to refer to an unbranched hydrocarbon as the *normal* form (*n*), and so structure (1) is called *n*-pentane, or simply pentane. The longest continuous carbon chain in structure (2) consists of four atoms; thus the second isomer is named as a derivative of butane. It appears that the chain may be numbered in any of four ways:

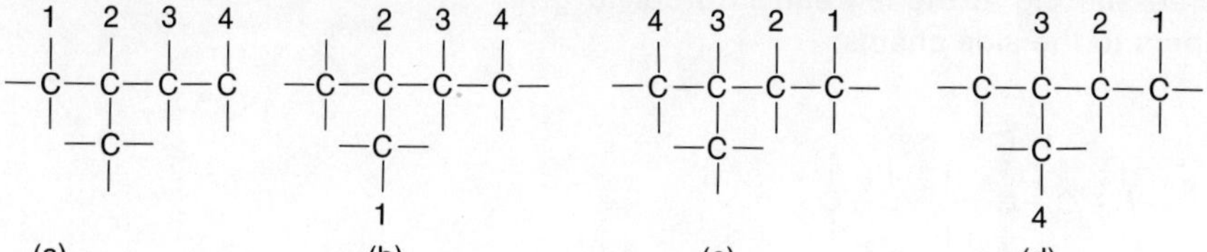

(a) (b) (c) (d)

However, both (c) and (d) violate the second rule because the side chain is not given the lowest possible number. The two remaining structures are actually identical and have the same name since they both have a methyl group (—CH_3) attached to the (2) carbon. The correct name for this isomer is 2-methylbutane.

The third isomer of pentane has a carbon chain made of three atoms, and thus is named as a derivative of propane. Two methyl groups (shown in circles) are attached to the center carbon. Numbering the chain from either end produces the same result:

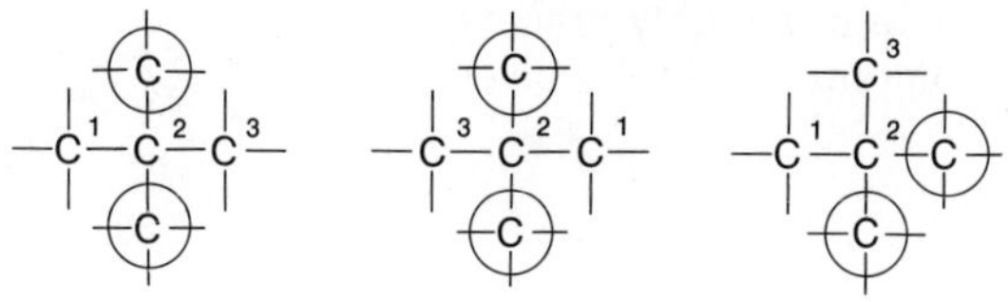

Since there are two methyl groups in this structure, its correct name is dimethylpropane.

The procedure for naming chlorinated alkanes is basically the same as for alkanes with side chains. Consider the two isomers of dichloroethane shown earlier. The structure in which both chlorine atoms are attached to the same carbon is called 1,1-dichloroethane, while the other isomer is 1,2-dichloroethane:

1,1-dichloroethane 1,2-dichloroethane

With these examples in mind, let us look at a more difficult example.

Sample problem 10.1 Name the compound whose structure is

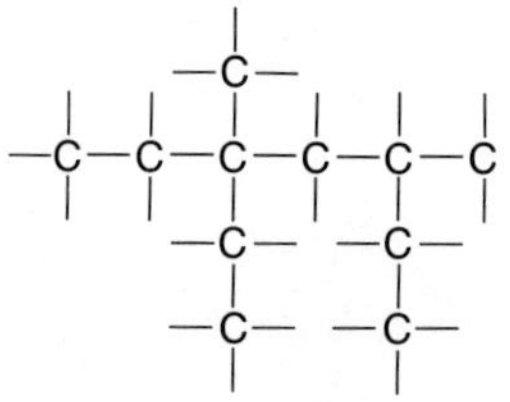

Solution The longest continuous carbon chain is seven atoms long, so the compound is named as a derivative of heptane. The chain is numbered starting at the left end in order to give the lowest numbers to the side chains:

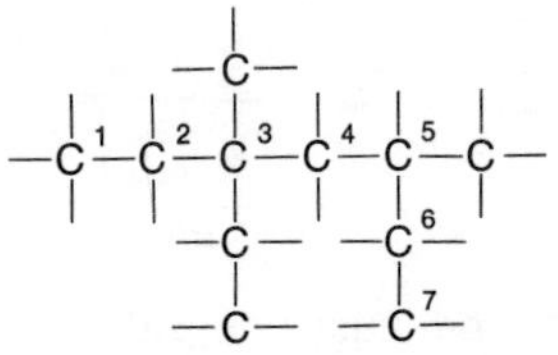

There is one methyl group attached to carbon 3 and another to carbon 5. The other side chain attached to carbon 3 is an ethyl ($—CH_2CH_3$) group. When there is more than one kind of substituent, they are named alphabetically. The correct name is 3,5-dimethyl-3-ethylheptane.

The rules for naming other organic compounds are basically the same as those governing the saturated hydrocarbons, except that special suffixes are used to indicate the presence of a highly reactive site within the molecule. Examples of such compounds will be given in the next chapter.

10.6 HETEROCYCLIC COMPOUNDS

Heterocyclic compounds are characterized by ring structures which include an atom or atoms other than carbon. The most important heterocyclic rings are those containing nitrogen. They are found in the nucleic acids (RNA and DNA), the oxygen-transporting portion of hemoglobin, chlorophyll, certain B vitamins, coenzyme A, ATP and ADP, proteins, and a diverse group of plant extracts called alkaloids, which includes nicotine, caffeine, quinine, and morphine. The pyrole ring

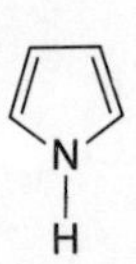

is common to hemoglobin and chlorophyll, the pyrimidine ring

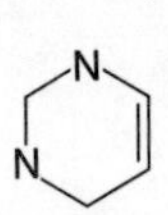

is found in RNA and DNA, and the purine structure

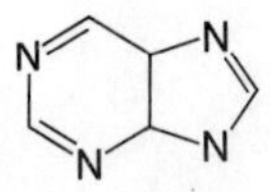

occurs in ATP, ADP, coenzyme A, NAD, and vitamin B_1. There is no one heterocyclic structure that is common to all of the alkaloids. The structures of nicotine, caffeine, and quinine give some idea of their diversity.

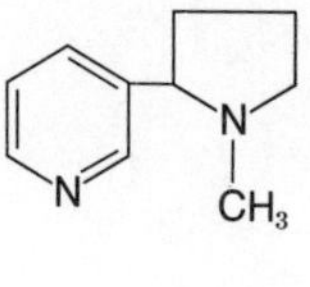

Nicotine 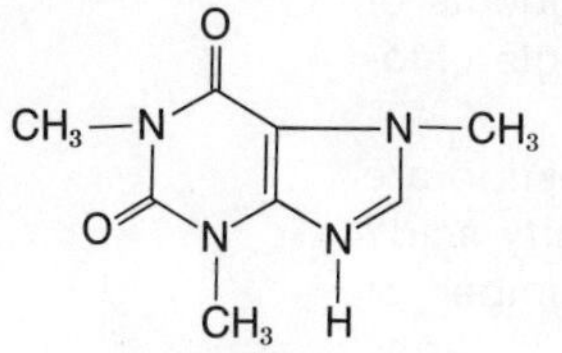   Caffeine 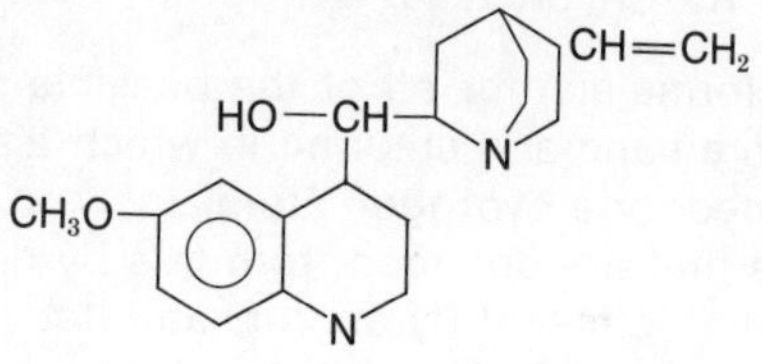   Quinine

APPLICATION OF PRINCIPLES

1. Draw structural formulas for all isomers of C_6H_{14}, and name each isomer according to the IUPAC system.
2. Draw a structural formula for each of these compounds:
   (a) 4-bromo-2,3-dimethylhexane
   (b) 2,2-dimethyl-4-ethylhexane
   (c) 2,2,3,3-tetramethylbutane
3. Write a balanced equation for the complete combustion of
   (a) benzene   (b) acetylene   (c) propane
4. Draw the structural formula for the product resulting from the complete hydrogenation of

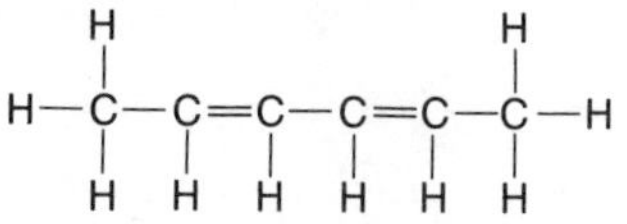

```
     H                   H
     |                   |
  H—C—C=C—C=C—C—H
     |   |   |   |   |   |
     H   H   H   H   H   H
```

5. What information concerning structure is revealed by the following experimental results? A dilute solution of $KMnO_4$ turns from purple to muddy brown shortly after being mixed with a hydrocarbon.
6. Which of the following compounds would you expect to have the higher boiling point? (a) 2-methylpentane (b) 2,2-dimethylbutane. Justify your choice.
7. Write an IUPAC name for each of the following compounds:

   (a)

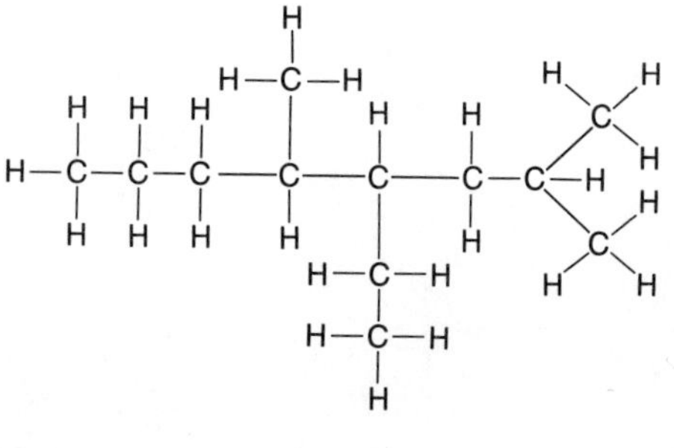

(b)

$$\begin{array}{ccccccccccccc} & & H & & I & & CH_3 & & H & & H & & H \\ & & | & & | & & | & & | & & | & & | \\ H & - & C & - & C & - & C & - & C & - & C & - & C & - & H \\ & & | & & | & & | & & | & & | & & | \\ & & H & & CH_3 & & CH_3 & & H & & H & & H \end{array}$$

8. Draw the structural formula(s) for all of the possible products of the reaction between ethane and chlorine in which a single chlorine atom is substituted for a hydrogen atom.
9. The carboxylic acids that are obtained from fats by digestion are called fatty acids. What is meant by an "unsaturated" fatty acid?
10. The reaction of benzene with chlorine produces a number of products, including HCl. What is the significance of this in terms of the structure of benzene?

Ingredients: Gelatin, Adipic Acid (for tartness), Potassium Citrate (controls acidity), Fumaric Acid (for tartness), Calcium Saccharin, Artificial and Natural Flavors with BHA added as a preservative, U.S. Certified Color. 17.7 gm.

Chemical additives play an important role in the food industry. (Photograph by J. Asdrubal Rivera)

11

FUNCTIONAL GROUPS AND THEIR PROPERTIES

LEARNING OBJECTIVES

1. Identify the functional group(s) of a compound from the structural formula or from the name of the compound.
2. Write the IUPAC names of simple organic compounds from their structural formulas, and the reverse.
3. Compare the structures and reactions of primary, secondary, and tertiary alcohols and amines.
4. Use structural formulas to demonstrate the formation of an ester linkage.
5. Summarize the reactions of the carbonyl groups of aldehydes and ketones.
6. Use structural formulas to demonstrate the formation of an amide linkage.

KEY TERMS AND CONCEPTS

alcohol
aldehyde
amide
amine
carbonyl group
chelating agent
esterification
fermentation
functional group
ketone
meta position
ortho position
para position
phenol coefficient
polyfunctional compound
reducing sugar

It was pointed out in the previous chapter that carbon atoms joined by a double bond react more readily than carbon atoms joined by single bonds in the same molecule. In other words, the behavior of the double bond is distinctive and is relatively unaffected by the length or complexity of the carbon chain. This is also true of the carboxyl group of an organic acid, which is the only part of the molecule to undergo ionization. Observations such as these simplify the study of carbon compounds, since this means that compounds having the same kind of reactive site will, in most cases, have the same chemical properties.

11.1 TYPES OF FUNCTIONAL GROUPS

Reactive sites such as the carbonyl group or the double bond are called *functional groups.* Each functional group is a specific configuration of atoms, and compounds having the same functional group are said to belong to the same class. For example, compounds whose structures include the carboxyl group ($-\overset{\overset{\displaystyle O}{\|}}{C}-OH$) belong to the class of organic acids. Other important functional groups are shown in Table 11.1.

The rules for naming compounds containing these functional groups are basically the same as those governing the saturated hydrocarbons, except that special endings are used to indicate the functional group in the molecule. Recall that the names of the olefins end in *-ene,* while the ending *-oic acid* designates a carboxyl group. The endings associated with the various functional groups are shown in Table 11.2.

The following procedure applies to the naming of compounds containing functional groups:

1. Pick out the longest continuous carbon chain that includes the functional group.
2. Number the carbon atoms in the chain so that the lowest numbers are assigned to atoms which are part of the functional group.
3. Name the compound as a derivative of the longest carbon chain, and indicate the location of the functional group by a prefix number.

TABLE 11.1 Important Functional Groups

Functional Group	Name of Group	General Name for Compounds Containing This Group
$-\overset{\vert}{C}=\overset{\vert}{C}-$	Double bond	Alkenes (olefins)
$-C\equiv C-$	Triple bond	Alkynes
$-\underset{\vert}{\overset{\vert}{C}}-X$ (where X = halogen)	Halide	Halides
$-\underset{\vert}{\overset{\vert}{C}}-OH$	Hydroxyl	Alcohols
$>C=O$	Carbonyl	Aldehydes and ketones
$-\overset{O}{\overset{\Vert}{C}}-OH$	Carboxyl	Organic acids and amino acids
$-\underset{\underset{H}{\vert}}{N}-H$	Amino	Amines and amino acids

4. Indicate the nature of the functional group by the appropriate ending.

By using this procedure it is possible to differentiate between the two isomers of butene which were described earlier:

$$\underset{(a)}{-\overset{1}{C}-\overset{2}{C}=\overset{3}{C}-\overset{4}{C}-} \qquad \underset{(b)}{-\overset{1}{C}=\overset{2}{C}-\overset{3}{C}-\overset{4}{C}-}$$

A chain of four carbon atoms is found in both isomers and so they are named as derivatives of butane. The functional group is a double bond which is indicated by the suffix *-ene.* In the first isomer, the double bond is attached to carbon 2; thus the name of the isomer is 2-butene.

TABLE 11.2 Endings Used in Naming Organic Compounds

Class of Compound	Ending
Alkene	-ene
Alkyne	-yne
Alcohol	-ol
Aldehyde	-al
Ketone	-one
Organic acid	-oic acid
Amine	-ine

(Note that numbering from the opposite end of the chain produces the same name). The second isomer is called 1-butene since the double bond involves carbon 1.

11.2 ALCOHOLS

Alcohols are organic compounds having one or more hydroxyl groups. Despite their apparent resemblance to metallic hydroxides, they do not ionize to form hydroxide ions, and thus do not have a bitter taste, a slippery feel, or a reaction with litmus. Since the hydroxyl group is covalently bonded, alcohols are chemically similar to water—although hydrogen bonding is less extensive in alcohols. The chemical names of alcohols end in *-ol.*

Important Aliphatic Alcohols

The simplest alcohol is structurally related to methane and is called methanol, although it also is known by its common name—methyl alcohol. Methanol is a colorless volatile liquid that is miscible in all proportions with water and has rather broad solvent properties. Large quantities of methanol are used in the manufacture of formaldehyde, and it is the solvent in shellac and duplicating fluid. A colloidal dispersion of methanol, sold under the trade name Sterno, is the fuel which keeps chafing dishes hot. Methanol is extremely poisonous and when ingested or inhaled may cause blindness or paralysis. Large doses of methanol are generally fatal.

The most widely used and best-known alcohol is structurally related to ethane; thus it is called ethanol, or ethyl alcohol. Beverages containing ethyl alcohol have been produced for thousands of years, and some kind of alcoholic drink is known in every region of the world.

Ethanol acts as a depressant on the central nervous system. When consumed in small amounts, it suppresses inhibitions, slows reaction times, and reduces coordination: in large amounts it may result in loss of consciousness. Most governments levy a tax based on the alcohol content of beverages, and in the United States the revenue from this source amounts to more than 4 billion dollars per year.

Ethanol is also used as an antiseptic and a preservative because of its ability to coagulate protoplasm. When applied to the skin, it acts as an astringent, but repeated application toughens the epidermis and helps to prevent bedsores. As alcohol evaporates, it absorbs heat, and so alcohol sponge baths are used to reduce the temperature of a person with a dangerously high fever. As a solvent it is found in elixirs, tinctures, and spirits. When alcohol is used for commercial purposes, it is usually rendered unsafe for drinking by the addition of very small amounts of benzene or methanol, both of which are poisonous. Neither of these additives can be completely removed by distillation. Ethanol treated in this way is said to be *denatured.*

Ethyl alcohol is produced industrially by the combination of ethene

Facts about Alcohol

Ethyl alcohol is a depressant. It does not have to be digested. Some of it is absorbed through the wall of the stomach, but most of it enters the bloodstream through the intestinal lining. The body can oxidize about 10 grams of alcohol per hour: if more than this is ingested, measurable amounts of alcohol appear in the blood. A blood alcohol concentration of 0.05–0.15% causes a loss of coordination in movement and a lessening of inhibitions. In many states a blood alcohol level of 0.15% is taken as evidence of intoxication. Concentrations of 0.5% can be fatal.

and water at high pressure. The reaction is

$$\begin{array}{c}\ \ \\ H-\underset{\displaystyle H}{\underset{|}{C}}=\underset{\displaystyle H}{\underset{|}{C}}-H + H_2O \xrightarrow[300°C]{70\ atm} H-\overset{\displaystyle H}{\overset{|}{\underset{\displaystyle H}{\underset{|}{C}}}}-\overset{\displaystyle H}{\overset{|}{\underset{\displaystyle H}{\underset{|}{C}}}}-OH\end{array}$$

It is also produced from most carbohydrates by the action of enzymes in a process commonly called *fermentation.* The fermentation of glucose by the enzyme zymase (found in yeast cells) is typical. The reaction actually involves several steps, but may be represented by the following equation:

$$C_6H_{12}O_6 \xrightarrow{zymase} 2\ CO_2 + 2\ C_2H_5OH$$

The concentration of ethyl alcohol in a solution is usually expressed in terms of *proof,* where the proof is equal to twice the volume percent. For example, a solution that is 50% alcohol is 100 proof. This rating scale had its origin in the practice of pouring alcohol on a pile of sawdust and igniting it: if the sawdust burned, it was taken as proof that there was very little water in the liquid.

Isopropanol (also known as rubbing alcohol) has astringent and coagulant properties similar to those of ethanol, but is more poisonous. Since it is also less expensive, it is generally used instead of ethanol for back rubs and sponge baths. Its formula is

$$H-\overset{\displaystyle H}{\overset{|}{\underset{\displaystyle H}{\underset{|}{C}}}}-\overset{\displaystyle H}{\overset{|}{\underset{\displaystyle OH}{\underset{|}{C}}}}-\overset{\displaystyle H}{\overset{|}{\underset{\displaystyle H}{\underset{|}{C}}}}-H$$

IUPAC Nomenclature of Aliphatic Alcohols

The name *isopropanol* suggests that the alcohol whose structure is shown above is an isomer of propanol, which is indeed the case. To differentiate between the two isomers, the molecule in which the —OH group is attached to a terminal carbon atom is designated the *normal* form, and is called *n*-propanol. You may recall from the example of pentane that this method of distinguishing between isomers does not work when there are more than two possibilities. Since there are four isomers of butyl alcohol, their names are assigned according to the IUPAC system.

In one of the isomers the functional group is located at one end of the carbon chain: $-\overset{|}{\underset{|}{C}}-\overset{|}{\underset{|}{C}}-\overset{|}{\underset{|}{C}}-\overset{|}{\underset{|}{C}}-OH$. This isomer is named as a derivative of butane, and the suffix *-ol* is used to show that it is an alcohol. The carbon chain is numbered so that the carbon atom bearing the hydroxyl group receives the lowest possible number, and this number is placed before the name: 1-butanol. The isomer in which the hydroxyl group is attached to the carbon atom next to the end is named 2-butanol.

The longest carbon chain in the remaining isomers is only three atoms long; thus they are both named as propanols. The two structures are shown below:

```
         |                  |
       —C—                —C—
    |    |    |        |    |    |
  —C—C—C—     —C—C—C—OH
    |    |    |        |    |    |
        OH
       (1)                 (2)
```

Notice that there is a methyl group attached to the second carbon in both molecules. The location and identity of this group is shown by the prefix 2-methyl. Since the hydroxyl group in structure (1) is also on the second carbon atom, the complete name of this isomer is 2-methyl-2-propanol. Structure (2) is named 2-methyl-1-propanol.

Classification of Aliphatic Alcohols

Unlike the carboxyl groups of organic acids which are always at the end of a carbon chain, the hydroxyl group may vary in its location within the alcohol molecule, and may be more active in one position than in another. The study of alcohols is simplified by subdividing them into groups having similar reactive tendencies. Alcohols are classified as *primary* (1°), *secondary* (2°), or *tertiary* (3°), depending upon the number of hydrogen atoms attached to the carbon atom bearing the hydroxyl group. A primary alcohol has at least two, while secondary alcohols have one, and tertiary alcohols have none.

```
                                                  |
     H                 H                        —C—
     |            |    |    |              |    |    |
   —C—OH        —C—C—C—                 —C—C—C—
     |            |    |    |              |    |    |
     H                OH                       OH
Primary alcohol   Secondary alcohol    Tertiary alcohol
```

As a general rule, primary alcohols are the most reactive, and tertiary alcohols the most stable of the three types. This order of activity is not unexpected, since the hydroxyl group of a primary alcohol is in an exposed position at the end of the chain, while access to the functional group in a tertiary alcohol is obstructed by neighboring atoms.

Aromatic Alcohols

The hydroxyl groups of aromatic alcohols are influenced by the delocalized electrons in the nearby ring, and tend to be more acidic than those of their aliphatic relatives. The electrons in the oxygen portion of the hydroxyl group are displaced toward the ring and away from the hydroxyl hydrogen, which weakens the bond and permits ionization of the hydrogen, as shown by the example of hydroxybenzene (phenol):

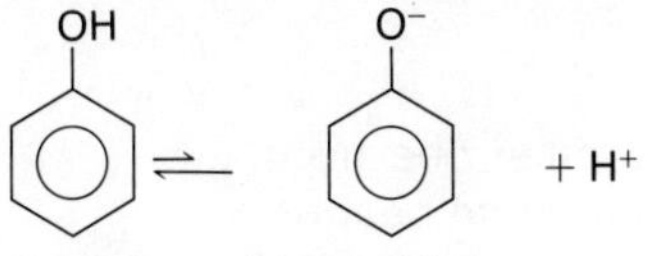

Phenol ⇌ Phenoxide ion + H^+

The acidic properties of phenol are so pronounced that it reacts with metallic hydroxides to form salts and water. For this reason phenol is also known as carbolic acid. Pure phenol looks very much like melting ice. It is poisonous and irritating to the skin, and may cause blistering. A phenol solution was used as a disinfectant during surgery by Joseph Lister in 1865. One way of measuring the effectiveness of a disinfectant is by comparison with phenol. The *phenol coefficient* is the disinfectant or antiseptic action of a chemical compared with that of phenol acting for the same length of time on the same organism under identical conditions.

Alcoholic derivatives of toluene, called *cresols,* are also effective as disinfectants. The hydroxyl group in a cresol can occupy any of three positions relative to the methyl group, as shown by the following structural formulas:

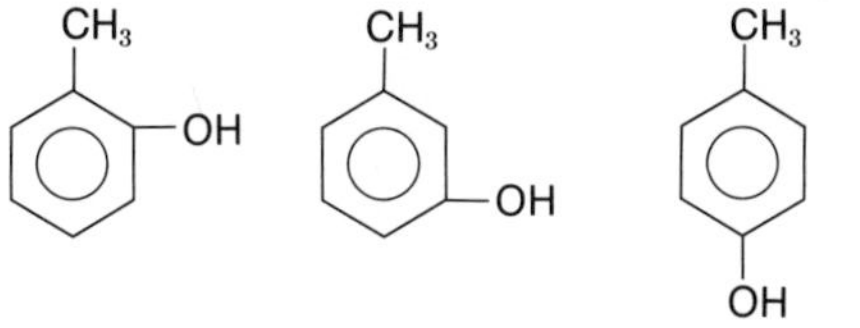

ortho-cresol *meta*-cresol *para*-cresol

A substituent on the carbon atom next to the methyl group is said to be in the *ortho* position, one on the second carbon from the methyl group is in the *meta* position, and the *para* position is directly across the ring. Lysol® is a combination of soap and *ortho, meta,* and *para* cresols.

Other important aromatic alcohols are adrenaline (a hormone), and resorcinol. Resorcinol is used in leather tanning and in the manufacture of dyes, perfumes, plastics, explosives, cosmetics, and adhesives. It is also an antifungal agent. The reason for the many uses for this compound can partially be found in the chemistry of the hydroxy or alcohol groups.

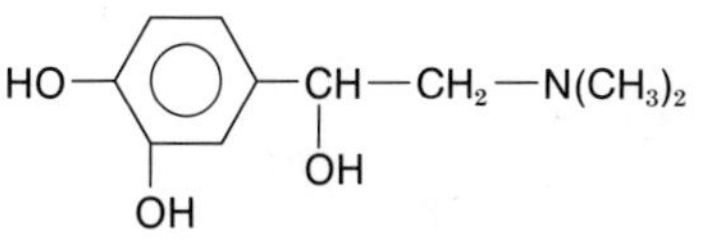

Adrenaline 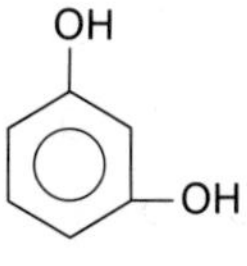 Resorcinol

Structural formulas of several antiseptics

Roccal (Zephiran)

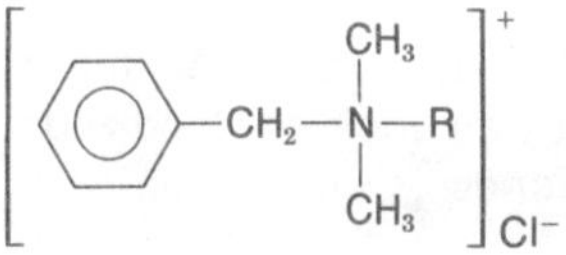

Merthiolate

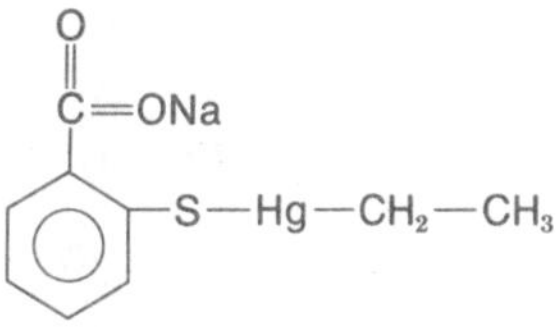

Iodoform

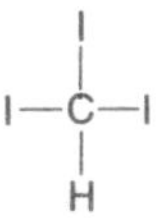

Hexachlorophene

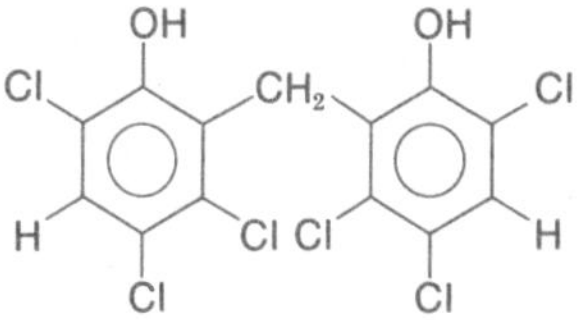

Chloramine-T

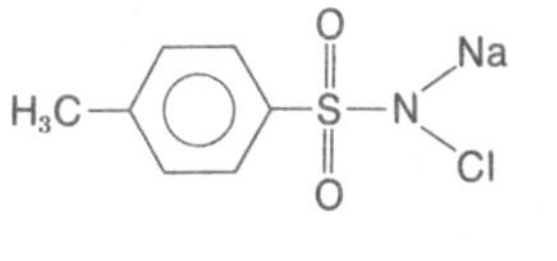

Reactions of Alcohols

Two important reactions of alcohols are oxidation and esterification.

Oxidation

Alcohols are oxidized by removal of two hydrogen atoms: one from the hydroxyl group itself, and the other from the carbon atom bearing the hydroxyl group. Primary alcohols have at least two hydrogen atoms on this carbon and are oxidized rather easily. Secondary alcohols react more slowly, and tertiary alcohols cannot be oxidized in this way.

Oxidation converts the hydroxyl group to a carbonyl group. Primary alcohols are oxidized to aldehydes, and secondary alcohols form ketones.

$$R\text{—}\overset{\displaystyle H}{\underset{\displaystyle H}{C}}\text{—}O\text{—}H \xrightarrow{[O]} R\text{—}\overset{\displaystyle H}{C}\text{=}O + H_2O$$

Primary alcohol   Aldehyde

$$R\text{—}\overset{\displaystyle \text{—}C\text{—}}{\underset{\displaystyle H}{C}}\text{—}O\text{—}H \xrightarrow{[O]} R\text{—}\overset{\displaystyle \text{—}C\text{—}}{C}\text{=}O + H_2O$$

Secondary alcohol   Ketone

$$R\text{—}\overset{\displaystyle \text{—}C\text{—}}{\underset{\displaystyle \text{—}C\text{—}}{C}}\text{—}OH \xrightarrow{[O]} \text{No reaction}$$

Tertiary alcohol

Esterification

Esterification is the reaction of an alcohol with either a carboxylic acid, or with an oxygen-containing acid such as sulfuric, nitric, or phosphoric. Each of these acids has an —OH group in its structure; thus the esterification reaction is a dehydration synthesis involving the —OH group of the alcohol and an —OH group from the acid. The synthesis yields water and an *ester,* as shown by the following example:

$$H\text{—}\overset{\displaystyle H}{\underset{\displaystyle H}{C}}\text{—}\overset{\displaystyle H}{\underset{\displaystyle H}{C}}\text{—}OH \;\; HO\text{—}\overset{\displaystyle O}{\overset{\|}{C}}\text{—}\overset{\displaystyle H}{\underset{\displaystyle H}{C}}\text{—}H \rightleftharpoons H\text{—}\overset{\displaystyle H}{\underset{\displaystyle H}{C}}\text{—}\overset{\displaystyle H}{\underset{\displaystyle H}{C}}\text{—}O\text{—}\overset{\displaystyle O}{\overset{\|}{C}}\text{—}\overset{\displaystyle H}{\underset{\displaystyle H}{C}}\text{—}H + H_2O$$

Ethanol + Acetic acid   Ethyl acetate + Water

Ethyl acetate is a solvent for plastics and is used in lacquers and varnishes, and in the manufacture of artificial leather, artificial silk, perfumes, photographic film, and smokeless powder.

Esters are also responsible for the distinctive flavors of many foods, chiefly fruits, and synthesis of these esters has made it possible to duplicate their flavors in beverages, ice cream, candy, and desserts. A partial list of esters and their flavors is contained in Table 11.3.

Neutral fats are esters of glycerol (also called glycerine) and long-chain carboxylic acids. Glycerol is a *triol;* that is, an alcohol containing three hydroxyl groups per molecule. Its structure is

$$\begin{array}{c} \mathrm{H} \\ | \\ \mathrm{H{-}C{-}OH} \\ | \\ \mathrm{H{-}C{-}OH} \\ | \\ \mathrm{H{-}C{-}OH} \\ | \\ \mathrm{H} \end{array}$$

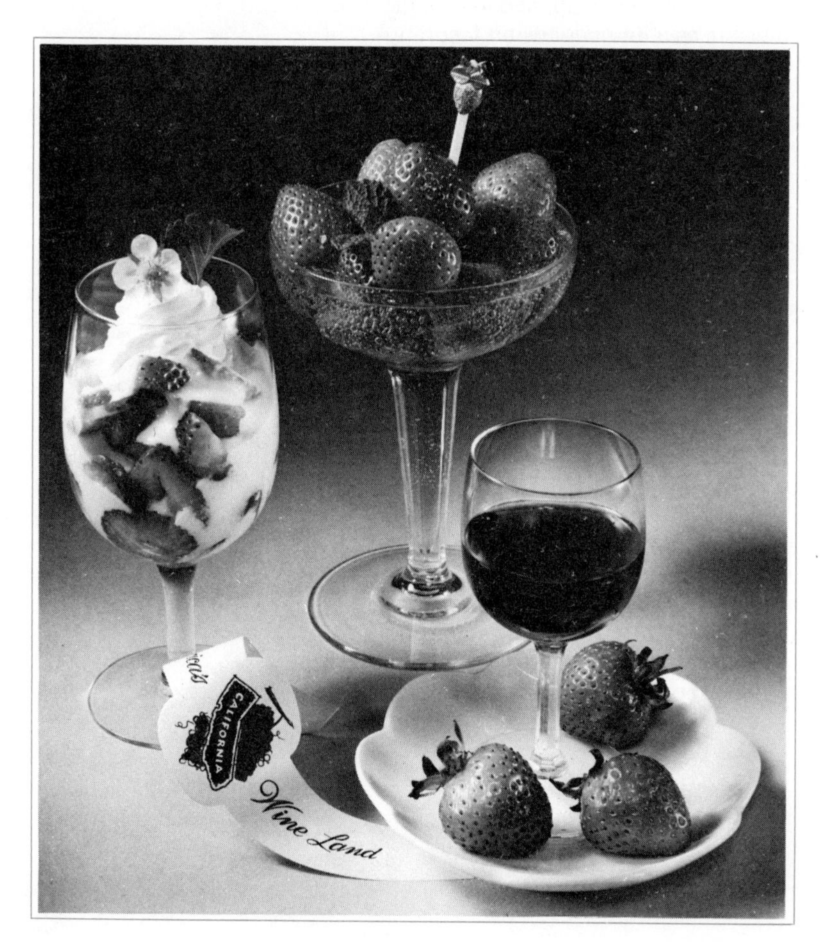

Esters contribute to the distinctive flavors of strawberries and wine. (Courtesy Wine Institute, San Francisco, California)

TABLE 11.3 The Chemical Names of Some Common Esters

Flavor	Name
Apple	Isoamyl isovalerate
Apricot	Amyl butyrate
Banana	Amyl acetate
Orange	Octyl acetate
Pear	Isoamyl acetate
Pineapple	Ethyl butyrate
Raspberry	Isobutyl formate
Rum	Ethyl formate
Wintergreen	Methyl salicylate

and its IUPAC name is 1,2,3-propanetriol. Glycerol is a colorless, viscous, sweet-tasting liquid that is miscible with water in all proportions. It is used in cosmetics, medicines, paints, tobaccos, and in the manufacture of photographic film. Its ability to absorb moisture from the air helps it to keep skin soft and moist when it is applied in lotions. In the laboratory its most common use is as a lubricant when inserting glass tubing or thermometers into rubber stoppers. It is a by-product of soap making.

When glycerol is completely esterified with nitric acid, the products are water and glyceryl trinitrate (better known as nitroglycerin):

$$\begin{array}{c} \text{H-C(H)-O(H} \quad \text{HO)-NO}_2 \\ \text{H-C(H)-O(H} \quad \text{HO)-NO}_2 \\ \text{H-C(H)-O(H} \quad \text{HO)-NO}_2 \end{array} \longrightarrow \begin{array}{c} \text{H-C(H)-O-NO}_2 \\ \text{H-C(H)-O-NO}_2 \\ \text{H-C(H)-O-NO}_2 \end{array} + 3\,H_2O$$

Glycerol + Nitric acid ⟶ Glycerol trinitrate + Water

Nitroglycerin dilates the coronary artery to increase the supply of blood to the heart, and is used to reduce pain in angina pectoris.

Another esterification reaction involving an oxyacid is used for production of a cleaning agent called sodium lauryl sulfate. The starting materials are sulfuric acid and lauryl alcohol (dodecanol), and there are two steps in the process. The first step is the esterification reaction itself:

$$C_{12}H_{25}(OH \quad H)O-S(OH)(=O)=O \longrightarrow C_{12}H_{25}-O-S(OH)(=O)=O + H_2O$$

Lauryl alcohol (dodecanol) + Sulfuric acid ⟶ Lauryl hydrogen sulfate + Water

The second step is the reaction of the ester with sodium hydroxide:

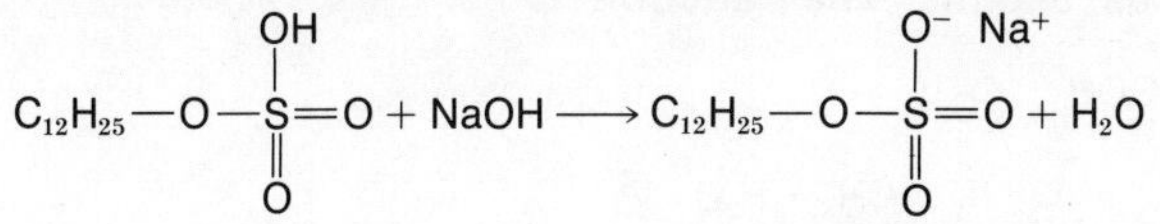

$$C_{12}H_{25}{-}O{-}\overset{\overset{\displaystyle OH}{|}}{\underset{\underset{\displaystyle O}{\|}}{S}}{=}O + NaOH \longrightarrow C_{12}H_{25}{-}O{-}\overset{\overset{\displaystyle O^- \; Na^+}{|}}{\underset{\underset{\displaystyle O}{\|}}{S}}{=}O + H_2O$$

Sodium lauryl sulfate

Sodium lauryl sulfate belongs to a class of materials, called *detergents,* which are used in place of soap in the kitchen and the laundry. The way in which a detergent works will be described in Chapter 13.

From a biological point of view the esters of greatest importance are those of phosphoric acid. They include the phosphate esters of adenosine (whose role in energy transfer was described in Chapter 9), a number of coenzymes which are essential for chemical transformations in the cell, and the nucleic acids which are responsible for the transmission of the genetic characteristics of living organisms.

11.3 CARBOXYLIC ACIDS

The names and structures of the monocarboxylic acids containing fewer than 10 carbon atoms were given in Chapter 5. The general properties and reactions of acids were discussed in Chapter 8, and esterification was described in the preceding section.

Carboxylic acids and their derivatives have both commercial and biological importance. The calcium salt of propionic acid retards the spoilage of bread, while the sodium salt of benzoic acid ($C_6H_5{-}COONa$) is used as a preservative in cider, ketchup, carbonated beverages, and a host of other foods. Benzoic acid itself acts as a diuretic and an antiseptic.

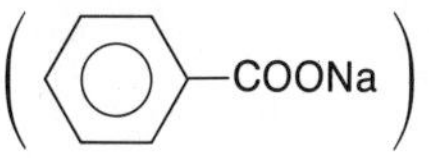

Oxalic acid $\left(\begin{array}{l} C{-}OOH \\ | \\ C{-}OOH \end{array}\right)$ is used as a bleach for straw and leather, and for the removal of rust stains from fabric and porcelain. Although oxalic acid is poisonous in quantity, small amounts of it are found in rhubarb and cranberries.

Polyfunctional Compounds

A number of organic acids are actually *polyfunctional compounds;* that is, they have two different functional groups. The polyfunctional acid of greatest commercial importance is salicylic acid (benzene ring bearing COOH and, on the adjacent carbon, OH).

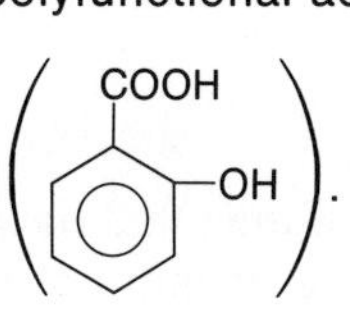

Its hydroxyl group can be esterified by reaction with acetic acid to for acetyl salicylic acid—better known as aspirin. The reaction is

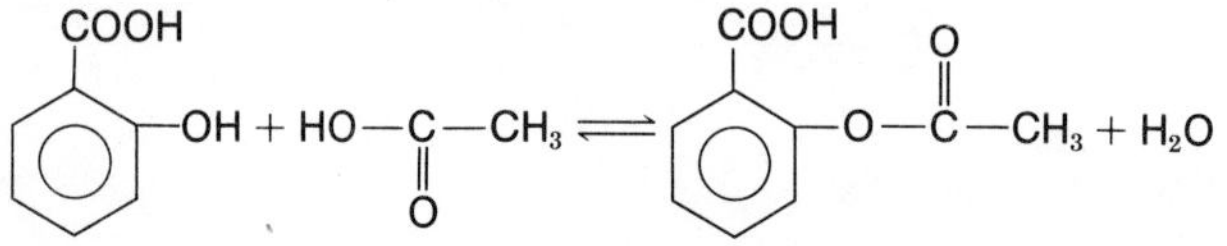

$$C_6H_4(COOH)\text{—}OH + HO\text{—}\overset{\overset{}{}}{\underset{\underset{O}{\|}}{C}}\text{—}CH_3 \rightleftharpoons C_6H_4(COOH)\text{—}O\text{—}\overset{\overset{O}{\|}}{C}\text{—}CH_3 + H_2O$$

Salicylic acid + Acetic acid ⇌ Acetyl salicylic acid + Water

Aspirin is a widely used *antipyretic* (fever reducer) and *analgesic* (pain reliever) whose mode of action is still largely unknown. Despite claims by manufacturers to the contrary, there is no real difference between various brands of aspirin unless one of them has decomposed. Decomposition is caused by moisture, which hydrolyzes the ester to salicylic and acetic acids. Aspirin that has hydrolyzed has the odor of vinegar because of the acetic acid.

Salicylic acid is used in the treatment of fungal infections and in the manufacture of "enteric-coated" pills. The enteric coating is a layer of phenyl salicylate

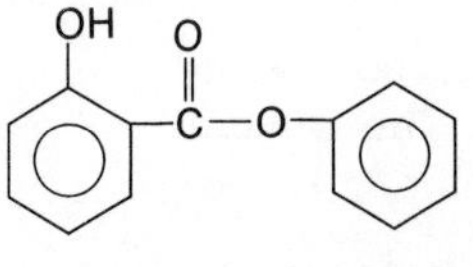

$$C_6H_4(OH)\text{—}\overset{\overset{O}{\|}}{C}\text{—}O\text{—}C_6H_5$$

which is unaffected by the acid of the stomach, but readily dissolves when it contacts the alkaline fluid in the intestine. A coating of this type insures that medication intended for the intestinal tract is not released by reaction with acid in the stomach.

The metabolic pathways of fats, proteins, and carbohydrates involve acidic intermediates and often yield polyfunctional acids such as citric, lactic, and pyruvic. As mentioned earlier, lactic acid is formed during muscle contraction. It is also produced by the bacterial fermentation of lactose, or milk sugar, and is responsible for the sour taste of buttermilk. A mixture of buttermilk and baking soda can be substituted for baking powder, since the reaction between them produces carbon dioxide gas, which causes dough to rise. The reaction is

$$CH_3\overset{\overset{OH}{|}}{CH}\text{—}\overset{\overset{O}{\|}}{C}\text{—}OH + NaHCO_3 \longrightarrow CO_2 + H_2O + CH_3\overset{\overset{OH}{|}}{CH}\text{—}\overset{\overset{O}{\|}}{C}\text{—}ONa$$

Lactic acid + Sodium bicarbonate ⟶ Carbon dioxide + Water + Sodium lactate

The metabolic relationships between pyruvic, lactic, and citric acids are discussed in Chapter 12.

11.4 AMINES

Amines can be thought of as compounds that have been formed from ammonia by the replacement of one or more of its hydrogen atoms.

Like alcohols, they are classified as primary, secondary, or tertiary. The nitrogen atom in a primary amine is bonded to two atoms of hydrogen and one of carbon, secondary amines have a single N—H bond, while tertiary amines have none:

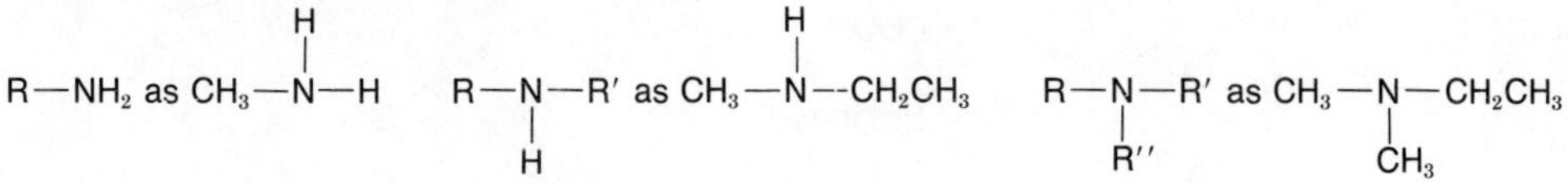

$$R{-}NH_2 \text{ as } CH_3{-}\underset{}{N}(H){-}H \qquad R{-}N(H){-}R' \text{ as } CH_3{-}N(H){-}CH_2CH_3 \qquad R{-}N(R''){-}R' \text{ as } CH_3{-}N(CH_3){-}CH_2CH_3$$

Primary Methylamine Secondary Ethyl methyl amine Tertiary Ethyl dimethyl amine

Amines resemble ammonia chemically as well as structurally. Recall that the lone pair of electrons on its nitrogen atom qualifies ammonia as a base, and that it reacts with acids to form salts. Amines react in a similar manner, as the following examples indicate:

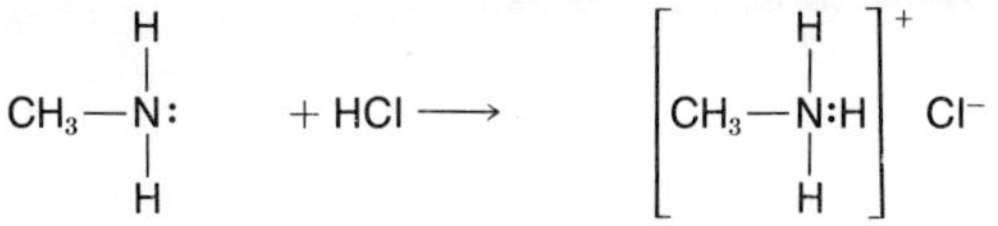

$$CH_3{-}\ddot{N}H_2 + HCl \longrightarrow [CH_3{-}NH_3]^+ \; Cl^-$$

Methyl amine — Methyl ammonium chloride

$$(CH_3)_3\ddot{N} + HCl \longrightarrow [(CH_3)_3N{:}H]^+ \; Cl^-$$

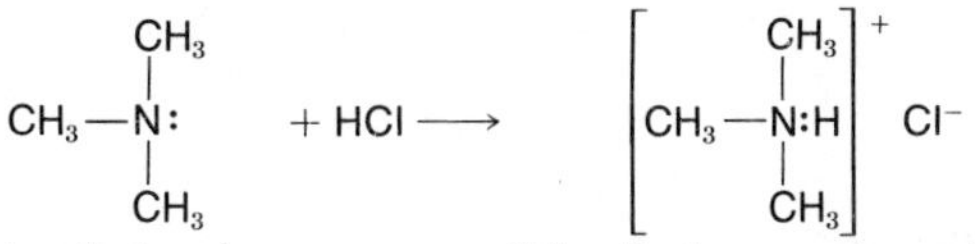

Trimethyl amine — Trimethyl ammonium chloride

Tertiary amines also react with alkyl halides to form quaternary ammonium salts. Some of these salts have detergent as well as disinfectant properties. Benzalkonium chloride, also called Zephiran® and Roccal,® is used at a dilution of 1:1000 for preoperative cleansing and disinfection of the skin. Its formula is

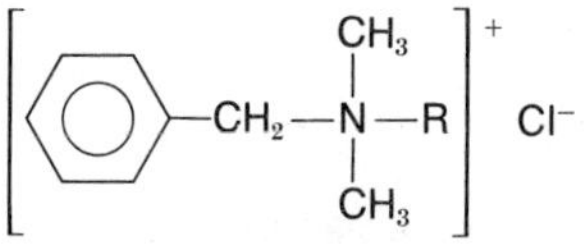

$$[C_6H_5{-}CH_2{-}N(CH_3)_2{-}R]^+ \; Cl^-$$

Benzalkonium chloride

In addition to their reactions with acids, primary and secondary amines also react with carboxylic acids to form *amides*. This reaction is a dehydration involving the removal of a hydrogen atom from the amino nitrogen:

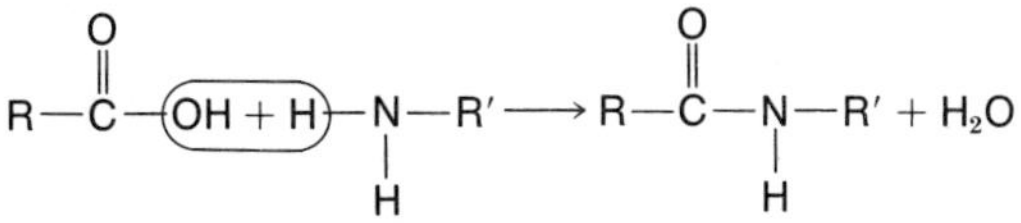

$$R{-}\overset{O}{\overset{\|}{C}}{-}\boxed{OH + H}{-}N(H){-}R' \longrightarrow R{-}\overset{O}{\overset{\|}{C}}{-}N(H){-}R' + H_2O$$

Tertiary amines cannot form amides because their amino hydrogens have all been replaced. Amide formation is an important biological process, since amino acids are joined by amide linkages to produce proteins. This process is discussed in Chapter 14.

One of the most unusual tertiary amines is a chelating agent called ethylenediaminetetraacetic acid (EDTA):

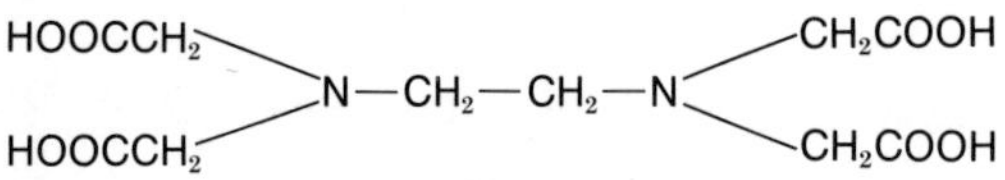

$$(HOOCCH_2)_2N{-}CH_2{-}CH_2{-}N(CH_2COOH)_2$$

EDTA

Chelating agents are organic compounds which form loose attachments to certain metal ions and remove them from solution. Because of this property, chelating agents may be administered intravenously to combat poisoning by lead, arsenic, or other heavy metals. EDTA is also administered to persons suffering from bone cancer: it chelates the calcium ions required by the cancerous cells for their growth.

Aromatic amines are related to aminobenzene (aniline), whose structure is

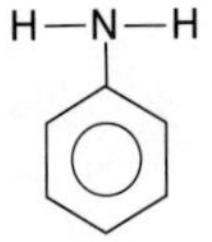

$$C_6H_5{-}NH_2$$

Aniline

Synthetic reactions with aniline produce dyes, perfumes, drugs, explosives, and a large variety of other useful compounds. For example, reaction of aniline with acetic acid produces acetanilide, whose physiological effects are similar to those of aspirin.

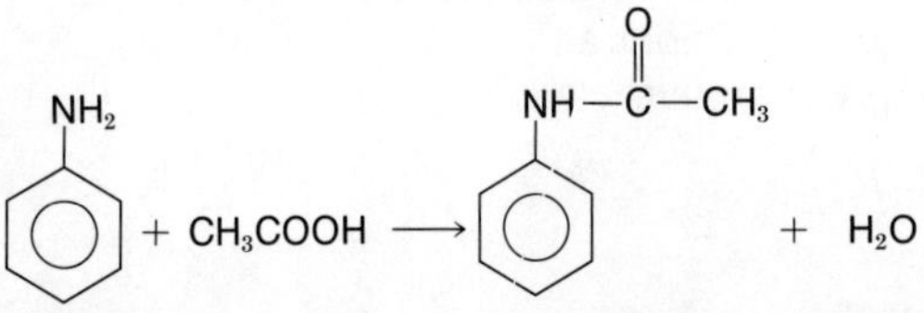

$$C_6H_5NH_2 + CH_3COOH \longrightarrow C_6H_5NH{-}\overset{O}{\overset{\|}{C}}{-}CH_3 + H_2O$$

Aniline + Acetic acid ⟶ Acetanilide + Water

Because of its undesirable side effects, acetanilide has been replaced by a related compound called phenacetin (or acetophenetidin), whose toxicity is much lower. The structure of phenacetin is

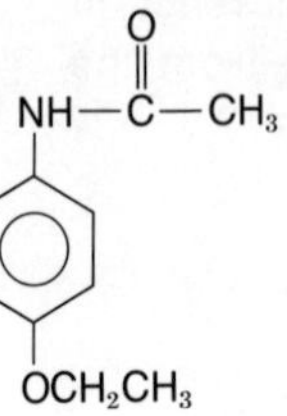

$$CH_3CH_2O{-}C_6H_4{-}NH{-}\overset{O}{\overset{\|}{C}}{-}CH_3$$

A class of antibacterial compounds called sulfa drugs are also derivatives of aniline. The first of these compounds to be synthesized was sulfanilamide:

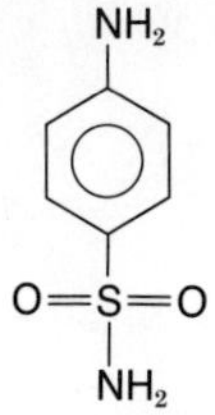

NH_2

$O{=}S{=}O$

NH_2

Toxic reactions in patients using sulfanilamide led to the synthesis of related structures with fewer adverse side effects, including sulfathiazole, sulfadiazine, and sulfaguanidine, which are sometimes used in the treatment of urinary tract infections, pneumonia, and diarrhea. The synthesis of drugs and related compounds is part of a new, highly empirical pharmacological endeavor called drug design. Because of the complexity of biochemical reactions, it is not possible to predict the effects of using related molecules in treating human illness.

11.5 ALDEHYDES

Aldehydes are compounds in which a carbonyl group is located at one end of the carbon chain. Numbering of the chain begins with the functional group, and the names of aldehydes end in *-al.* Methanal is the IUPAC name for the aldehyde whose molecular formula is HCHO. Methanal—a colorless gas with a sharp odor—is also known by the common name formaldehyde. It is a polar compound and extremely soluble in water. A 40% solution is called formalin, and is used to sterilize surgical instruments and to harden and preserve biological specimens, and as embalming fluid.

Paraformaldehyde $[HCHO]_3$, a polymer of methanal, is a solid that decomposes into formaldehyde when heated. Candles containing paraformaldehyde are sometimes used for disinfecting rooms.

The aldehyde resembling ethane in structure is ethanal (acetaldehyde). A polymer of ethanal, called paraldehyde $[CH_3CHO]_3$, is a potent sleep inducer (*hypnotic*) and is an effective sedative for persons who are extremely agitated and suffering from hallucinations following excess consumption of alcohol. Because of its foul taste and odor, it is administered rectally.

Reactions of Aldehydes

Aldehydes are readily oxidized to the corresponding carboxylic acids, as shown by the following examples:

$$C_6H_5CHO \xrightarrow[\Delta]{[O]} C_6H_5COOH$$

Benzaldehyde Benzoic acid

$$CH_3CHO \xrightarrow{[O]} CH_3COOH$$

Acetaldehyde Acetic acid

When the oxidation takes place in an alkaline solution containing Cu^{2+} ion, a precipitate of red Cu_2O is produced. The tendency of copper ion to precipitate in alkaline solution as $Cu(OH)_2$ is prevented by the addition of a complexing agent such as citrate ion (used in Benedict's reagent) or tartrate ion (Fehling's reagent). The oxidation of an aldehyde by either reagent produces the same result, as shown by the following equation (not balanced):

$$\underset{\text{Aldehyde}}{RCHO} + \underset{\substack{\text{Fehling's or Benedict's} \\ \text{reagent (blue)}}}{2\ Cu^{2+}\ (\text{complex})} + NaOH \longrightarrow \underset{\text{Salt}}{RCOONa} + \underset{\text{Red precipitate}}{Cu_2O}$$

Aldehydes may also be oxidized by Tollen's reagent, which is an alkaline solution of silver ions. The silver is stabilized in solution by the formation of a silver-ammonia complex. The overall reaction is

$$\underset{\text{Aldehyde}}{RCHO} + \underset{\substack{\text{Tollen's reagent} \\ \text{(colorless)}}}{2\ Ag(NH_3)_2^+} + OH^- \longrightarrow \underset{\text{Salt}}{RCOONH_4} + 2\ Ag + 3\ NH_3 + H_2O$$

The metallic silver resulting from the reduction of silver ion is deposited on the walls of the vessel, forming a silver mirror.

Most of the simple sugars contain an aldehyde group and are oxidized by the reagents just described. Since the oxidation of the carbonyl group is accompanied by reduction of copper ion, sugars containing a free carbonyl group are called *reducing sugars.* Glucose is a reducing sugar whose presence in the urine generally indicates a disease condition. Inasmuch as Fehling's solution is affected by high concentrations of NaCl, Benedict's solution is the reagent of choice for the detection of sugar in urine.

11.6 KETONES

The carbonyl group of a ketone is linked to two carbon atoms, and thus is located within the molecule rather than at one end. The names of ketones end in *-one,* and the position of the carbonyl group is designated by a numerical prefix. Ketones are also named according to the groups attached to the carbonyl carbon. For example, the structure

```
          O
  |   |   ‖   |   |
—C—C—C—C—C—
  |   |       |   |
```

is called either 2-pentanone or diethyl ketone. Propanone also has two common names: dimethyl ketone and acetone. It is a volatile, water-soluble liquid which is found in the breath of persons suffering from uncontrolled diabetes. It is also the solvent in lacquers and in fingernail polish remover.

As mentioned earlier, ketones are formed from secondary alcohols by oxidation. They are less active than aldehydes because their carbonyl group is less exposed. Ketones are not normally oxidized by Benedict's, Fehling's, or Tollen's reagents. Compounds that contain a primary alcohol group adjacent to the carbonyl group are notable exceptions to this rule. One such compound is the ketone sugar, fructose, which is oxidized by Benedict's but not by Tollen's reagent.

```
      H
      |
  H—C—OH
      |
      C=O
      |
 HO—C—H
      |
  H—C—OH
      |
  H—C—OH
      |
  H—C—OH
      |
      H
```

Fructose

Ketones produce two molecules of organic acid when oxidized, while aldehydes produce only one.

Several ketone acids are formed during the aerobic breakdown of glucose. They are pyruvic acid, oxaloacetic acid, α-ketoglutaric acid, and oxalosuccinic acid. The process by which they are formed is described in the next chapter.

11.7 A POSTSCRIPT

Compounds containing hybridized carbon atoms number in the millions, and this discussion has been intended only to focus attention on those types of compounds which have special significance for living creatures. Each of the individual functional groups introduced in this chapter will be considered further in the remaining chapters.

APPLICATION OF PRINCIPLES

1. Draw the structural formulas for all isomers of butanol and name each isomer according to the IUPAC system.
2. Draw a structural formula for each of these compounds:
   (a) 5-bromo-3-methylpentanal (b) o-chlorophenol
   (c) 1,2-dihydroxybutane (d) 3-methyl-1-hexene
3. Write balanced equations for the complete combustion of
   (a) acetone (b) ethanol
4. Draw the structural formula(s) for the product(s) of the following reaction:

   COOH (on benzene ring, ortho to OH)

   $$\text{C}_6\text{H}_4(\text{COOH})\text{—OH} + \text{CH}_3\text{OH} \xrightarrow{\text{H}_2\text{SO}_4}$$

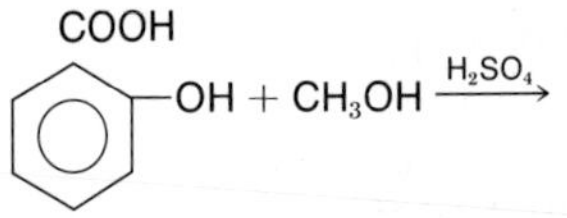

5. What information concerning structure is revealed by each of the following results?
   (a) Benedict's solution turns green upon being heated with an unidentified carbon compound.
   (b) An alcohol upon mild oxidation is converted into an aldehyde.
6. Which of the following structures contains an amide linkage?
   (a) $\text{C}_6\text{H}_5\text{—C(=O)—O—CH}_2\text{CH}_3$ (b) $\text{CH}_3\text{—NH—CH}_2\text{CH}_3$
   (c) $\text{CH}_3\text{CH}_2\text{—O—NHCH}_3$ (d) $\text{CH}_3\text{CH}_2\text{C(=O)—NH—CH}_3$
7. Identify the functional group(s) in the following structures:
   (a) $\text{CH}_3\text{CHOHCOOH}$ (b) $\text{CH}_3\text{—NH—CH}_2\text{CH}_3$
   (c) COOH

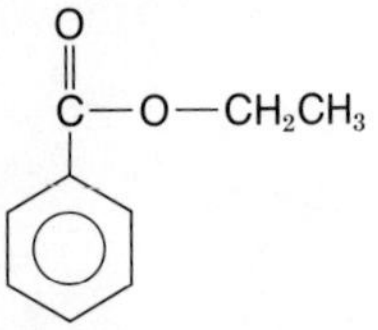

   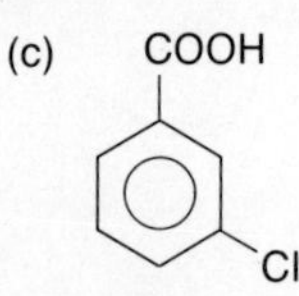

   Cl

   (d) $\text{CH}_3\text{C(=O)CH}_2\text{CH}_3$
8. What is the concentration of ethanol in 90-proof whiskey?
9. Hexachlorophene is a disinfectant formerly used in Dial® soap and pHisoHex.® It has a phenol coefficient of about 1200. What does this mean?
10. Draw the structural formula for the product of the following reaction:

$$\text{CH}_3\text{CH}_2\text{—C(H)(CH}_3\text{)—OH} \xrightarrow{[\text{O}]}$$

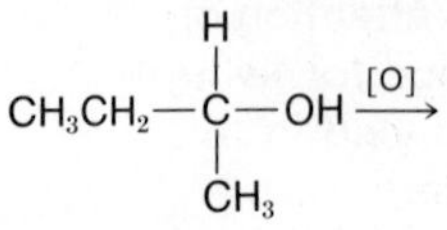

11. Draw the structural formula for the product of the following reaction:

$$CH_3CH_2CHO \xrightarrow{[O]}$$

12. Draw the structural formula for the product of the reaction between butanoic acid and dimethyl amine.

Carbohydrates provide a large portion of our energy requirements. (Courtesy Betty Crocker Kitchens, General Mills, Inc., Minneapolis, Minnesota)

12

CARBOHYDRATES

LEARNING OBJECTIVES

1. Compare the composition, structures, and properties of the common monosaccharides, disaccharides, and polysaccharides.
2. Derive the Haworth configuration of a monosaccharide from its Fischer projection formula.
3. Name and describe the several qualitative tests for carbohydrates, and explain the significance of each test.
4. Trace the digestion of a carbohydrate, including the name and specific chemical action of each digestive enzyme.
5. Compare and explain the results of the glucose tolerance test for normal, diabetic, and hypoglycemic individuals.
6. Distinguish between glycogenesis and glycogenolysis.
7. Compare the changes that occur during glycolysis with those taking place in the Krebs cycle.
8. Describe the actions of the various hormones that are involved in the regulation of glucose metabolism.

KEY TERMS AND CONCEPTS

aldonic acid
aldose
alduronic acid
asymmetric carbon atom
carbohydrate
citric acid (Krebs) cycle
disaccharide
Embden-Meyerhof pathway
enediol
fermentation
glucose tolerance test
glycogenesis
glycogenolysis
glycolysis
Haworth configuration
hypoglycemia
ketose
monosaccharide
oligosaccharide
optical isomerism
perspective formula
photosynthesis
polysaccharide
reducing sugar

One significant early event in the history of chemistry was the discovery that the element carbon always occurs in compounds manufactured by cells. Early research indicated that a number of these compounds contained carbon, hydrogen, and oxygen in the ratio CH_2O. This ratio was interpreted as a molecule of water attached to an atom of carbon, suggesting that these compounds were hydrates of carbon. Such compounds were called carbohydrates. Later research showed that the carbon was not actually bonded to water molecules; however, as often happens in science, the name persisted even though it was not correct. The current view of carbohydrates is that they are polyhydroxyaldehydes or polyhydroxyketones, or substances that yield these compounds when they are hydrolyzed.

12.1 CLASSIFICATION

Individual carbohydrates fall into one of three major categories: monosaccharides (simple sugars), oligosaccharides, and polysaccharides. A carbohydrate is classified as a *monosaccharide* if it cannot be hydrolyzed to smaller polyhydroxyaldehydes or ketones. Although there is some disagreement among chemists regarding the upper limit, a carbohydrate that yields from two to ten monosaccharide units when hydrolyzed is considered to be an *oligosaccharide* [<L. *oligo,* few]. Most of the important oligosaccharides are *disaccharides,* i.e., composed of two monosaccharide units. The *polysaccharides* are polymers formed from monosaccharides and may contain as many as 500 units.

Monosaccharides may be further subdivided according to the nature of their carbonyl group, and according to the number of carbon atoms they contain. When the carbonyl group is on the terminal carbon atom, the monosaccharides display some of the properties of aldehydes and are called *aldoses.* Simple sugars that have a ketone configuration are called *ketoses* (not to be confused with *ketosis,* which is a physiological condition described in Chapter 8).

The number of carbon atoms in a simple sugar is indicated by inserting a numerical stem between the *keto-* prefix and the *-ose* ending. Thus a sugar that contains six carbons and a ketone group would be called a *ketohexose,* and an aldehyde sugar having five carbon atoms would be an *aldopentose.*

12.2 STRUCTURES

For the sake of simplicity the structures of the monosaccharides are often represented by Fischer projection formulas which show them as open-chain compounds. For example, the structure of glucose (an aldohexose) may be represented by the formula

$$\begin{array}{rcl} H- & \overset{1}{C} & =O \\ & | & \\ H- & \overset{2}{C} & -OH \\ & | & \\ HO- & \overset{3}{C} & -H \\ & | & \\ H- & \overset{4}{C} & -OH \\ & | & \\ H- & \overset{5}{C} & -OH \\ & |\,{\scriptstyle 6} & \\ & CH_2OH & \end{array}$$

while fructose (a ketohexose, and a structural isomer of glucose) is shown as

$$\begin{array}{rcl} & \overset{1}{CH_2OH} & \\ & | & \\ & \overset{2}{C} & =O \\ & | & \\ HO- & \overset{3}{C} & -H \\ & | & \\ H- & \overset{4}{C} & -OH \\ & | & \\ H- & \overset{5}{C} & -OH \\ & |\,{\scriptstyle 6} & \\ & CH_2OH & \end{array}$$

Such formulas are useful because they clearly indicate the various functional groups in each monosaccharide, yet they fail to account for some of the properties of the various sugars. For instance, solid glucose is quite inert with oxygen, yet aldehydes are generally oxidized with ease. And while aldehydes react with fuchsin to give a positive Schiff test, glucose and other aldoses do not. Although these observations do not fit the open-chain structure, they *can* be explained in terms of a ring structure formed by the natural folding of the carbon chain. The formation of such a ring can be visualized in two steps, with only the carbon chain and the participating functional groups being shown for clarity. First, the chain folds so that the carbonyl group on carbon 1 is adjacent to the OH group on carbon 5:

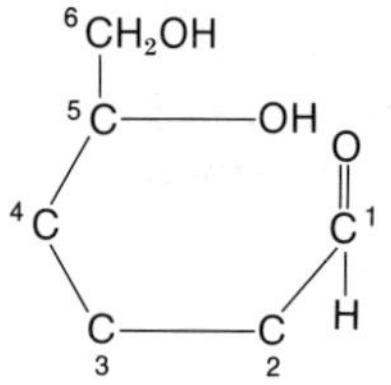

Then the ring is closed when the two carbon atoms bond through an oxygen bridge:

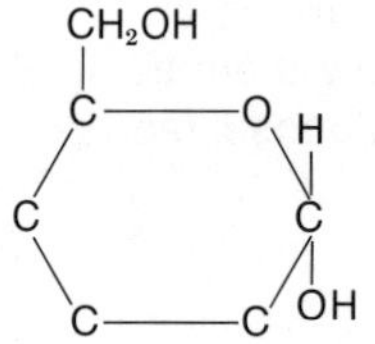

The English chemist W. H. Haworth suggested that the ring structure might be more clearly represented as a hexagonal ring lying in a plane perpendicular to the plane of the paper, with the substituents projecting either above or below the plane of the ring. To show this perspective, the side of the ring that is closer to you is indicated by a thickened line. The *Haworth configuration* for glucose is

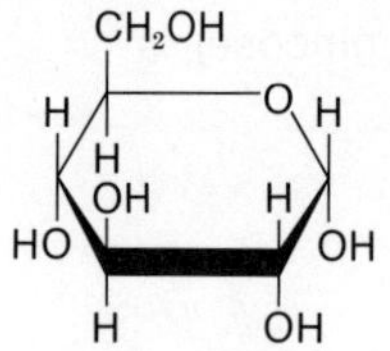

Notice that closure of the ring can result in either of two different configurations at carbon 1: the one in which the —OH group projects below the plane of the ring, and another in which the —OH group is above the ring. The two configurations are designated α and β, respectively.

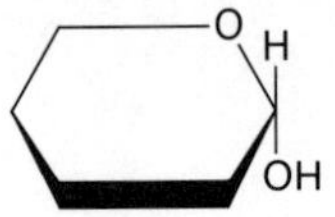

α configuration of glucose 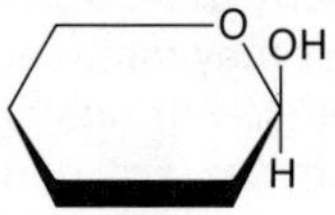 β configuration of glucose

Ketoses, such as fructose, also exist predominantly in the ring structure. Closure of the ring involves the carbonyl group on carbon 2 and the hydroxyl group on carbon 6. Once again two configurations, α and β, are possible:

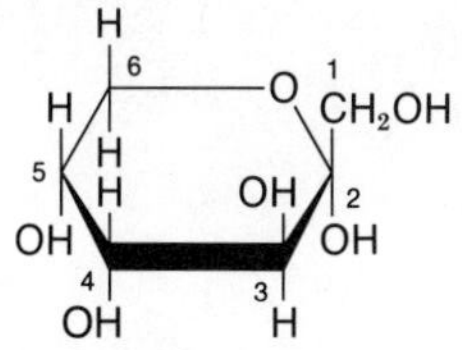

α configuration of fructose

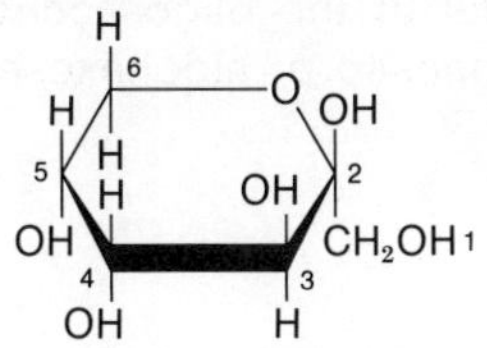

β configuration of fructose

However, when fructose bonds to another molecule of fructose, or to a different monosaccharide, it assumes a ring structure that is composed of only five members and links carbon 2 to carbon 5:

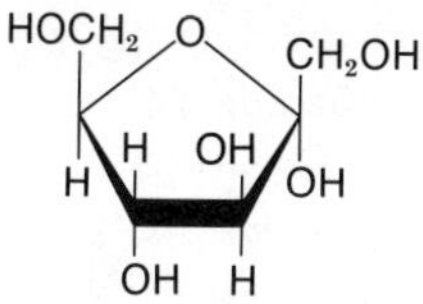

α configuration

Galactose, like fructose, is an isomer of glucose. However, fructose is a structural isomer while galactose differs from glucose only with respect to the configuration around the #4 carbon atom. In glucose the hydroxyl group on this carbon is below the plane of the ring, while in galactose it is above the ring:

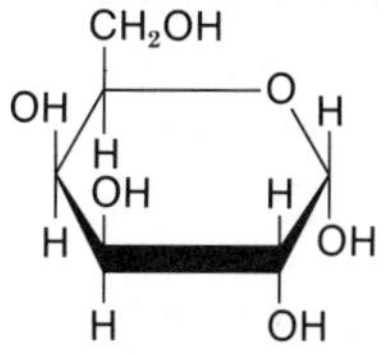

Galactose

The most important pentose is probably ribose—the aldopentose component of RNA, ATP, NAD, and coenzyme A. The closely related 2-deoxyribose is found in DNA. The Fischer projection and Haworth formulas for ribose are

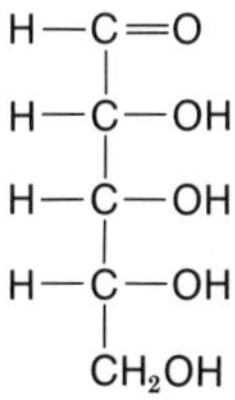

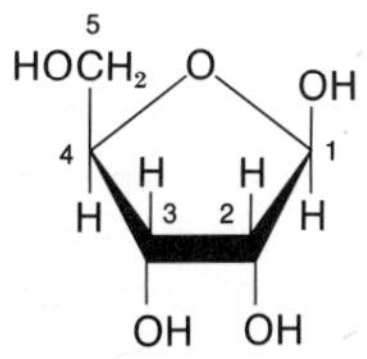

$$\begin{array}{c} H-C=O \\ | \\ H-C-OH \\ | \\ H-C-OH \\ | \\ H-C-OH \\ | \\ CH_2OH \end{array}$$

Both the α- and β-isomers can exist in solution, but the β-isomer alone is found in the compounds just mentioned.

Oligosaccharides and Polysaccharides

Oligosaccharides and polysaccharides are formed from monosaccharides by dehydration synthesis. Several of the oligosaccharides, and most of the polysaccharides are composed of aldohexose units linked either α-1,4 or α-1,6.

The α-1,4 Linkage

Maltose is a disaccharide consisting of two molecules of α-glucose joined through the hydroxyl group attached to carbon 1 of one molecule to the hydroxyl group on carbon 4 of the second molecule. The formation of maltose may be represented as

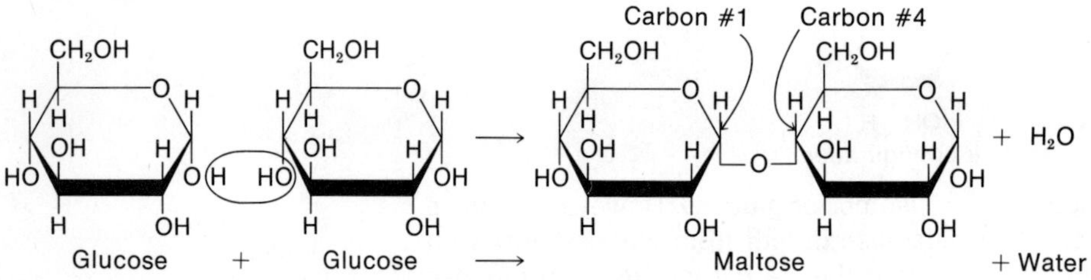

Carbon #1 Carbon #4

Glucose + Glucose $\longrightarrow$ Maltose + Water ($+ H_2O$)

A significant amount of the glucose that a plant produces may be polymerized in this fashion to form amylose. Amylose is one of the two components of starch. The other component is a branched polymer of glucose called amylopectin. The number of glucose units in amylose ranges from several dozen to several hundred.

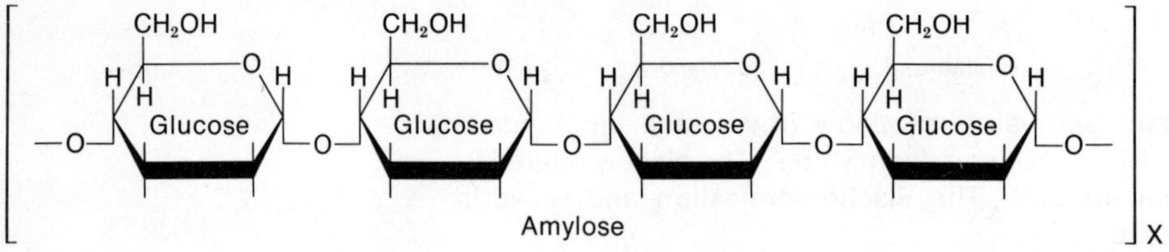

$[\text{—O—Glucose—O—Glucose—O—Glucose—O—Glucose—O—}]_X$

Amylose

The α-1,6 Linkage

The difference between amylose and amylopectin is that amylose is a continuous chain of glucose molecules, while in amylopectin shorter chains (about 25-30 units) of α-1,4-linked glucose units are joined by α-1,6 linkages.

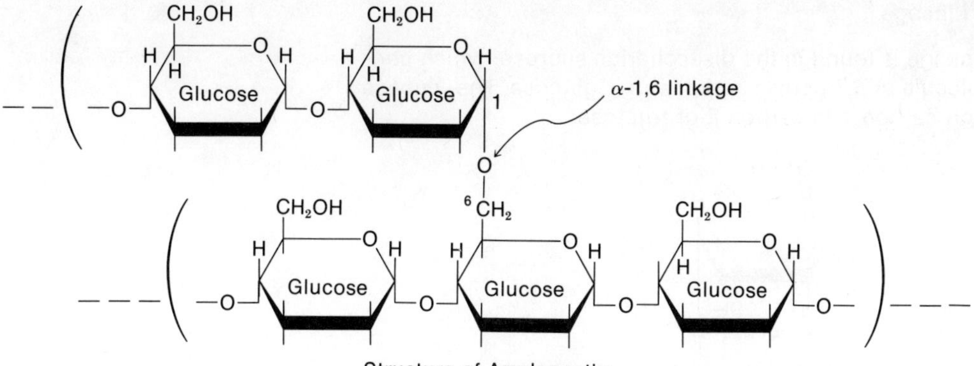

Structure of Amylopectin

Animals store carbohydrates in the form of a compound called glycogen, which is similar in structure to amylopectin, but more highly branched. Glycogen has an average of one branch point for every 8-10 glucose units.

The β-1,4 Linkage

The structure of cellulose is identical with that of amylose in all respects except one: the glucose molecules in cellulose all have the β configuration. Because of this, the linkage between glucose units in cellulose is β-1,4:

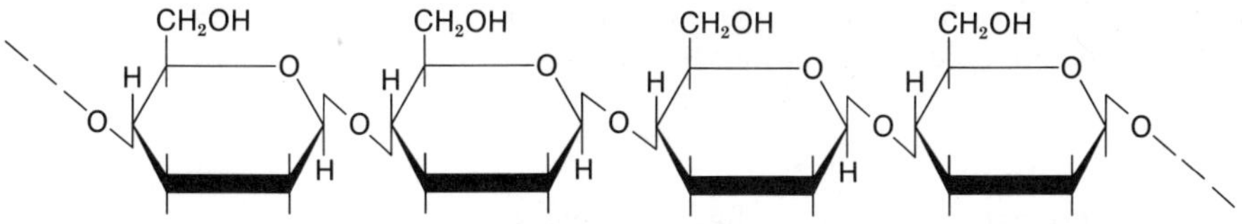

Cellulose

The disaccharide lactose also contains the β-1,4 linkage. Lactose is formed from one molecule of β-galactose and one molecule of α-glucose. The bond is through the C-1 hydroxyl group of galactose to the C-4 hydroxyl group of glucose:

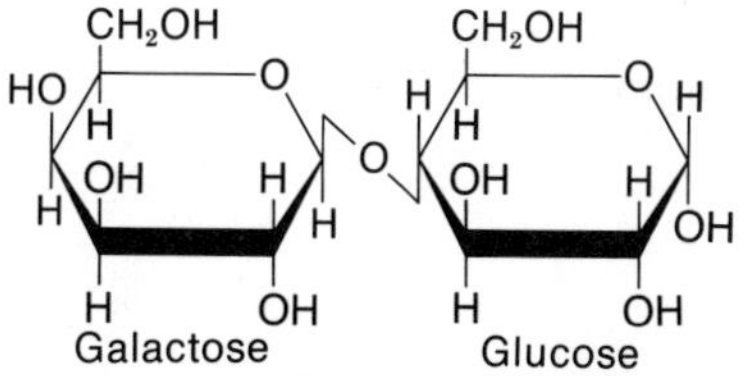

Lactose

The α-1 to β-2 Linkage

This unusual linkage is found in the disaccharide sucrose, which consists of one molecule of β-fructose and one of α-glucose. The glucose is joined through carbon 1 to carbon 2 of fructose:

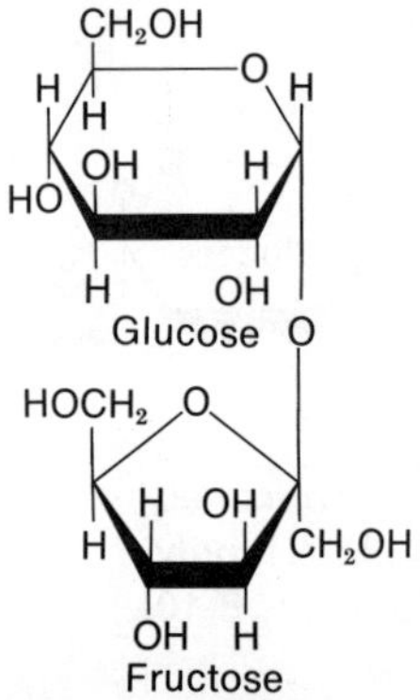

Glucose

Fructose

Sucrose

One important aspect of the chemistry of sugars is that all of the various types of linkages between monosaccharides can be ruptured by treatment with acid, yet each linkage type can be broken enzymatically only by a specific enzyme. Thus the enzyme that hydrolyzes starch cannot hydrolyze cellulose, and vice versa. Most animals lack the enzyme needed for the breakdown of cellulose, and so cellulose passes through their digestive tracts unaltered. Horses, cows, and other ruminants obtain energy from cellulose because bacteria and protozoa in their digestive systems are able to attack the β-1,4 linkage.

12.3 OPTICAL ISOMERISM

Compounds such as glucose and galactose, which have the same molecular and structural formulas but which differ in the arrangement of their atoms in space (configuration), are called *optical isomers.* Optical isomerism is possible when a molecule contains one or more *asymmetric carbon atoms*—that is, carbon atoms to which four different groups are attached. A carbon atom having either a double bond or two identical attached groups cannot be asymmetric.

The reference compound for optical isomers is glyceraldehyde, an aldotriose. Glyceraldehyde was chosen because it is the simplest monosaccharide containing an asymmetric carbon atom.

H—C=O
|
H—C—OH
|
H—C—OH
|
H

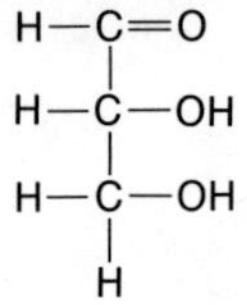

Glyceraldehyde

Optical isomers. Each isomer contains one asymmetric carbon atom. (Photograph by J. Asdrubal Rivera)

The configuration of glyceraldehyde can be represented by a perspective formula in which wedges, dashes, and dotted lines are used to show the spatial relationship between the asymmetric atom and an attached group. The bonds of groups lying in the same plane as the carbon atom are indicated by solid lines. Groups that project toward the viewer are attached by wedges to the central atom, and dotted lines are used for groups that lie behind the plane of the carbon atom. Figure 12.1 illustrates the relationship between a three-dimensional model and its perspective formula.

There are two optical isomers of glyceraldehyde:

D-glyceraldehyde	L-glyceraldehyde
CHO	CHO
H—C—OH	HO—C—H
CH_2OH	CH_2OH

When they are represented by perspective formulas, the configuration in which the —OH group on the center carbon atom is shown on the

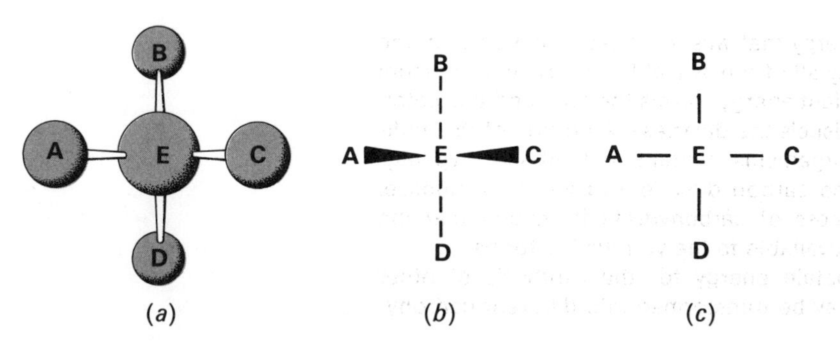

Figure 12.1
Representations of a molecule containing an asymmetric carbon atom. (a) Ball-and-stick model. (b) Perspective formula. (c) Projection formula.

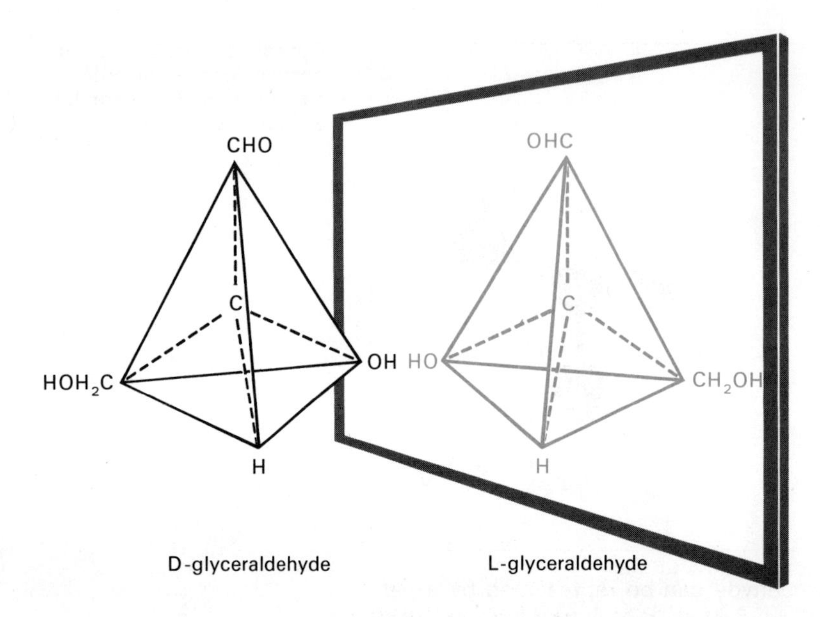

Figure 12.2
The relationship between D-glyceraldehyde and L-glyceraldehyde. Optical isomers are mirror images.

right is arbitrarily designated the D form. The configuration of a monosaccharide containing more than three carbon atoms is determined by the asymmetric carbon farthest from the carbonyl group.

A pair of optical isomers are, in effect, mirror images. Each isomer is related to the other in the same way as a right hand is related to a left hand: they cannot be superimposed on each other (Figure 12.2).

Optical isomers have identical physical properties (density, boiling point, vapor pressure, etc.), but may undergo reaction with the same reagent to produce compounds whose physical and chemical properties are different.

For many compounds there is a correlation between configuration and biological activity. For example, D-glucose is readily fermented by enzymes found in most microorganisms, but L-glucose is not; and only L forms of the amino acids are incorporated into proteins.

12.4 IMPORTANT CARBOHYDRATES AND THEIR PROPERTIES

With the exception of the energy that was imparted to the earth at the time of its formation, virtually all of the available energy on our planet has come from the sun. Radiant energy warms the land and the water, sets the air in motion, and dispels the darkness. A portion of this radiant energy is trapped by organisms containing light-sensitive pigments and is used to reduce carbon dioxide and water to glucose. Thus it is through the process of carbohydrate formation that the energy of the sun becomes available to the various life forms.

Glucose is oxidized to obtain energy for the synthesis of other cellular components, or it may be transformed into different carbohydrates, some of which may serve as energy storage forms. The carbohydrates of major importance are discussed in the following sections.

Glucose

Glucose is found in the juices (sap) of plants, in fruits, and in the blood and tissues of animals. The molecular structures of lactose, maltose, sucrose, starch, glycogen, cellulose, and the dextrins also include glucose. *Dextrins* are products of the partial hydrolysis of starch. Glucose is also known as dextrose, grape sugar, and blood sugar. Isotonic solutions of glucose are administered intravenously to provide a source of energy for persons who cannot take nourishment by mouth. The concentration of glucose in the blood of a normal fasting individual ranges from about 80 to 120 mg%.

Fructose

The relative sweetness of the common mono- and disaccharides appears to be a function of solubility. Fructose is the most soluble common sugar and is also the sweetest (Table 12.1). This fact accounts for the extreme sweetness of honey, which contains roughly equal amounts of fructose and glucose.

TABLE 12.1 Water Solubility and the Relative Sweetness of Common Sugars

	Solubility (g/100 g of water)	Relative Sweetness
Monosaccharides		
Fructose	200	173
Glucose	83	74
Galactose	10	32
Disaccharides		
Sucrose	179	100[1]
Maltose	108	33
Lactose	17	16

[1] Sweetness is expressed on a scale relative to sucrose, which is assigned a rating of 100.

For many persons who either cannot tolerate carbohydrates or must restrict their sugar intake, the desire for sweets creates a problem. Chemists have synthesized a substance called saccharin which has the ability to duplicate the sensation of sweetness, yet is not a carbohydrate.

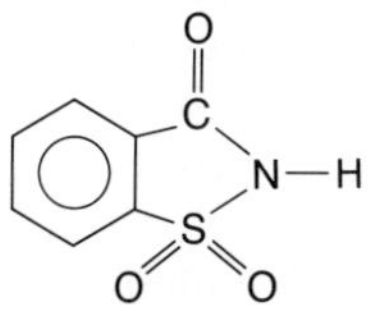

Saccharin

On the relative scale shown in Table 12.1, saccharin has a rating of approximately 50,000!

Because of its presence in many fruits, fructose is also known as fruit sugar. Like glucose, fructose is found in the sap of a number of plants, and a polymer of fructose (called inulin) occurs in Jerusalem artichokes. Inulin is not hydrolyzed by the enzymes in man's digestive tract.

Galactose

One of the unique features of galactose is that it does not occur in the free state, but is found in combination with other sugars or with long-chain carboxylic acids and amino alcohols. For example, it combines with glucose to form lactose, the sugar that is present in the milk of most mammals, and it is a component of fat-like substances called glycolipids that are distributed throughout brain and nervous tissue. Agar-agar, a polymeric form of galactose, is extracted from red algae and is used as a gelling agent in the preparation of solid media for culturing bacteria. Like inulin, agar-agar has no food value for man because it is not digested.

Sucrose

Sucrose, also known as table sugar, or cane sugar, is prepared from the juice of sugar cane or from sugar beets. The juice, which contains from 14 to 18% sucrose, is treated with lime to remove a protein-like substance, and then evaporated at low temperature and reduced pressure to a semi-solid mass from which raw sugar is separated by centrifugation, leaving behind a dark liquid called molasses. The raw sugar is filtered through bone or charcoal—a process that further reduces its impurities—and then evaporated to form the familiar translucent crystals.

In recent years the consumption of sucrose in the United States has risen to a yearly average of 100 pounds per person. A number of doctors and nutritionists, apparently alarmed by the poor eating habits suggested by these statistics, have raised their voices against the excessive intake of carbohydrates. Some health food hucksters have gone them one better by claiming that sucrose is just "empty calories," or that consumption of sucrose will lead to all sorts of ailments. Taking advantage of the evidence that many foods have a higher vitamin content when they are raw than after they have been processed, the hucksters argue that raw sugar has a greater nutritional value than completely refined sucrose—a claim that has not been substantiated. It is important to realize that the word "raw" when applied to sugar does not mean the same as when it is used to describe unprocessed or uncooked fruit, vegetables, or meat. Truly "raw" sugar is found only in the juices of plants.

When a concentrated solution of sucrose is subjected to prolonged heating, as in candy- or jelly-making, an equimolar mixture of glucose and fructose, called invert sugar, is produced. The conversion of sucrose to invert sugar is desirable because candy or jelly is less likely to crystallize when it contains a mixture of sugars.

Maltose

Maltose is obtained by the incomplete hydrolysis of starches, dextrins, and glycogen. It is also found in germinating grains. For example, barley that has been sprouted under controlled conditions contains large amounts of maltose, and is sometimes called malt. In the manufacture of beer, this maltose is hydrolyzed to glucose, which is then fermented by yeasts to produce ethanol.

Lactose

Lactose is synthesized in the mammary glands from glucose and galactose, and is found in milk at a concentration of 4-8%. Human milk contains about 30% more lactose than cow's milk; therefore, infant feeding formulas must be supplemented with sugar in order to provide an equivalent amount of carbohydrate.

Cellulose

The cells of plants, unlike those of animals, are separated from each other by partitions called cell walls. The chief constituent of these walls is cellulose, a linear polymer of glucose. The strength and rigidity of cell walls results from widespread hydrogen bonding between adjacent cellulose molecules. The strength of cellulose is readily apparent in wood and in plant fibers such as cotton and hemp.

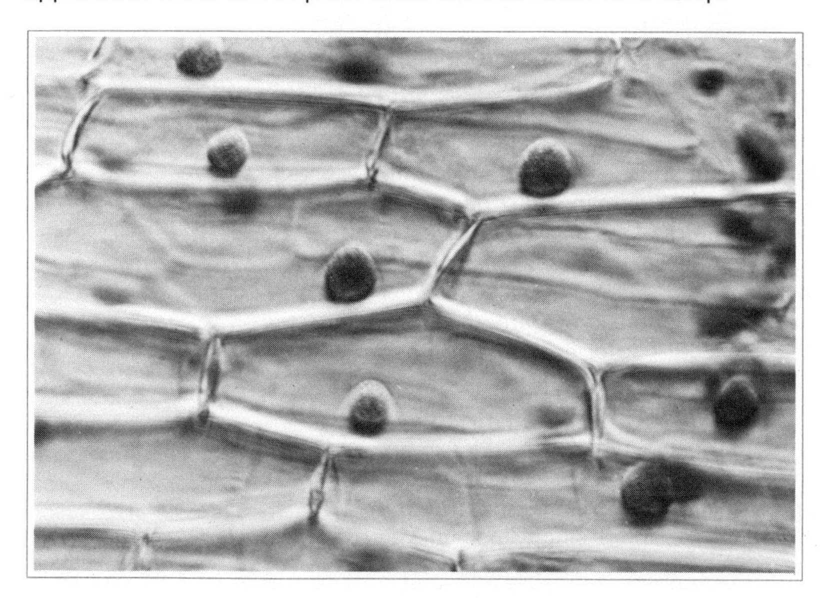

Cells of *Elodea*. The cell walls are clearly evident. (Photograph by A. M. Winchester, University of Northern Colorado)

Estimates of the molecular weight of cellulose range from 150,000 to about one million. Despite its potential for hydrogen bonding with water, cellulose is virtually insoluble because of its size and structure.

Many useful products result from the chemical modification of cellulose. Rayon threads, for example, are produced by dissolving cellulose in a mixture of sodium hydroxide and carbon disulfide and then forcing the solution through tiny holes into dilute sulfuric acid. Cellophane and cellulose acetate are two other familiar examples of modified cellulose.

Starch

As mentioned earlier, starches are mixtures of two glucose polymers—amylose, a linear polysaccharide containing 250-500 monosaccharide units, and amylopectin, a branched structure composed of about 1000 units. Starches are found in protein-covered granules in the leaves, stems, and seeds of plants. When the granules are heated in water, they rupture, forming a colloidal dispersion whose viscosity increases with additional heating. This property of starch makes it useful as a thickening agent in sauces, puddings, and gravies. A coating of starch also improves the writing qualities of paper, and strengthens textile fibers so that they can be woven without breaking.

The ratio of amylose to amylopectin varies with the source of the starch. Corn starch is composed of approximately 75% amylopectin and 25% amylose. Amylose is less soluble than amylopectin and tends to separate upon standing. It is also precipitated by 1-butanol, while amylopectin is not.

Dextrins

Partial hydrolysis of starch by either acid or enzymes converts it into a mixture of smaller polysaccharides called dextrins. Dextrins produced by the partial hydrolysis of amylose are linear, while those obtained from amylopectin are branched. Starch is also converted into dextrins by dry heat, and the golden color of bread crust is due to the presence of dextrins that have been formed during baking. Since dextrins represent partially hydrolyzed starch, toasting bread makes it easier to digest.

Dextrins are readily dispersed in water, but their concentrated solutions are extremely sticky. For this reason dextrins are useful as adhesives where great strength is not required; for example, on postage stamps and envelopes.

Glycogen

Glycogen is the animal counterpart of starch. It is similar in structure to amylopectin but more highly branched, and a single molecule of glycogen may contain more than 5000 glucose units. Glycogen comprises roughly 5% of the liver and about 0.5% of the weight of muscle cells; thus the liver stores approximately 110 g of glycogen, and the muscles account for another 250 g.

Agar

Agar is a polymer of galactose that occurs in certain types of seaweed. A typical molecule contains 53 galactose units; thus agar is a rather small polysaccharide. Despite the fact that it is indigestible, agar is widely used by the Japanese and is a common ingredient in sauces and ice creams and in several kinds of soft, fruit-flavored candy. It is also used in microbiology to thicken nutrient broths to a gel so that bacteria and fungi can be grown on their surfaces. Agar is sometimes prescribed for the relief of constipation because it swells and adds bulk to the feces.

Dextrans

Dextrans and dextrins have little in common besides the similarity in names and the fact that both are polymers of glucose. Dextrans are produced from sucrose by a bacterium—*Leuconostoc mesenteroides*—and differ from dextrins both in size and structure. Dextrans are immense molecules having molecular weights in excess of one million and held together by α-1,6 -1,4, and -1,3 linkages.

Dextrans which have been hydrolyzed to an average molecular weight of about 70,000 are used as plasma substitutes in the treatment of shock in cases where low plasma volume is the cause. The dextran remains in the blood for several days and is gradually eliminated in the urine.

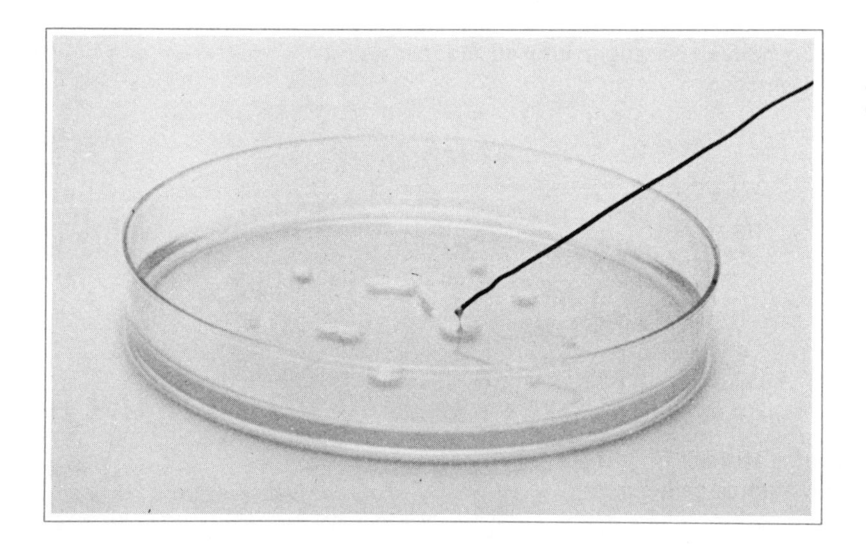

Bacterial colonies growing on a nutrient medium containing agar. (Photograph by J. Asdrubal Rivera)

12.5 REACTIONS OF CARBOHYDRATES

Oxidation

The oxidation of sugars containing either a free or potential carbonyl group by an alkaline solution of Cu^{2+} ion was described in Chapter 11. The sugar is oxidized to the corresponding sugar acid while the metal ion is reduced.

Since the linkage between two monosaccharide molecules generally involves the carbonyl group of at least one of them, it follows that a given amount of a polysaccharide will contain fewer of these groups than an equal quantity of mono-, di-, or oligosaccharide. The observation that Benedict's solution is reduced by all of the common monosaccharides and disaccharides except sucrose, and is unaffected by starch, cellulose, or glycogen shows that this reasoning is correct. The behavior of sucrose is explained by the fact that the carbonyl group of glucose is joined to that of fructose; thus sucrose is unable to act as a reducing sugar. Table 12.2 shows which of the common carbohydrates have the ability to reduce Benedict's solution.

TABLE 12.2 Reactions of Common Carbohydrates with Benedict's Solution

Carbohydrate	Reaction	Carbohydrate	Reaction
Ribose	+	Sucrose	−
Glucose	+	Glycogen	−
Fructose	+	Starch	−
Galactose	+	Dextrins	−
Lactose	+	Cellulose	−
Maltose	+	Agar	−

Oxidation of its aldehyde group converts a sugar into an *aldonic acid,* as shown by the example of glucose:

$$
\begin{array}{c}
CHO \\
| \\
H-C-OH \\
| \\
HO-C-H \\
| \\
H-C-OH \\
| \\
H-C-OH \\
| \\
CH_2OH \\
\text{D-glucose}
\end{array}
\xrightarrow{[O]}
\begin{array}{c}
COOH \\
| \\
H-C-OH \\
| \\
HO-C-H \\
| \\
H-C-OH \\
| \\
H-C-OH \\
| \\
CH_2OH \\
\text{D-gluconic acid}
\end{array}
$$

In the body, the end of the molecule farthest from the aldehyde group may be oxidized, resulting in compounds called *alduronic acids.*

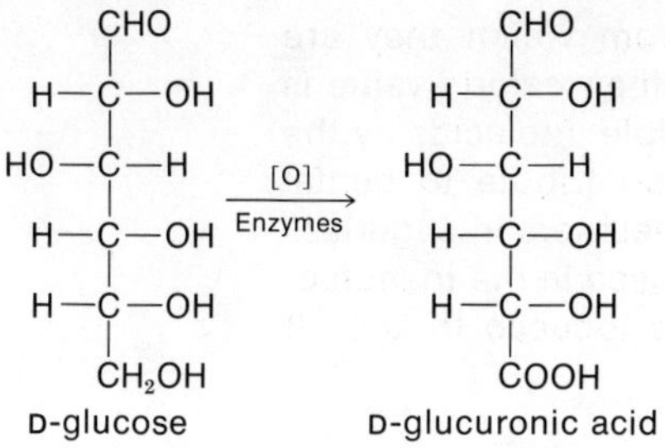

$$\begin{array}{c} CHO \\ H-C-OH \\ HO-C-H \\ H-C-OH \\ H-C-OH \\ CH_2OH \end{array} \xrightarrow[\text{Enzymes}]{[O]} \begin{array}{c} CHO \\ H-C-OH \\ HO-C-H \\ H-C-OH \\ H-C-OH \\ COOH \end{array}$$

D-glucose D-glucuronic acid

The alduronic acids, especially glucuronic acid, combine with such toxic by-products of metabolism as phenol and benzoic acid to render them less toxic. The detoxified substances, in the form of glucuronates, are then excreted. The detoxification of phenol illustrates glucuronate formation:

Heparin is a polysaccharide which prevents blood from clotting by inhibiting the conversion of prothrombin to thrombin. It consists of glucuronic acid linked α-1,3 with glucosamine. Some of the –OH and $-NH_2$ groups are modified to sulfates.

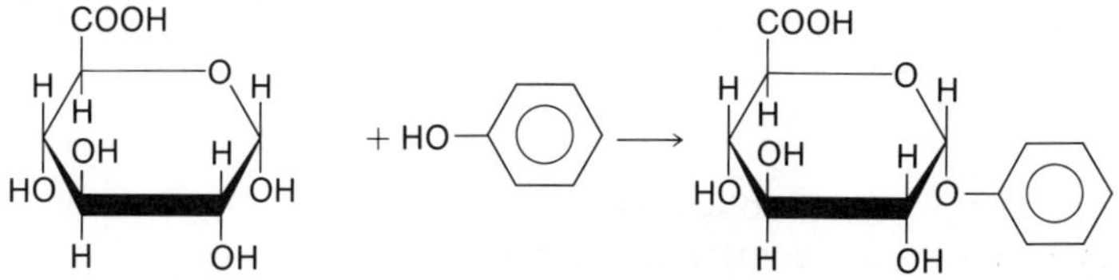

α-form of glucuronic acid + Phenol ⟶ Phenyl glucuronic acid

The complete oxidation of a carbohydrate, either enzymatically or by combustion, yields carbon dioxide and water as products and liberates approximately 4 kcal of energy per gram. It is worth noting that, although carbohydrates may be completely oxidized in the body, much of the energy that is released during the oxidation is given off in the form of heat, and thus performs no work.

Reduction

The carbonyl groups of aldoses and ketoses can be reduced to hydroxyl groups, forming compounds called sugar alcohols in which there is an —OH group on every carbon atom. Glucose, for example, can be reduced to sorbitol:

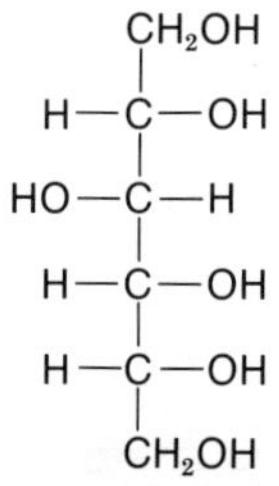

$$\begin{array}{c} CH_2OH \\ H-C-OH \\ HO-C-H \\ H-C-OH \\ H-C-OH \\ CH_2OH \end{array}$$

Sorbitol

Sugar alcohols are sweet, like the sugars from which they are derived, but they are metabolized so slowly that their caloric value is negligible. Sugar alcohols are not readily convertible into acids by the microorganisms in the mouth, and so do not contribute to dental decay. These properties make them useful as sweeteners in sugarless chewing gum and dietetic candy. Sorbitol is also used in the manufacture of ascorbic acid (vitamic C) and is added to tobacco to keep it from becoming dry and brittle.

Ester Formation

In addition to changes involving their carbonyl groups, monosaccharides take part in esterification reactions with phosphoric acid. As will be seen presently, the conversion of a sugar into its phosphate ester is the first step in the chain of reactions leading to its oxidation. Phosphate esters of ribose are also found in the structures of ATP, DPN, coenzyme A, and the nucleic acids (see Section 9.6).

Fermentation

Fermentation is an anaerobic process whereby a carbohydrate is enzymatically converted into a product which occurs somewhere in the oxidation pathway between the sugar and its completely oxidized form, carbon dioxide. The major product of fermentation may be an alcohol, a ketone, or an acid. One of the best-known fermentation reactions is catalyzed by enzymes from yeast cells, and changes glucose into ethanol. The overall reaction is

$$\underset{\text{Glucose}}{C_6H_{12}O_6} \xrightarrow{\text{zymase}} \underset{\text{Ethanol}}{2\ C_2H_5OH} + \underset{\text{Carbon dioxide}}{2\ CO_2}$$

Table 12.3 shows which of the common carbohydrates are acted on by zymase from yeast.

TABLE 12.3 Alcoholic Fermentation of Carbohydrates

Fermented by Zymase	Not Fermented
Glucose	Ribose
Fructose	Lactose
Sucrose	Galactose
Maltose	Glycogen
	Starch
	Dextrins
	Cellulose
	Agar

Although production of ethanol does not occur in our bodies, there *is* a similar series of reactions, called *glycolysis,* in which glucose is anaerobically changed into lactic acid. Glycolysis is described later in this chapter.

Reactions with Alkali

When either glucose or fructose is exposed to dilute alkali for several hours, a mixture of glucose, fructose, and mannose (another isomer of glucose) is produced. This reaction, known as the Lobry de Bruyn-von Ekenstein transformation, results from the increased activity in basic solution of the hydrogen alpha to the carbonyl group. Migration of the alpha hydrogen introduces a double bond into the chain between the first and second carbon atoms and converts the carbonyl oxygen into an alcohol group. This intermediate structure has a double bond between two hydroxyl groups and is thus called an *enediol.*

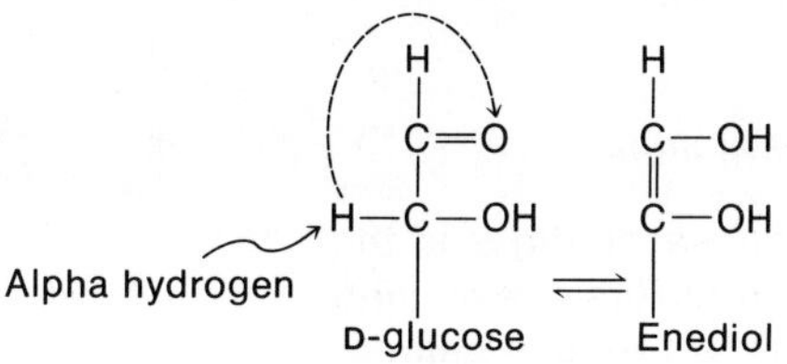

H
C=O
H—C—OH
Alpha hydrogen
D-glucose ⇌ Enediol
H
C—OH
C—OH

The enediol is unstable and the active hydrogen may either resume its original position (forming D-glucose), move to the opposite side of the #2 carbon atom (forming D-mannose), or jump to carbon #1—in which case D-fructose is the product.

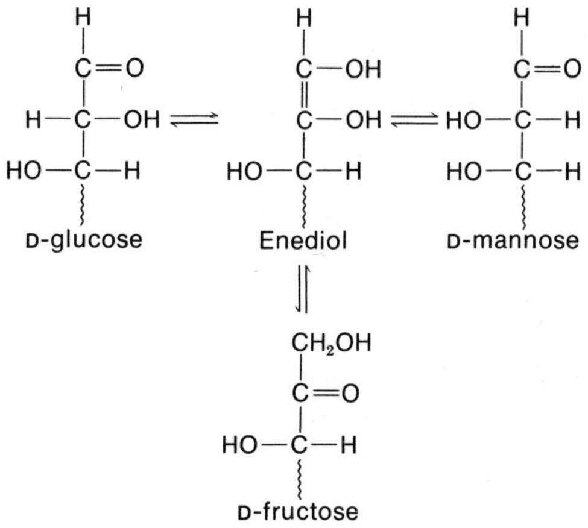

H | C=O | H—C—OH | HO—C—H (D-glucose) ⇌ H | C—OH | C—OH | HO—C—H (Enediol) ⇌ H | C=O | HO—C—H | HO—C—H (D-mannose)

Enediol ⇌ CH_2OH | C=O | HO—C—H (D-fructose)

As you might expect, the action of dilute alkali on any one of the three sugars produces the same mixture.

In higher concentrations of alkali the monosaccharides are unstable and undergo oxidation, degradation, and polymerization.

Reactions with Acids

Monosaccharides are generally unaffected by either hot or cold dilute mineral acids, although treatment with hot dilute acids causes hydrolysis of oligo- and polysaccharides.

The action of hot concentrated acids such as 5 *M* HCl on pentoses dehydrates them to furfural, while hexoses are dehydrated to 5-hydroxymethylfurfural:

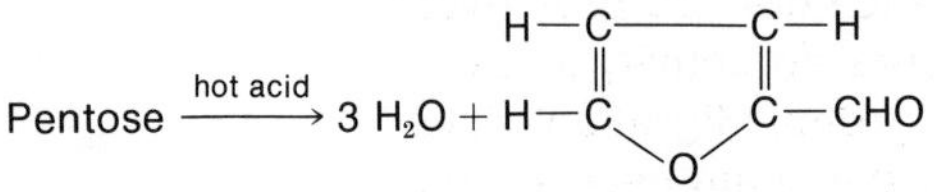

$$\text{Pentose} \xrightarrow{\text{hot acid}} 3\ H_2O + \text{Furfural}$$

H—C——C—H
H—C C—CHO
O

Furfural

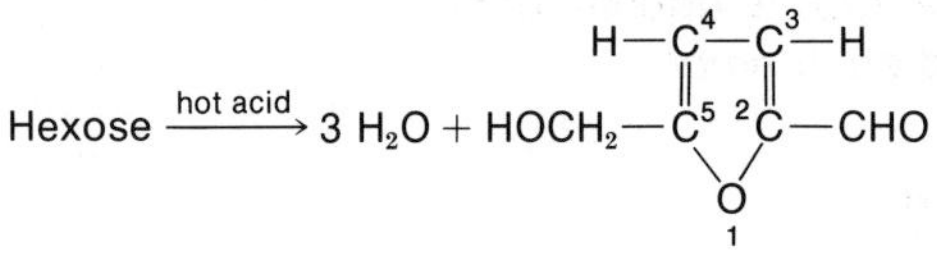

$$\text{Hexose} \xrightarrow{\text{hot acid}} 3\ H_2O + HOCH_2\text{—5-hydroxymethylfurfural ring}$$

H—C4—C3—H
HOCH$_2$—C5 2C—CHO
O
1

5-hydroxymethylfurfural

The formation of furfurals by dehydration of monosaccharides is the basis for the Molisch test for carbohydrates, since furfurals react with α-naphthol or certain other aromatic compounds to produce characteristic colored products. This test is described in the following section.

12.6 TESTS FOR CARBOHYDRATES

Simple qualitative tests make it possible to detect the presence of a carbohydrate and, in some cases, to differentiate between types of carbohydrates.

Tests Involving Dehydration by Strong Acids

The Molisch Test

This test is based on the conversion of a monosaccharide to furfural or its derivatives by reaction with concentrated sulfuric acid as described in the preceding section. The products of the reaction then combine with α-naphthol to give a purple complex.

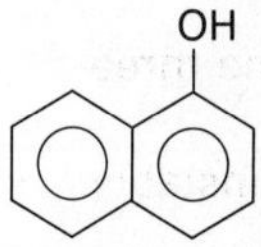

OH

α-naphthol
(α-hydroxynaphthalene)

Since oligosaccharides and polysaccharides are completely hydrolyzed by concentrated sulfuric acid, most carbohydrates will give a positive Molisch test.

Seliwanoff's Test

This is another general test for carbohydrates, but it may also be used to distinguish between hexoses, since the rate of dehydration with hot HCl is higher for ketohexoses than for aldohexoses. The dehydrated products react with resorcinol, producing a bright red color with ketoses and a pink color with aldoses after an equal period of time.

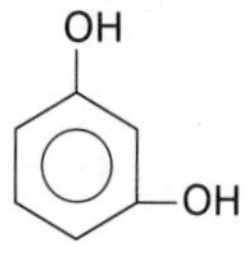

OH

OH

Resorcinol
(*m*-dihydroxybenzene)

Bial's Test

When pentoses are heated with concentrated HCl, furfural is formed. A condensation reaction between furfural and orcinol in the presence of Fe^{3+} ion produces a blue-green color.

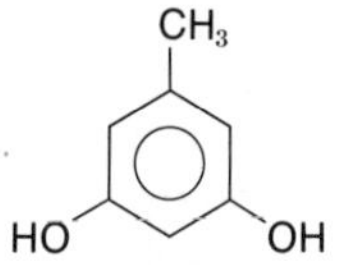

CH_3

HO OH

Orcinol
(3,5-dihydroxytoluene)

The reaction is not absolutely specific for pentoses, since prolonged heating of some hexoses in HCl yields hydroxymethylfurfural, which also reacts with orcinol to give colored complexes.

Tests Based on the Reducing Properties of Carbohydrates

Carbohydrates containing potential aldehyde or ketone groups are readily oxidized by Cu^{2+} ion in a basic solution (see Section 11.5). In addition, monosaccharides act as reducing agents in weakly acidic solutions.

Benedict's Test

Benedict's reagent is a solution of copper(II) citrate in sodium carbonate. Sodium carbonate causes the solution to be basic, while the citrate ions offset the tendency of Cu^{2+} ion to precipitate in an alkaline environment. Reducing sugars bring about the formation of insoluble copper(I) oxide. The reaction is quantitative—a high concentration of reducing sugar will produce a thick rust-brown precipitate, while extremely dilute solutions may cause a small amount of olive-green sediment to be formed.

Formic acid, phenols, and uric acid will also reduce Benedict's solution and may give false positive reactions.

Fehling's Test

Fehling's solution is prepared just before use by mixing equal volumes of copper(II) sulfate solution and a solution of sodium potassium tartrate. Fehling's reagent is more sensitive and less stable than Benedict's solution. Also, excess ammonia and ammonium salts interfere with the test, and urates give false positive reactions; therefore, Fehling's reagent has been generally replaced by Benedict's solution.

Barfoed's Test

Barfoed's reagent is weakly acidic and is reduced only by monosaccharides; thus it provides a means of distinguishing monosaccharides from other carbohydrates. However, prolonged heating may hydrolyze disaccharides, and misleading results may be obtained. Barfoed's reagent is a solution of copper(II) acetate in dilute acetic acid.

An Additional Test: The Iodine Reaction

The reaction between starch and a reagent composed of iodine in a solution of potassium iodide produces a brilliant blue-black complex. This color appears to be associated with the amylose portion of starch since amylopectin by itself forms a reddish complex with iodine.

The blue color fades if the starch is heated almost to boiling, but reappears when it cools. The accepted explanation of this behavior is that amylose assumes a coiled configuration at room temperature, and that the iodine is held in the interior of the coil. At higher temperatures the coil straightens out, the iodine is released, and the color disappears. Glycogen and amylopectin are branched structures and do not coil; thus they are incapable of trapping any appreciable amount of iodine.

Partial hydrolysis of starch splits amylose and amylopectin into dextrins, which are too small to form the blue-colored complex, but large enough to give a reddish color with iodine. More complete hydrolysis yields products that are colorless with iodine.

12.7 PHOTOSYNTHESIS

One of the characteristics which distinguishes living creatures from inanimate matter is their inability to maintain their molecular organization without a constant supply of chemical energy. Although a few types of primitive microorganisms obtain energy by oxidizing inorganic substances such as sulfur or iron, the majority of plants and animals depend—directly or indirectly—upon energy-rich compounds produced during photosynthesis.

Photosynthesis is an endergonic process in which radiant energy causes the conversion of carbon dioxide and water into glucose and molecular oxygen. Although the net result of photosynthesis is usually represented by the equation

$$6\ CO_2 + 6\ H_2O \xrightarrow[\text{light}]{\text{chlorophyll}} C_6H_{12}O_6 + 6\ O_2$$

the process is considerably more complex. Photons of light are absorbed by molecules of a green pigment called chlorophyll, causing an electron in the molecule to be raised to a higher energy level. The excited electron is captured by an acceptor molecule called substance Z (not yet identified). Following its capture, the electron is passed from substance Z to a succession of other acceptor molecules until it is either trapped by nicotinamide adenine dinucleotide phosphate (NADP) or returns to the molecule of chlorophyll from which it came.

Although the electron in each case is moving from an excited to a more stable condition, the net result is quite different. Electrons returning to the chlorophyll molecules donate their energy to the synthesis of ATP, while electrons captured by NADP cause it to have such a strong affinity for protons that they are removed from adjacent water molecules, and NADP becomes reduced $NADPH_2$. Oxygen atoms and free electrons are formed from water at the same time, and the oxygen atoms from two water molecules combine to form a molecule of oxygen. The reactions are

$$2\ H_2O \longrightarrow 4\ H^+ + 4\ e^- + O_2$$
$$2\ NADP + 4\ H^+ \longrightarrow 2\ NADPH_2$$

The transfer of electrons from chlorophyll to NADP creates electron vacancies in the chlorophyll molecules. These vacancies are promptly

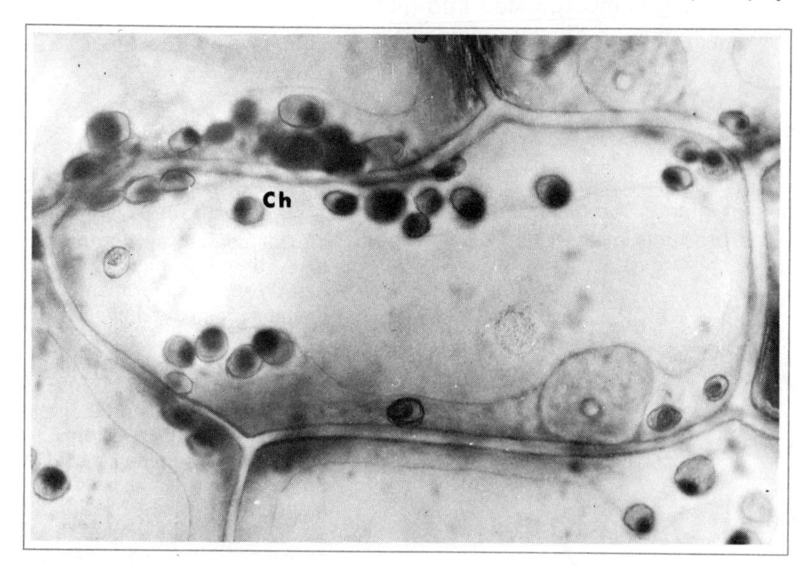

Cells of *Elodea* showing chloroplasts (Ch). (Photograph by A. M. Winchester, University of Northern Colorado)

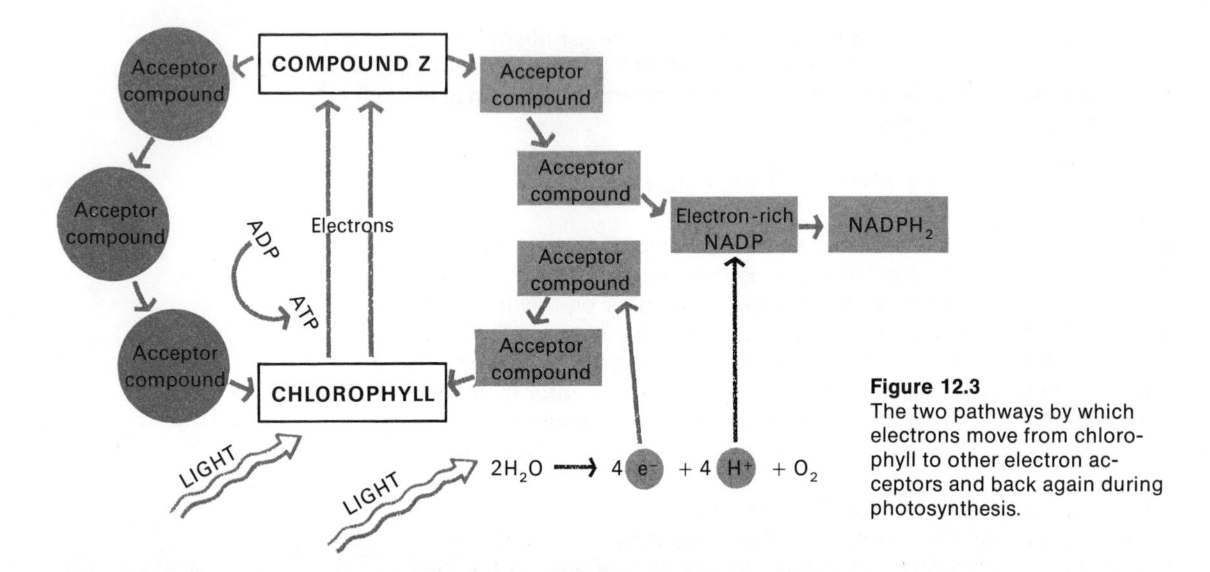

Figure 12.3
The two pathways by which electrons move from chlorophyll to other electron acceptors and back again during photosynthesis.

filled by electrons which were freed by the splitting of water. Figure 12.3 summarizes the movement of electrons from chlorophyll along the two pathways. The relevance of these electron pathways to the production of glucose is that $NADPH_2$ can act as a hydrogen donor in the reduction of carbon dioxide to carbohydrate, as shown in Figure 12.4.

It should be noted that only those reactions leading to the synthesis of ATP and $NADPH_2$ require light energy. The remaining reactions leading to the production of glucose can take place in the dark.

It should also be noted that both water and carbon dioxide are split, and that some of the oxygen from the CO_2 is incorporated into glucose while the remainder is used in the formation of water molecules, as shown in Figure 12.5.

12.8 DIGESTION AND ABSORPTION OF CARBOHYDRATES

The chief nutritional carbohydrates are starch, the disaccharides—lactose, sucrose, and maltose—and the products of their hydrolysis. The

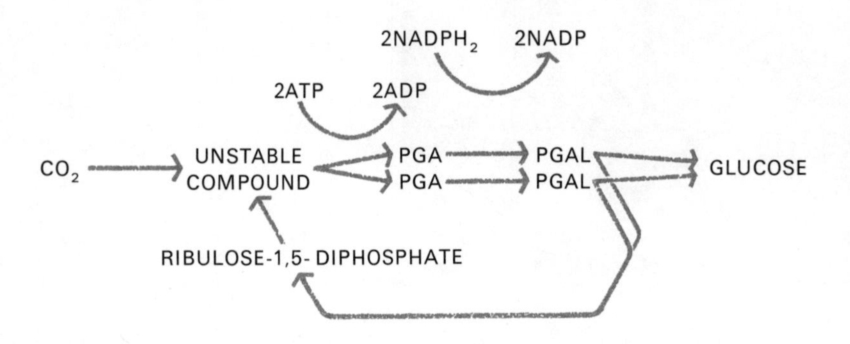

Figure 12.4
Reduction of carbon dioxide to glucose by $NADPH_2$. Carbon dioxide combines with ribulose-1,5-diphosphate (a five-carbon sugar) to form an unstable compound which immediately decomposes into two molecules of 3-phosphoglyceric acid (PGA). PGA is phosphorylated by ATP and then reduced by $NADPH_2$ to form 3-phosphoglyceraldehyde (PGAL). Some of the PGAL is used to regenerate ribulose-1,5-diphosphate through a complex series of reactions, and the remainder is converted into glucose.

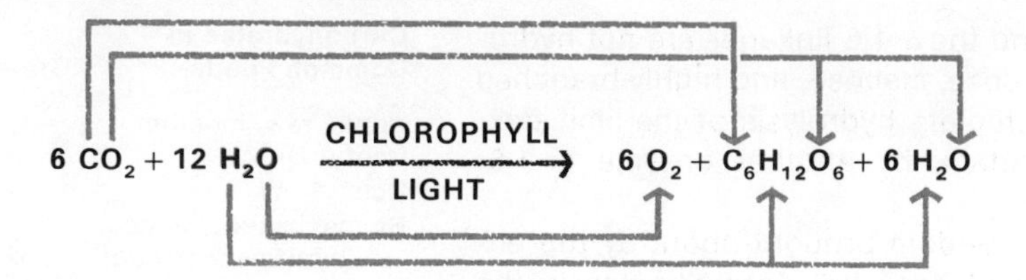

Figure 12.5
The fate of the atoms from the reactants in photosynthesis. The water produced during photosynthesis contains atoms that were not present in the original water molecules.

dietary carbohydrates are obtained primarily from plant sources, although lactose comes from milk, and traces of glycogen are found in animal tissues.

As mentioned previously, starch is enclosed in protein granules. If these granules have been ruptured, either by cooking or grinding, digestion of starch begins in the mouth.

Saliva contains the enzyme α-amylase (formerly called ptyalin), whose effect on amylose and amylopectin is depicted in Figure 12.6. α-amylase randomly attacks the α-1,4 linkages in amylose, producing a mixture of dextrins, maltose, and glucose. If the action of α-amylase is not interrupted, the amylose will eventually be completely converted into glucose.

Amylopectin is also attacked by α-amylase, but the α-1,4 linkages in

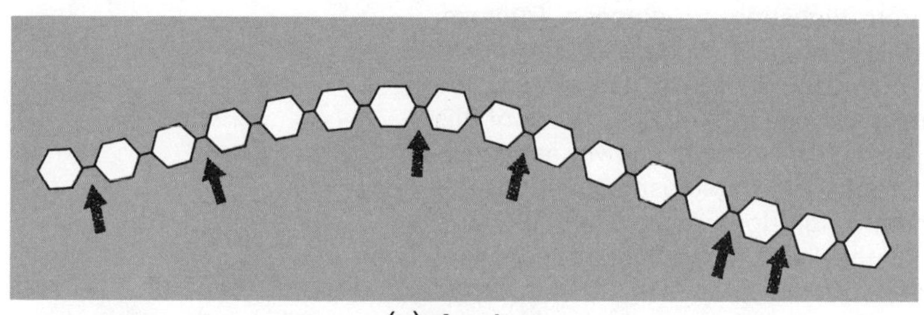

(*a*) **Amylose**

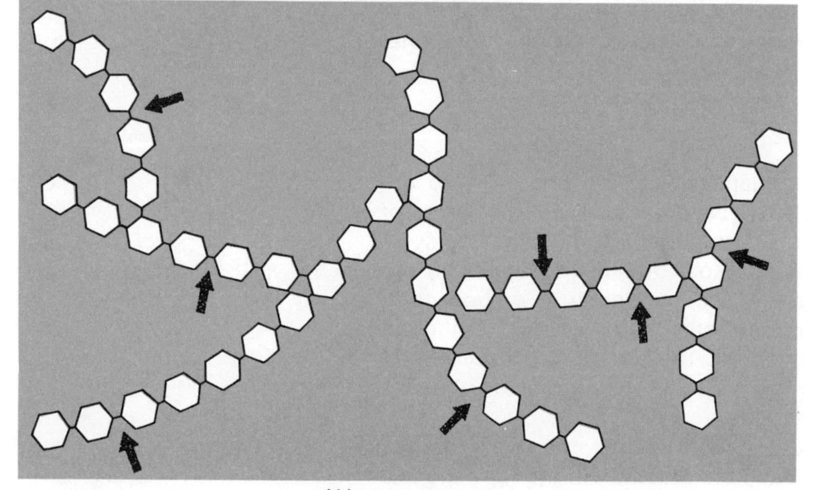

(*b*) **Amylopectin**

Figure 12.6
Hydrolysis of starch by α-amylase. (a) The attack on amylose is random, producing dextrins, maltose, and glucose. (b) Amylopectin is reduced to small branched structures (called limit dextrins), maltose, and glucose.

the vicinity of the branch points, and the α-1,6 linkages are not hydrolyzed. The result is a mixture of glucose, maltose, and highly branched structures called limit dextrins. Complete hydrolysis of the limit dextrins takes place following the attack by another enzyme (α-1,6-glycosidase) at the branch points.

Complete hydrolysis of starch is seldom brought about by the enzymes in saliva simply because food does not normally stay in the mouth long enough. The activity of the salivary enzymes ceases shortly after the food reaches the stomach since their optimum pH is about 6.6 and the pH of gastric juice is less than 2.0. The enzymes that are present in gastric juice do not attack carbohydrates; thus, except for a negligible amount of hydrolysis caused by hydrochloric acid, breakdown of carbohydrates does not occur in the stomach.

Digestion of carbohydrates is completed in the small intestine. Disaccharides are split into their constituent monosaccharides by pancreatic and intestinal secretions containing maltase, sucrase (invertase), and lactase; and the breakdown of starch is completed by the action of amylopsin, which is similar in its action to ptyalin.

There is some evidence that adult humans (and possibly other mammals as well) have lower levels of lactase than infants. This observation may have evolutionary significance in that the consumption of milk does not normally continue in mammals beyond the infant stage.

The end products of carbohydrate digestion are glucose, fructose, and galactose. These sugars are absorbed through the villi of the small intestine and enter the venous circulation, where an enzyme present in red blood cells converts galactose into glucose. Lack of this enzyme causes a metabolic disorder of infants known as galactosemia, or galactose diabetes. Infants with galactosemia cannot tolerate milk and suffer from diarrhea, weight loss, lack of appetite, and jaundice. Cirrhosis (hardening) of the liver, mental retardation, and death may follow if the diet is not corrected.

Fructose is also converted into glucose, but at a different location. The enzymes which bring about this change are present in cells of the liver. Thus the dietary carbohydrates share a common metabolic fate.

Carbohydrates in Common Foods	Grams
Apple, raw, medium	17
Beef (oz)	trace
Beer (8 oz)	11
Beans, baked, canned, without pork (1 cup)	52
Beans, green, canned (1 cup)	9
Bread, white (1 slice)	12
Cheese, cheddar (1 oz)	0.6
Coffee, black (8 oz)	0.6
Milk, whole (8 oz)	12
Orange juice, frozen, diluted, unsweetened (8 oz)	24
Potato, french fried (10 pc)	20
Tomato, fresh (medium)	6
Whiskey (1 oz)	0.0

12.9 METABOLISM OF GLUCOSE

The functions for which living organisms require energy were described in Chapter 9: they include muscle contraction, transmission of nerve impulses, synthesis of cellular components, and active transport of materials across cell membranes. Most organisms obtain the energy for these processes by the stepwise oxidation of carbon compounds—specifically carbohydrates, fats, and proteins. Although these three classes of compounds are quite different structurally, their metabolic pathways are so closely interwoven that except for the initial steps they are essentially the same.

The initial stage in the oxidation of glucose does not require molecular oxygen. It is a series of 13 reactions resulting in the formation of

8 moles of ATP and 2 moles of lactic acid[1] from each mole of glucose.

$$\underset{\text{Glucose}}{C_6H_{12}O_6} \xrightarrow{8\,ADP \;\to\; 8\,ATP} \underset{\text{Lactic acid}}{2\,CH_3\overset{\overset{\large OH}{|}}{C}HCOOH}$$

The production of lactic acid from a carbohydrate is called *glycolysis,* and the series of reactions is referred to as the Embden-Meyerhof pathway.

The conversion of glucose to lactic acid is significant because it provides a way in which tissues can obtain energy even when the partial pressure of oxygen is very low. Thus the muscles are able to respond immediately to demands for increased activity without waiting for additional oxygen to be brought to them. However, prolonged muscle activity results in an accumulation of lactic acid and is accompanied by a sensation of muscle fatigue. Since lactic acid cannot be further oxidized without oxygen, the accumulation of lactic acid constitutes an oxygen debt. Heavy breathing following exercise repays the debt by bringing in oxygen so that the oxidation of lactic acid can proceed.

In the presence of molecular oxygen, lactic acid is changed into pyruvic acid, which is oxidized to carbon dioxide and water by a series of reactions known as the Krebs, or citric acid, cycle. Approximately 25% of the lactic acid generated in the muscles is acted on in this way—the remainder is transported to the liver where it is synthesized into glycogen.

The details of the Krebs cycle are shown in Figure 12.7. Pyruvic acid enters the cycle as acetyl CoA following the loss of one of its carbon atoms and two atoms of oxygen. Acetyl CoA condenses with oxaloacetic acid to form citric acid, which undergoes a number of molecular rearrangements, including the loss of two additional molecules of carbon dioxide, and completes the cycle as oxaloacetic acid. Thus oxaloacetic acid is regenerated and can condense with another molecule of acetyl CoA to repeat the cycle. Notice that keto acids formed by the breakdown of proteins, and the carboxylic acids and glycerol from the hydrolysis of simple fats also enter the Krebs cycle and are oxidized to carbon dioxide and water.

The complete oxidation of one mole of glucose yields 38 moles of ATP—8 from glycolysis and 30 from the Krebs cycle. The conversion of 1 mole of ADP into ATP stores 7.4 kilocalories; thus the total usable energy obtained from the oxidation of a mole of glucose is 38×7.4, or roughly 280, kilocalories. Since the theoretical yield of energy from the combustion of glucose is about 690 kcal/mole, the conversion process is only $280/690 \times 100\%$, or 40%, efficient.

---

[1] It should be remembered that carboxylic acids are generally weak acids and only slightly ionized in dilute solutions; however, at the pH of the body they exist primarily as their corresponding anions. Lactic acid, for example, is found as lactate ion. It is an accepted practice among biochemists to use the name of the acid when referring to the anionic species that are involved in metabolic reactions.

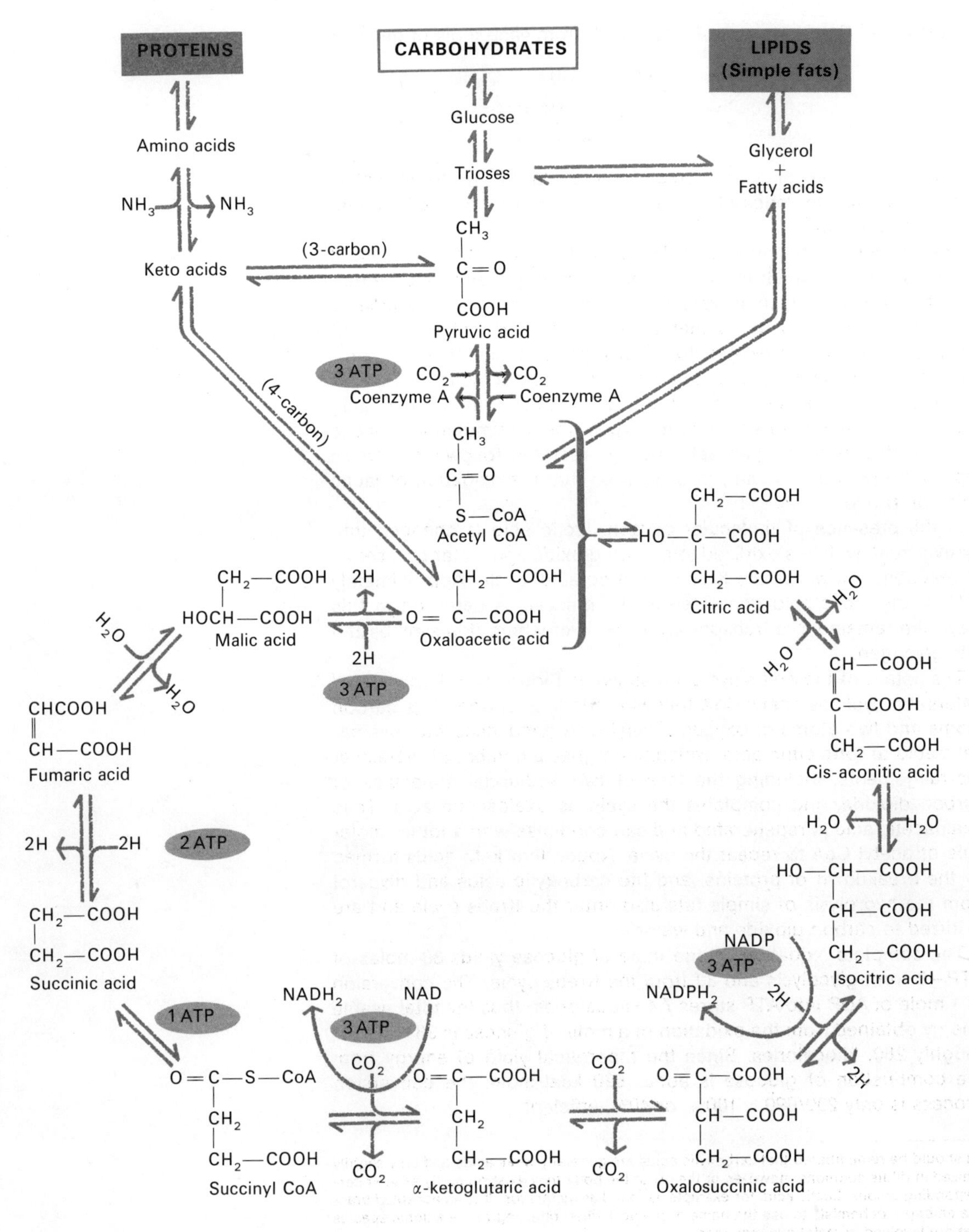

Figure 12.7
The Krebs (citric acid) cycle.

12.10 REGULATION OF CARBOHYDRATE METABOLISM

As pointed out in Chapter 9, living organisms require a constant supply of energy in order to offset the adverse effects of entropy. Multicellular organisms obtain this energy primarily by the oxidation of glucose, and adverse physiological changes are observed when the supply of glucose is less than adequate. The condition in which the concentration of blood sugar is inadequate is termed *hypoglycemia,* and may be characterized by any or all of the following symptoms: weakness, trembling, profuse perspiration, rapid heart beat (tachycardia), delirium, and loss of consciousness. Because of the similarity of symptoms, hypoglycemia may be mistaken for conditions such as drunkenness, allergic reactions, epilepsy, brain disease, and heart trouble. For most human beings an adequate supply of glucose is provided when its concentration in the blood is between 80 and 120 mg%.

There is abundant evidence that the amount of sugar in the blood is regulated by secretions (hormones) from several different glands, including epinephrine (adrenalin) from the adrenal medulla, insulin and glucagon from the pancreas, and thyroxin from the thyroid. In general, it can be said that hormones which bring about a reduction in the blood sugar level do so either by increasing the rate of glucose oxidation or by causing an increased conversion of glucose into glycogen (*glycogenesis*), and hormones that raise the blood sugar level stimulate the breakdown of glycogen to glucose (*glycogenolysis*). The specific action of each of the hormones mentioned above is shown in Figure 12.8.

Note that the action of insulin is opposite to that of epinephrine and glucagon. Under most conditions these secretions are balanced in

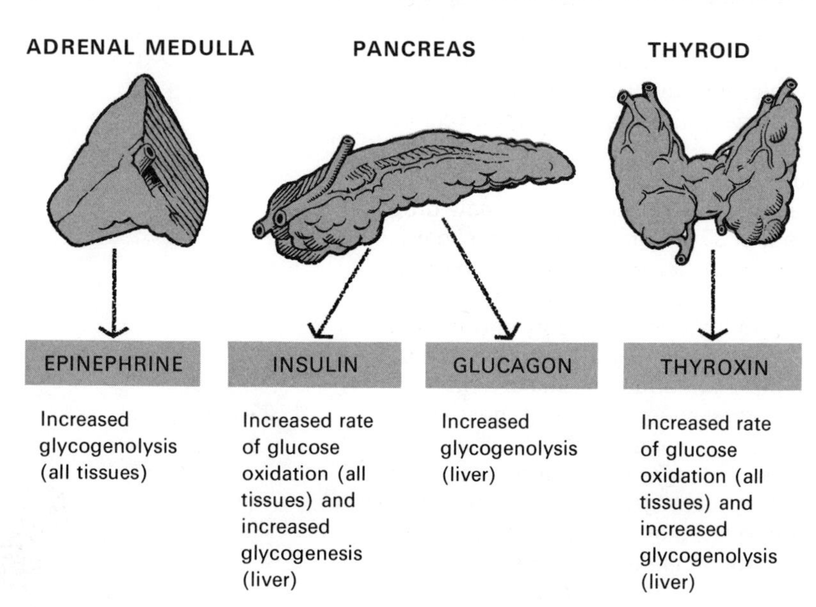

Figure 12.8
Hormones that influence glucose metabolism.

such a way that the glucose level is maintained within normal limits; but when a person becomes angry or frightened, additional epinephrine is released, there is increased glycogenolysis, and the blood sugar level rises. These changes prepare the body in case it is called upon for a sudden burst of activity.

Insulin production is stimulated by an elevation of the blood sugar level, such as that which follows the digestion and absorption of carbohydrates. The additional insulin causes increased glycogenesis, thereby reducing the amount of glucose in the blood. Insulin also facilitates the transport of glucose across the membranes of muscle cells and adipose (fatty) tissue, but appears not to be needed for transport into the cells of the blood, liver, intestines, or nervous system.

The condition called hypoglycemia, which was described earlier, can be caused by an excess of insulin, which in turn may result from a proliferation of cells that make insulin (a tumor of the beta cells of the pancreas), or overproduction of insulin in response to carbohydrate stimulus. So powerful is this stimulus that many individuals who are otherwise normal experience symptoms of hypoglycemia 2-3 hours after ingesting relatively small quantities of carbohydrate. These individuals suffer from functional hypoglycemia, and the reason for their low tolerance of carbohydrates is not yet understood. The condition is alleviated by a low-carbohydrate, high-protein diet.

Hypoglycemia is not the only disease condition associated with insulin production: a deficiency of insulin causes diabetes mellitus (sugar diabetes). The deficiency may come about because the beta cells of the pancreas stop producing insulin, or because there are substances present in the body which inhibit or interfere with insulin activity. In any case, the blood sugar level remains elevated, and may increase by as much as 300-400% during a glucose tolerance test. This test begins with the ingestion of 50-100 grams of glucose by an individual who has been fasting for at least 8 hours. Blood and urine samples are taken at timed intervals and analyzed for their glucose content. Typical test results for normal, diabetic, and hypoglycemic subjects are shown in Figure 12.9.

The diabetic curve is identified by a high fasting level, an extreme rise in blood sugar following the intake of glucose, and the slow drop toward the normal range. After the first 30 minutes of the test, sugar is usually detected in the urine. This is because the concentration of glucose in the blood is so great that the kidneys are unable to recover all of it, and the excess spills over into the urine. The concentration at which spillover occurs is called the *renal threshold,* and in normal individuals is approximately 175 mg%. A condition called *renal diabetes* may also be responsible for the presence of sugar in the urine. A person with renal diabetes has a normal blood sugar level, but a lower than normal renal threshold. *Glycosuria* (sugar in the urine) may also be the result of liver damage, kidney damage, or emotional upset.

As a consequence of their failure to store and utilize glucose properly, persons with uncontrolled diabetes mellitus must derive a large

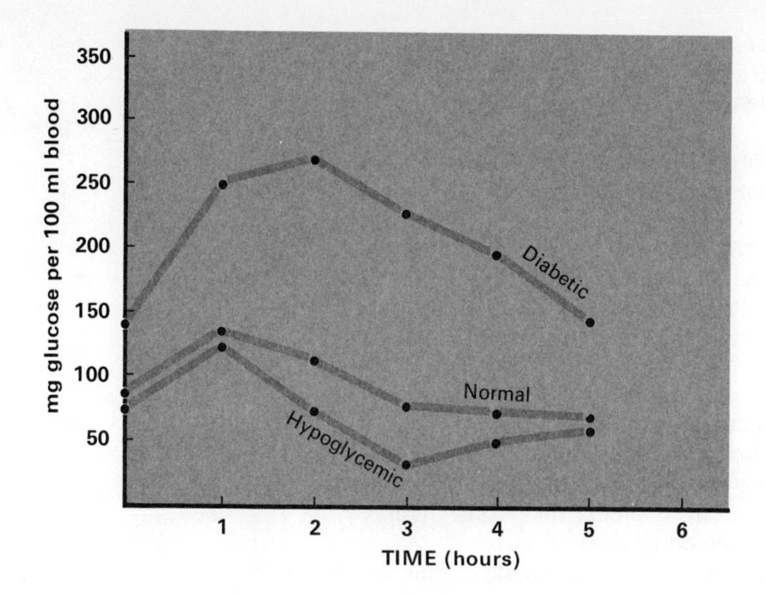

Figure 12.9
Results of glucose tolerance tests.

portion of their energy from the metabolism of fats, and thus produce abnormal amounts of acetoacetic acid, acetone, and β-hydroxybutyric acid, all of which contribute to metabolic acidosis (see Section 8.13). They may also have polyuria (an increased output of urine) because a greater volume of water is required for the excretion of the ketone bodies and the excess glucose.

Diabetes mellitus can usually be controlled by daily injection of insulin, and in some mild cases by oral administration of tolbutamide (Orinase®) a synthetic compound whose action mimics that of insulin.

$$CH_3-C_6H_4-SO_2-NH-\overset{\overset{\displaystyle O}{\|}}{C}-NH-(CH_2)_3-CH_3$$

Tolbutamide

APPLICATION OF PRINCIPLES

1. Explain why a 2-hour glucose tolerance test will detect diabetes mellitus, but will not detect functional hypoglycemia.
2. If an overdose of insulin is administered to a diabetic, it may cause a condition called "insulin shock." What is this condition and why does it occur?
3. Assume that you are given samples of fructose, glucose, and galactose identified only by the letters A, B, and C. Explain how you could tell which sample was which.

4. Draw a projection formula for L-glucose and a Haworth formula for the β-form of D-galactose.
5. Which of the following substances is capable of forming optical isomers?
   (a) 2-butanol (b) $CH_3CHOHCOOH$ (c) CH_3—NH—CH_2CH_3
6. Explain why maltose will give a positive Benedict's test while sucrose will not.
7. Keto acids are derived from amino acids by the removal of the amino group. Keto acids containing three carbon atoms enter the Krebs cycle as pyruvic acid, while those with four carbons enter as oxaloacetic acid. Where would you expect 5-carbon keto acids to enter the cycle?
8. A current theory suggests that functional hypoglycemia may lead to diabetes if it is not counteracted by diet. Why might this be so?
9. It has been suggested that the cereal grains now used in the production of beer and whiskey should be used as food, and that alcoholic beverages be produced from lumber mill wastes. Is this possible? Suggest how it might be done.
10. Criticize the statement: "Complete oxidation of a mole of glucose releases more energy than oxidation of a mole of lactose."

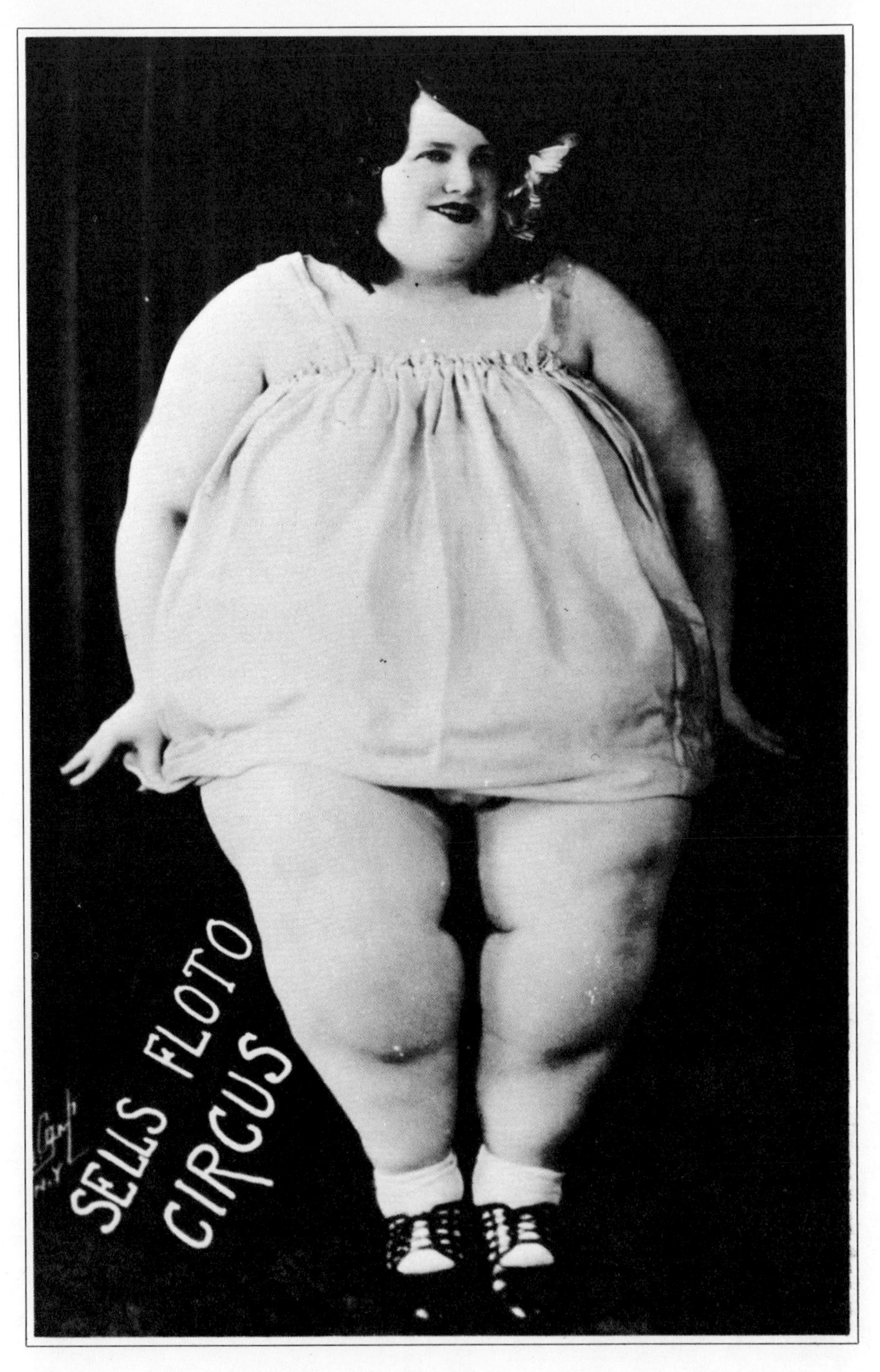

Obesity is a condition in which the body contains excessive amounts of fats. (Courtesy Circus World Museum, Baraboo, Wisconsin)

13

LIPIDS

LEARNING OBJECTIVES

1. Compare the composition and structures of the various classes of lipids.
2. Discuss the importance of steroids to the body.
3. Summarize the main chemical reactions of neutral fats.
4. Distinguish between hydrolytic and oxidative rancidity.
5. Describe the mechanism by which a soap aids in the removal of dirt.
6. Trace the digestion of a neutral fat, including the specific chemical action of the digestive enzymes.
7. Describe the role of the bile salts in lipid digestion and absorption.
8. Compare the absorption of a digested fat with that of a digested carbohydrate.
9. Outline the process by which fatty acids are oxidized.

KEY TERMS AND CONCEPTS

acrolein test
atherosclerosis
arteriosclerosis
beta oxidation
compound lipid
fatty acid
iodine number
ketosis
lipase
polyunsaturation
rancidity
saponification
saponification number
simple lipid
soap
steroid
triglyceride

The tissues of plants and animals contain a variety of compounds that are insoluble in water but soluble in nonpolar liquids such as ether, chloroform, and benzene. Compounds of this kind are known as *lipids.*

13.1 LIPID COMPOSITION AND CLASSIFICATION

Most lipids are composed entirely of carbon, hydrogen, and oxygen, but a few contain nitrogen, or nitrogen and phosphorus, as well. Lipids have neither a common functional group nor a common structure, although the majority of them are esters derived from long-chain carboxylic acids (called fatty acids).

Fatty acids generally contain an even number of carbon atoms and are characterized by a lack of branching. The most abundant fatty acids are those having either 18 or 20 carbon atoms per molecule, although carbon chains ranging in length from 4 to more than 40 atoms are also found. This means that a typical fatty acid consists of a polar carboxyl group that has an affinity for water, and a nonpolar hydrocarbon chain that does not. More often than not, the carbon chain is unsaturated, and multiple double bonds are common. Fatty acids having multiple double bonds are said to be *polyunsaturated.* The common names and condensed structural formulas of some of the more important fatty acids are given in Table 13.1. The IUPAC names of several of the acids are included for comparison.

Lipids are classified according to the products that result from their hydrolysis. The following classifications are generally accepted:

I. Simple Lipids. Esters derived from fatty acids and aliphatic alcohols.
   A. Neutral Fats. Neutral fats are simple lipids that yield three molecules of fatty acid and one of glycerol per molecule of fat. Because of their composition, they are also referred to as *triglycerides.* Unlike some of the other types of lipids, triglycerides are not charged—hence the name. The structure of a neutral fat may be represented in this way:

TABLE 13.1 Names and Structures of Important Fatty Acids

Common Name	Number of Carbons	Condensed Structural Formula
Saturated Fatty Acids		
Butyric	4	$CH_3(CH_2)_2COOH$
Caproic	6	$CH_3(CH_2)_4COOH$
Caprylic	8	$CH_3(CH_2)_6COOH$
Capric	10	$CH_3(CH_2)_8COOH$
Lauric	12	$CH_3(CH_2)_{10}COOH$
Myristic	14	$CH_3(CH_2)_{12}COOH$
Palmitic (hexadecanoic)	16	$CH_3(CH_2)_{14}COOH$
Stearic (octadecanoic)	18	$CH_3(CH_2)_{16}COOH$
Unsaturated Fatty Acids		
Palmitoleic (9-hexadecenoic)	16	$CH_3(CH_2)_5CH{=}CH(CH_2)_7COOH$
Oleic (9-octadecenoic)	18	$(CH_3(CH_2)_7CH{=}CH(CH_2)_7COOH$
Linoleic[1]	18	$CH_3(CH_2)_4CH{=}CHCH_2CH{=}CH(CH_2)_7COOH$
Linolenic[1]	18	$CH_3CH_2CH{=}CHCH_2CH{=}CHCH_2CH{=}CH(CH_2)_7COOH$
Arachidonic[1]	20	$CH_3(CH_2)_4CH{=}CHCH_2CH{=}CHCH_2CH{=}CHCH_2CH{=}CH(CH_2)_3COOH$

[1] Fatty acids that are considered to be essential to helath. The absence of these substances from the diet of an infant causes eczema and loss of weight.

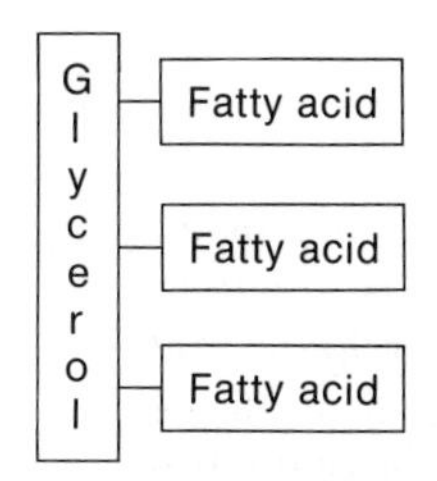

B. Waxes. Waxes are complex mixtures that contain—in addition to their lipid components—ketones, alkanes ranging from 25 to 35 carbons, and secondary alcohols. The lipid components yield one molecule of an alcohol other than glycerol and one of a long-chain fatty acid.

II. Compound Lipids. Compound lipids yield fatty acids and alcohols, plus one or more additional compounds.

A. Phospholipids. These are composed of phosphoric acid, an alcohol, fatty acids, and a fourth component—generally a nitrogenous compound. Phospholipids may be further subdivided according to the identity of the alcohol that they contain.

1. Phospholipids That Contain Glycerol. In almost every phospholipid of this type one of the fatty acid molecules that is linked to glycerol is saturated and the other fatty acid is not. The general structure of this kind of phospholipid is

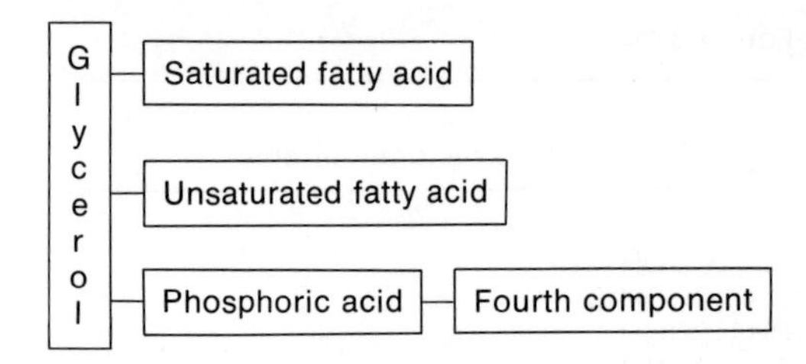

a. Lecithins. Lecithins are charged lipids because the fourth component is the quaternary ammonium salt choline.

$$HO—CH_2CH_2—\overset{+}{N}(CH_3)_3$$

b. Cephalins. The fourth component in cephalins is serine, inositol, or ethanolamine.

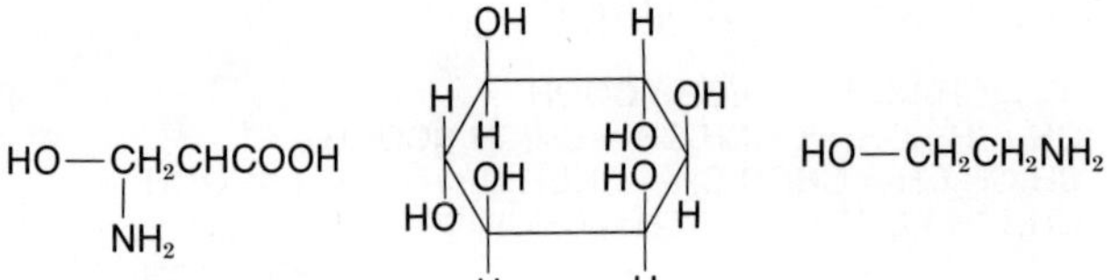

$$HO—CH_2CH(NH_2)COOH$$

Serine

(Inositol ring: substituents OH, H; H, OH; H, OH; HO, H; HO, H; HO, H; H, H)

Inositol

$$HO—CH_2CH_2NH_2$$

Ethanolamine

2. Phospholipids That Contain Sphingosine. Sphingosine is an amino alcohol whose formula is

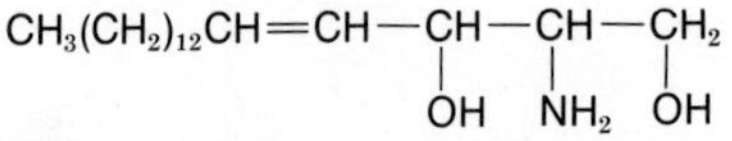

$$CH_3(CH_2)_{12}CH{=}CH—\underset{OH}{CH}—\underset{NH_2}{CH}—\underset{OH}{CH_2}$$

The terminal —OH group in sphingosine is involved in an ester linkage with phosphoric acid which, in turn, is joined to the fourth component—usually choline. Phospholipids that contain sphingosine are called sphingomyelins. The single molecule of fatty acid that is found in this type of phospholipid is attached to the amino group of sphingosine by an amide linkage. The typical structure is

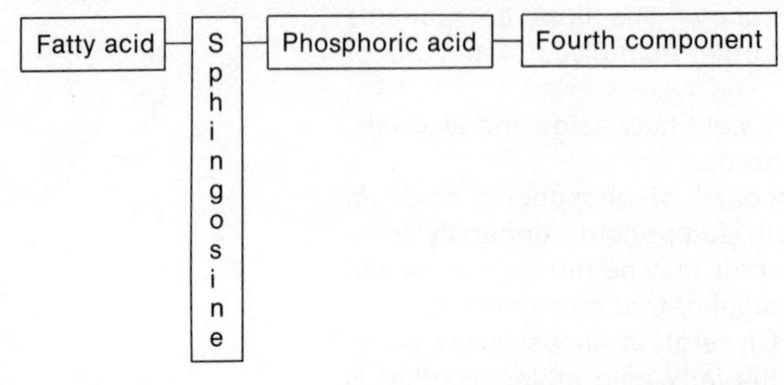

B. Glycolipids. The major differences between a phospholipid and a glycolipid are that the latter contains a sugar instead of the

phosphate group, and there is no fourth component. As in the phospholipids, the alcohol in a glycolipid may be either glycerol or sphingosine; thus there are two types of glycolipids:

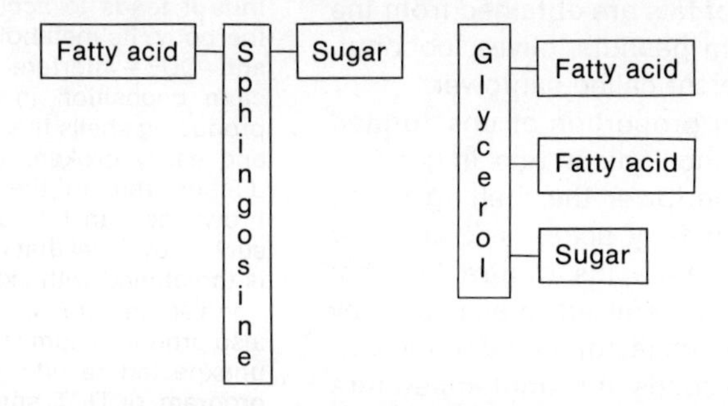

III. Nonsaponifiable Lipids. *Saponification* is the alkaline hydrolysis of an ester. The lipids discussed thus far are said to be saponifiable since they may be hydrolyzed by heating with alkali. Cells also contain small amounts of other lipids that are not esters and thus cannot be saponified. The main types of nonsaponifiable lipids are the steroids and the terpenes.

A. Steroids. Steroids are derivatives of a complex ring structure called perhydrocyclopentanophenanthrene:

B. Terpenes. These are hydrocarbons constructed by joining several molecules of 2-methyl-1,3-butadiene (isoprene):

$$CH_2{=}\overset{\overset{\displaystyle CH_3}{|}}{C}{-}CH{=}CH_2$$

The resulting substances may be either linear or cyclic, or both.

13.2 OCCURRENCE AND FUNCTIONS OF LIPIDS

Neutral Fats

Neutral fats that contain three molecules of the same fatty acid are called *simple triglycerides,* while those in which there are two or more different fatty acids are *mixed triglycerides.* Triglycerides are the major components of depot or storage fats in plant and animal cells, especially in the cells comprising the adipose tissues of vertebrates. Adipose tissue in an animal body serves as a source of reserve energy and cushions and protects the kidneys, heart, and other vital organs, and a layer of subcutaneous fat insulates the body against excessive heat loss or gain. In plants the highest concentrations of fats are found in specialized tissues such as those of seeds or fruits, where they act as an energy reserve. High yields of fats are obtained from the seeds of cotton and flax (linseed oil), from peanuts, olives, coconut, and castor beans, and from a thistle-like plant called safflower.

Fats from plants usually contain a higher proportion of unsaturated fatty acids, and have lower melting points, than animal fats. In general, the greater the degree of unsaturation, the lower the melting point; and fats that are formed from short-chain fatty acids melt at lower temperatures than those containing long-chain acids. Neutral fats that are liquids at room temperature are normally referred to as *oils.* They should not be confused with mineral oil, which is not a lipid but a mixture of hydrocarbons. Neutral fats that are solids at room temperature are simply called fats.

Pure fats are colorless (or white), odorless, tasteless, and less dense than water. The colors that are associated with certain fats are usually due to fat-soluble pigments. Butter, for example, owes its color to traces of the carotene that was present in grass eaten by the cow.

Butter is a stable emulsion of water in fat. Hydrolysis of butter produces mostly palmitic and oleic acids, and small amounts of stearic, myristic, butyric, capric, and caproic. Its relatively low melting point stems from the presence of double bonds as well as from the fact that it contains many short-chain fatty acids.

Another source of dietary fat is the adipose tissue of hogs, cattle, and sheep. Fat obtained from hogs is called lard, and is predominantly a mixture of three simple triglycerides—triolein, tristearin, and tripalmitin. However, diet influences the composition of the depot fat, and when hogs are raised on feeds containing large amounts of unsaturated triglycerides, their depot fat is softer and has a lower melting point. Fat obtained from sheep or cattle is called tallow.

The DDT Problem

The insecticide DDT has a marked affinity for depot fats, thus it tends to accumulate in the body. Its metabolites—DDD and DDE—interfere with calcium deposition in eggshells, producing shells that are fragile and easily broken. The reproduction rate of the California brown pelican has dropped to such a low level that the species is threatened with extinction.

Indiscriminate use of DDT also produces some bizarre and unexpected results. A massive program of DDT spraying was carried out in Borneo to eliminate mosquitos and the malaria that they carried. Cockroaches contaminated with DDT were eaten by lizards, which were then eaten by cats. The DDT had become so concentrated by this time that the cats died, and the area was plagued by an increase in rats.

Waxes

An ideal protective coating for a plant or animal would be waterproof, flexible, and unreactive chemically. In general, these are the characteristics of waxes. They are found on the leaves and fruits of higher plants, on the skin, fur, or feathers of many animals, and on the hard exoskeletons of insects. Wax coatings on fruits protect them against insects and prevent water losses. On animal bodies they keep the skin soft and pliable, and the wax coatings on their feathers help birds that live in an aquatic environment to float on the surface of the water.

In recent years there have been a number of industrial accidents in which unrefined petroleum was spilled into waters inhabited by marine birds. Birds whose feathers became coated with crude oil attempted to clean themselves by preening, and large numbers of them died from ingesting the toxic hydrocarbons. In an attempt to save these birds, rescuers removed the oil with detergents—which, unfortunately, also removed the wax from their feathers. Deprived of this

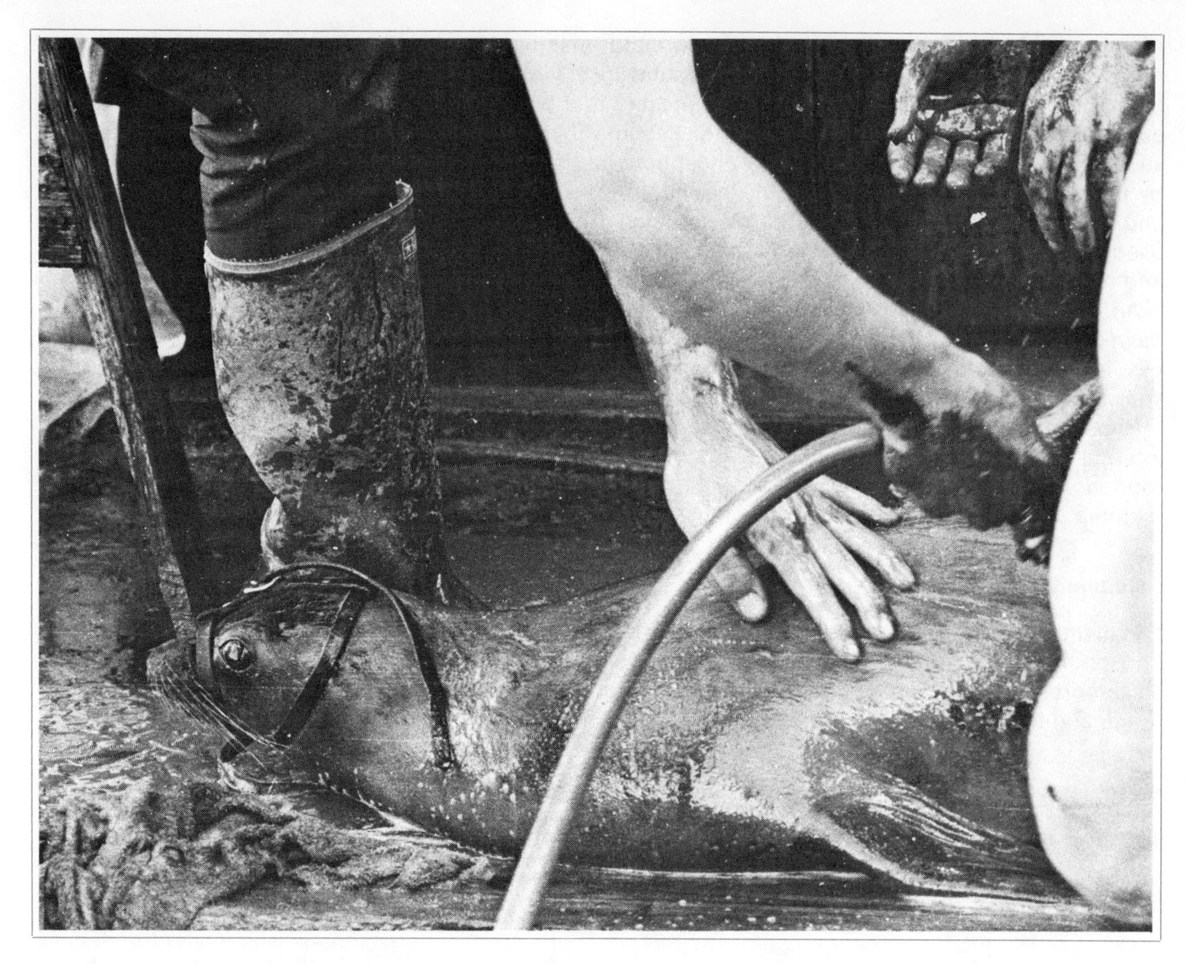

A California sea lion undergoes oil-removal treatment at a wildlife rescue station. (Photograph by Dick Smith, Santa Barbara News-Press)

coating, many of the birds were unable to swim and subsequently died because they were accustomed to feeding only in the water.

Among the commercially important waxes are beeswax, carnauba wax, and lanolin. Beeswax is mostly myricyl palmitate—an ester of myricyl alcohol ($C_{30}H_{61}OH$) and palmitic acid. It is used in making better grades of candles. Cheaper candles are often made of paraffin wax, which is not a lipid but a mixture of solid hydrocarbons. Carnauba wax comes from the carnauba palm and is used extensively in polishes for leather and furniture. Lanolin is the wax coating found on wool, and it is a common ingredient in ointments, lotions, and salves.

Phospholipids

As mentioned earlier, the end of a phospholipid closest to the phosphate group is markedly polar in character, while the opposite end is nonpolar. The nonpolar segment has an affinity for lipids, and the polar segment associates with water: thus phospholipids are capable of bringing together the water-soluble and water-insoluble components of cells. Since tiny fat droplets are found in the blood a short time after fats are ingested, it has been suggested that phospholipids, especially the lecithins, are responsible for keeping these globules in suspension. The emulsifying ability of phospholipids is demonstrated by the use of egg yolk, which is rich in lecithin, to emulsify vinegar and salad oil in mayonnaise. Lecithins obtained from soybeans are used as emulsifying agents in the manufacture of chocolate candies and margarine.

An enzyme (lecithinase A) that is present in rattlesnake venom removes one of the fatty acids from lecithin and produces lysolecithin. Lysolecithin causes hemolysis, damages capillary walls, and produces spasmodic muscle contractions.

Phospholipids are also associated with membranes, especially the internal membranes of cells, and with the material (myelin) that surrounds some nerve fibers. Defective myelin sheaths are observed in multiple sclerosis and in a number of other diseases.

Glycolipids

The membranes of nerve and brain cells contain glycolipids as well as phospholipids. For this reason glycolipids are sometimes referred to as cerebrosides. Glycolipids have also been isolated from seeds, from fungi, and from the woody parts of plants.

Steroids

The most abundant steroid is cholesterol—a steroid alcohol, or sterol.

Cholesterol

It comprises more than 10% of the dry weight of the brain and 10-15% of the spinal cord. It is abundant in egg yolks, and is found in virtually every type of cell. The membranes of cells and the external covering of the skin contain either cholesterol or one of its derivatives. The skin covering prevents excessive loss of water and of water-soluble components.

The body manufactures 3-5 grams of cholesterol per day. This is in addition to the approximately 0.3 gram of cholesterol present in a normal diet. Cholesterol is synthesized from acetyl CoA, primarily in the cells of the liver. The normal fasting level of serum cholesterol is 130-250 mg%.

"It is not fully understood why some people are more susceptible than others to the serious results of atherosclerosis, but scientists have identified some of the factors that increase the risks: high levels of cholesterol in the blood; overweight; cigarette smoking; lack of exercise; high blood pressure; diabetes; and family inheritance of a tendency to heart disease.

There is much encouraging evidence that most people—including those who inherit a *tendency* to heart disease—can substantially reduce their risk of having a heart attack if they follow a diet to control blood cholesterol levels; avoid cigarette smoking; maintain a normal weight; exercise regularly; and get medical treatment if they have high blood pressure or diabetes." (From "The Way to a Man's Heart" ©1972 American Heart Association. Reprinted with permission.)

A high serum cholesterol level appears to be a major factor in *atherosclerosis*—a disease in which lipids are deposited on the inner walls of large blood vessels. These deposits, called plaques, contain large amounts of cholesterol. As the disease advances, the plaques are invaded by fibroblasts, and they become calcified. The plaques may obstruct the flow of blood, or may cause clots to form. If the plaques cover a large area, the vessels lose their elasticity and may rupture. Hardening of the arteries is called *arteriosclerosis.*

Although the cause of atherosclerosis is unknown, statistics support the view that a high-fat diet—especially one containing cholesterol and saturated fats—increases the likelihood of contracting the disease.

Approximately 80% of the cholesterol manufactured by the body is converted into cholic acid, the precursor of the bile salts (discussed in Section 13.5). Other important derivatives of cholesterol include vitamin D and the steroid hormones of the adrenal cortex, ovaries, and testes.

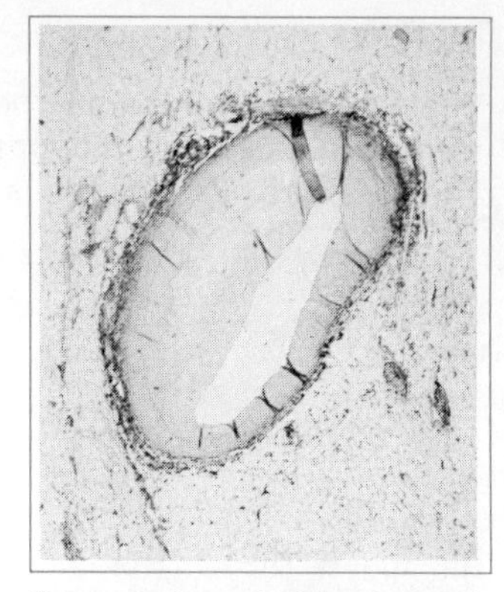

Cross-section of a blood vessel showing atherosclerotic deposits. (Courtesy Pathology Department, Santa Barbara Cottage Hospital)

Ergosterol. Ultraviolet irradiation changes this into vitamin D_2.

Progesterone. A female sex hormone—essential for gestation.

Cholic acid. Precursor of bile salts.

Testosterone. Male sex hormone that regulates development of reproductive organs.

Terpenes

Vitamins A, E, and K are terpenes, along with carotene (a reddish pigment of plants), and rhodopsin (a pigment in the retina of the eye). The structure of vitamin A_1 is

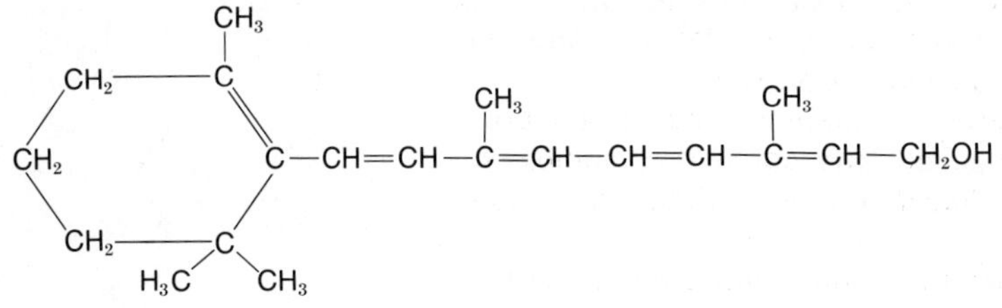

$$\text{(cyclohexene ring: } CH_2, CH_2, CH_2, C(CH_3), C(CH_3)_2\text{)}-C-CH{=}CH-\underset{}{C}(CH_3){=}CH-CH{=}CH-C(CH_3){=}CH-CH_2OH$$

13.3 REACTIONS OF NEUTRAL FATS

The reactions of neutral fats deserve special attention not only because they are the most important dietary lipids, but also because a number of commercially valuable products are derived from them. The principal reactions are hydrolysis, oxidation, and hydrogenation.

Hydrolysis

The ester linkages in a fat or oil may be hydrolyzed by solutions of sodium or potassium hydroxide, by hot mineral acids, or by the action of enzymes.

Alkaline hydrolysis, or saponification, produces salts of the fatty acids rather than the acids themselves. The effect of each type of hydrolytic agent on a typical mixed triglyceride is illustrated by the following example:

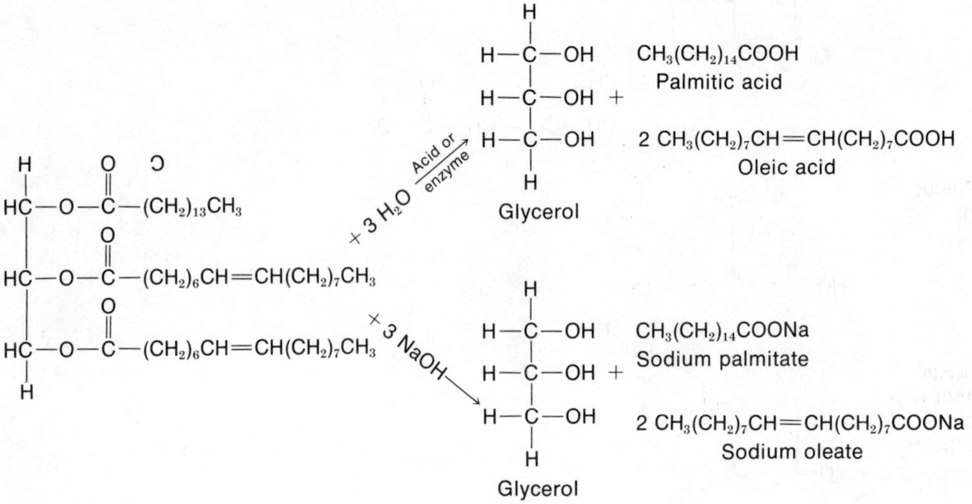

Triglyceride:

$HC{-}O{-}C(=O){-}(CH_2)_{13}CH_3$

$HC{-}O{-}C(=O){-}(CH_2)_6CH{=}CH(CH_2)_7CH_3$

$HC{-}O{-}C(=O){-}(CH_2)_6CH{=}CH(CH_2)_7CH_3$

$\xrightarrow[\text{enzyme}]{+3\,H_2O,\ \text{Acid or}}$ Glycerol ($H{-}C(H){-}OH$, $H{-}C{-}OH$, $H{-}C(H){-}OH$) + $CH_3(CH_2)_{14}COOH$ (Palmitic acid) + $2\ CH_3(CH_2)_7CH{=}CH(CH_2)_7COOH$ (Oleic acid)

$\xrightarrow{+3\,NaOH}$ Glycerol ($H{-}C(H){-}OH$, $H{-}C{-}OH$, $H{-}C(H){-}OH$) + $CH_3(CH_2)_{14}COONa$ (Sodium palmitate) + $2\ CH_3(CH_2)_7CH{=}CH(CH_2)_7COONa$ (Sodium oleate)

Salts of long-chain fatty acids are known as *soaps.* Soaps that ionize in water are valuable cleansing agents because the opposite ends of a soap ion are quite different in character. The nonpolar hydrocarbon end dissolves in lipids and in hydrocarbon greases that tend to trap dirt, while the polar salt group has an affinity for water. Thus a soap ion acts as a link between water and water-insoluble substances, thereby making it possible for the nonpolar material to be brought into "solution" and washed away, carrying the dirt with it (Figure 13.1).

Bar soaps are made from the sodium salts of fatty acids, while liquid soaps are made from potassium salts. Hard soaps are salts of saturated fatty acids. Softer soaps are produced from unsaturated fats.

Soaps are hydrolyzed by acids to form free fatty acids, as shown by the example of sodium stearate:

$$\underset{\text{Sodium stearate}}{CH_3(CH_2)_{16}COONa} + HCl \longrightarrow \underset{\text{Stearic acid}}{CH_3(CH_2)_{16}COOH} + NaCl$$

Fatty acids containing more than 10 carbons are insoluble in water; thus they cannot act as a link between grease and water. For this reason the cleansing power of soap is affected by the pH of the water: the lower the pH, the less the cleansing power.

Soaps are also affected by the presence of certain metal ions in the water (see Section 6.11). These metal ions (especially Mg^{2+} and Ca^{2+}) form soaps that are virtually insoluble and are therefore ineffective in the removal of grease and dirt.

Because of their shortcomings, soaps have been losing in popularity to detergents. A detergent is a substance whose cleansing action is similar to that of a soap, but whose structure is quite different (Section 11.2). Detergents are generally unaffected by either low pH or metal ions. However, despite their unsatisfactory behavior in water that is excessively hard or acidic, soaps have a distinct advantage over

A Recipe for Making Soap

Mix together in a bucket a quantity of hardwood ash and some hot water. Allow the ash to settle and pour off the liquid, straining it through cheesecloth. Add a quart of this liquid to a quarter cup of bacon grease or other fat in a heavy pot. Cook the mixture for at least an hour, then allow it to cool. Add a quarter cup of table salt (or rock salt) and stir. The soap is insoluble in salt solution: it will precipitate and rise to the surface. Skim it off and press it into a cake. Soap made in this fashion tends to be rather alkaline and may irritate sensitive skin.

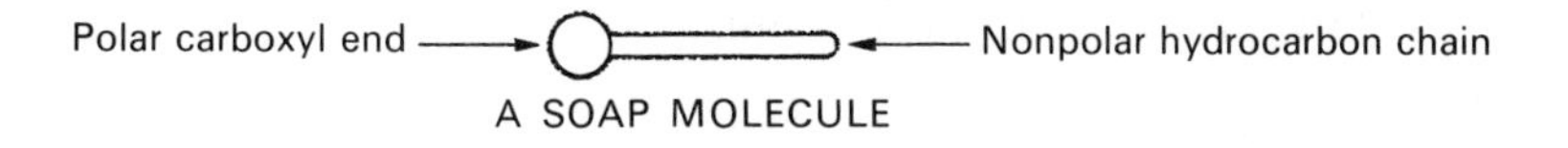

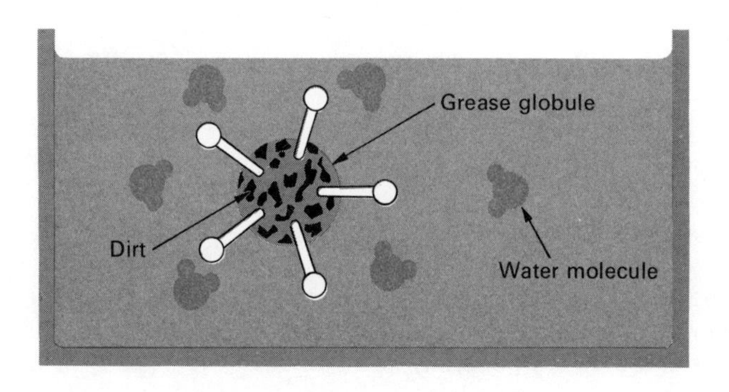

Figure 13.1
The cleansing action of soap. The nonpolar segment of the soap ion dissolves in dirt-carrying grease. The polar group is attracted by water molecules; thus the soap is able to lift the grease and carry it away.

most detergents in that they are biodegradable—that is, they are metabolized by microorganisms in the soil and by those present in a sewage treatment plant. Thus the continued use of soaps creates no obstacles as far as the reuse of water is concerned.

The saponification reaction can be used to obtain information concerning the length of the fatty acid chains in a neutral fat. A weighed quantity of fat is hydrolyzed and the free fatty acids are titrated with standard KOH. The shorter the carbon chains, the more acid groups there are per gram of fat, and the larger the quantity of KOH required for the titration. The *saponification number* is the number of milligrams of KOH required to neutralize the fatty acids resulting from complete hydrolysis of 1 gram of neutral fat.

Oxidation

Complete oxidation of a neutral fat produces CO_2 and water and liberates 9 kcal of energy per gram of fat. Note that this is more than twice the amount of energy that is released when a gram of carbohydrate is completely oxidized. This is because fats have less oxygen per molecule than sugars.

Fats that contain unsaturated fatty acids are subject to oxidation at the double bonds. This oxidation is promoted by heat, light, air, and moisture, and is generally followed by cleavage of the molecule at the double bonds, resulting in the formation of various aldehydes, ketones, and carboxylic acids, many of which have unpleasant odors. A fat containing these odoriferous compounds is said to be rancid. Rancidity is often observed in cooking fats that have been heated either too long or at too high a temperature. The tendency of a fat to become rancid can be counteracted by the addition of antioxidants such as vitamin E, vitamin C, and hydroquinone, because they are more readily oxidized than the double bonds.

Rancidity may also result from hydrolysis of fats containing short-chain fatty acids such as butyric, caproic, and caprylic—which all have disagreeable odors. Hydrolytic rancidity is often caused by enzymes from microorganisms. Keeping fats tightly covered helps to prevent both oxidative and hydrolytic rancidity.

Hydrogenation

The addition of hydrogen to a double bond was described in Section 10.2. This reaction changes an unsaturated compound into the corresponding saturated structure. Both pressure and a catalyst are required, as shown in the following example:

$$\text{Oleic acid} \xrightarrow[\text{2000 psi}]{\text{Ni}} \text{Stearic acid}$$

As noted earlier, the melting point of a triglyceride corresponds to its degree of saturation. Triglycerides from plant sources contain only a small percentage of saturated fatty acids and thus are generally liquids at room temperature. Partial hydrogenation of a plant lipid converts it into a soft white solid. Spry®, Crisco®, and Fluffo® are solid shortenings that have been formed in this way.

Since iodine reacts more readily than hydrogen with double bonds, it is possible to approximate the degree of unsaturation in a fat by simply measuring the amount of iodine that combines with a specific amount of the fat. This is done by placing the weighed sample of fat in a standard iodine solution, allowing it to remain for a period of time, and then titrating the remaining iodine. The number of grams of iodine absorbed by 100 grams of a neutral fat is called the *iodine number* of the fat. Liquid fats (oils) usually have iodine numbers above 70. The iodine and saponification numbers of some common fats and oils are given in Table 13.2.

TABLE 13.2 A Comparison of Solid and Liquid Fats. Median Values for Saponification and Iodine Numbers

	Saponification Number	Iodine Number
Solid Fats		
Butter	220	27
Lard	200	53
Tallow	195	40
Liquid Fats		
Corn oil	192	121
Cottonseed oil	194	110
Olive oil	192	85
Peanut oil	192	93
Safflower oil	192	148
Soybean oil	192	133

13.4 DETECTION AND IDENTIFICATION OF LIPIDS

The Acrolein Test

When glycerol is heated in the presence of $KHSO_4$, it is dehydrated to acryl aldehyde, or acrolein. The reaction is

$$\begin{array}{c} H \\ | \\ H-C-OH \\ | \\ H-C-OH \\ | \\ H-C-OH \\ | \\ H \end{array} \xrightarrow[\Delta]{KHSO_4} \begin{array}{c} H-C=O \\ | \\ H-C \\ \| \\ H-C \\ | \\ H \end{array} + 2\,H_2O$$

Acrolein

This is a standard test for neutral fats since the reaction takes place regardless of whether the glycerol is free or esterified.

Acrolein, which has a most unpleasant odor, is also formed when fats are heated to more than 300°C or when they are maintained at a high temperature for a long time, as in a deep-fat fryer. Since acrolein is irritating to the digestive tract, fried foods may sometimes cause digestive upsets.

The Liebermann-Burchard Test

This is a specific test for cholesterol. Acetic anhydride reacts with a solution of cholesterol in chloroform to produce a blue-green color. The exact nature of the product is not known.

13.5 DIGESTION AND ABSORPTION OF LIPIDS

Many of the substances classified as lipids are neither digested nor absorbed by the body. The chief dietary lipids are neutral fats. The ester linkages in neutral fats are hydrolyzed by digestive enzymes called *lipases,* producing glycerol and fatty acids, as well as mono- and diglycerides. In adult humans the digestion of lipids occurs primarily in the small intestine, although a lipase is also found in gastric fluid. Gastric lipase has an optimum pH of about 7, and in the mild acidity of an infant's stomach it is an active enzyme. However, as the acidity of the stomach increases with age, the activity of gastric lipase diminishes until it is virtually inactive. The principal lipase—steapsin—is produced by the pancreas, and the small intestine contributes cholesterol esterase and lecithinase.

Digestion of a neutral fat is complicated by the fact that the digestive enzymes are soluble in water but fats are not. Fat molecules are repelled by water and tend to cluster together, forming globules. Since digestive enzymes cannot penetrate the globules, only those fat molecules that are on the surface of the globule can be acted upon. The surface area of the fat can be increased by breaking it up into smaller droplets—a process called emulsification. This occurs when the fat enters the intestine and is mixed with an alkaline fluid called bile.

Bile is a secretion of the liver that is stored in the gallbladder and enters the intestine via the bile duct. The chief constituents of bile are the bile salts (derivatives of cholic acid) and two waste products—cholesterol and bile pigments (mostly bilirubin). The bile salts are structurally similar to cholesterol, yet they behave like soaps in that they disperse fat globules into numerous tiny fat droplets. The droplets do not come together again because the bile salts on their surfaces cause them to repel each other. The increased surface area permits greater contact between enzyme and fat, thus increasing the activity of the enzyme.

Bile salts also aid in the absorption of digested fats. Glycerol is miscible with water and is readily absorbed, but the fatty acids are not. They combine with bile salts to form more soluble complexes. Once

Lipids in Common Foods

	grams
Avocado, Fuerte, 4" diameter	52.8
Butter, 1 pat	5.6
Cottage cheese, 1 cup	2.6
Egg, 1 medium	5.7
Ham, smoked, baked, $4 \times 2\frac{1}{2} \times \frac{3}{4}$"	18.5
Hamburger, lean, $3 \times 3 \times \frac{1}{4}$"	8.5
Hershey bar, with nuts, $1\frac{5}{8}$ oz.	16.8
Ice cream, chocolate, $\frac{1}{2}$ cup	12.8
Milk, whole, 8 oz.	7.8
Sirloin steak, medium, $4 \times 3 \times \frac{1}{2}$"	13.0
Soybeans, cooked, $\frac{3}{4}$ cup	9.0
Swiss cheese, $4 \times 4 \times \frac{1}{8}$ slice	7.0
Tuna fish, canned, drained, $\frac{3}{4}$ cup	13.5
Walnuts, English, 10–12 meats	9.7

they have passed through the intestinal wall, the bile salts are split off and returned to the liver for reuse.

The bile pigments are colored substances resulting from the breakdown of the heme portion of hemoglobin. They are further degraded in the intestines and are responsible for the characteristic color of feces.

Cholesterol, the other waste product in the bile, is excreted when the supply exceeds the body's requirement. Sometimes cholesterol precipitates in the gallbladder, forming gallstones. The stones may obstruct the bile duct, preventing the excretion of bile pigments and causing a condition known as obstructive jaundice in which the pigments are diverted into the bloodstream. The resulting yellowish pigmentation of the skin is one of the diagnostic signs of jaundice.

Absorption

As mentioned earlier, digestion of a neutral fat produces a mixture of glycerol and fatty acids as well as mono- and diglycerides. These products attach themselves to bile salts and pass into the epithelial cells that line the intestine, where they are apparently reassembled into triglycerides. The triglycerides leave the epithelial cells and enter

Gallstones. (Photograph by J. Asdrubal Rivera)

the lymphatic fluid as components of fat droplets, called chylomicrons, that also contain cholesterol and phospholipids. Since the lymphatic system joins the venous circulation, the chylomicrons soon appear in the blood, giving it a milky appearance that may persist for several hours. The fasting level of serum triglycerides normally ranges from 40 to 175 mg%.

13.6 METABOLISM OF LIPIDS

The serum triglycerides represent an even greater energy potential than that of the blood sugar. Although their concentrations are roughly equivalent, oxidation of triglycerides yields more energy because they contain more bonds. And, whereas the carbohydrate reserves of the body are depleted after one day without food, the fat stored in adipose tissues can sustain a person for 30-40 days.

The release of energy from serum triglycerides takes place mainly in the liver, although partially degraded fats are utilized by other tissues. The triglycerides are once again hydrolyzed to glycerol and fatty acids, following which the two components are oxidized by different pathways.

Glycerol is converted into PGAL (see Section 12.7) and either enters the glycolytic pathway or is used to make glycogen. The oxidation of a fatty acid proceeds by a stepwise process that begins at the carboxyl end of the molecule. With each step the carbon chain is shortened by two atoms. Since the fatty acid chain is split between the second (α) and third (β) carbon atoms, the process is called beta oxidation. The 2-carbon fragments combine with molecules of coenzyme A to form acetyl CoA, which can be further oxidized via the Krebs cycle, used in the synthesis of steroids and fatty acids, or condensed with another molecule of acetyl CoA to form acetoacetyl CoA.

Recall that one of the steps in the oxidation of a carbohydrate is its conversion to acetyl CoA. When carbohydrate oxidation produces more acetyl CoA than the body needs, the excess is generally synthesized into triglycerides. Amino acids can also be converted into acetyl CoA: thus both carbohydrates and proteins can be turned into fats (see Figure 12.7). Persons who persistently overeat are proof of the ability of the body to make this transformation.

The reactions that take place during beta oxidation are illustrated by the example of hexanoic acid. Note that the first step is unusual because ATP is cleaved to produce the diphosphate (pyrophosphate), symbolized here by PP_1.

1. The fatty acid combines with coenzyme A.

$$CH_3(CH_2)_2\overset{\beta}{C}H_2-\overset{\alpha}{C}H_2-\overset{O}{\overset{\|}{C}}-OH + H-S-CoA \xrightarrow[Mg^{2+}]{ATP \rightarrow AMP + PP_i} CH_3(CH_2)_2\overset{\beta}{C}H_2-\overset{\alpha}{C}H_2-\overset{O}{\overset{\|}{C}}-S-CoA + H_2O$$

2. The α- and β-carbons each lose one hydrogen atom and form a double bond.

$$CH_3(CH_2)_2\overset{H}{\underset{H}{C}}-\overset{H}{\underset{H}{C}}-\overset{O}{\overset{\|}{C}}-S-CoA \xrightarrow{NAD \quad NADH_2} CH_3(CH_2)_2\underset{H}{C}=\underset{H}{C}-\overset{O}{\overset{\|}{C}}-S-CoA$$

3. Water is added to the double bond.

$$CH_3(CH_2)_2\underset{H}{C}=\underset{H}{C}-\overset{O}{\overset{\|}{C}}-S-CoA + H_2O \longrightarrow CH_3(CH_2)_2\overset{OH}{\underset{H}{C}}-\overset{H}{\underset{H}{C}}-\overset{O}{\overset{\|}{C}}-S-CoA$$

4. The secondary alcohol group is oxidized to a carbonyl group.

$$CH_3(CH_2)_2\overset{OH}{\underset{H}{C}}-\overset{H}{\underset{H}{C}}-\overset{O}{\overset{\|}{C}}-S-CoA \longrightarrow CH_3(CH_2)_2\overset{O}{\overset{\|}{C}}-\overset{H}{\underset{H}{C}}-\overset{O}{\overset{\|}{C}}-S-CoA$$

5. The chain is broken, freeing acetyl CoA.

$$CH_3(CH_2)_2\overset{O}{\overset{\|}{C}}-\overset{H}{\underset{H}{C}}-\overset{O}{\overset{\|}{C}}-S-CoA + H_2O \longrightarrow CH_3\overset{O}{\overset{\|}{C}}-S-CoA + \underset{\text{Butanoic acid}}{CH_3CH_2CH_2\overset{O}{\overset{\|}{C}}-OH}$$

The acetyl CoA thus produced enters the Krebs cycle while the butanoic acid (activated by H—S—CoA) undergoes additional oxidation by repeating the sequence of steps just shown. Since the majority of the fatty acids found in nature have an even number of carbon atoms, beta oxidation yields acetyl CoA as a final product. Beta oxidation of a fatty acid containing an odd number of carbon atoms ceases when there are three carbon atoms left, and the propionyl CoA eventually enters the Krebs cycle as succinyl CoA.

The oxidation of a fatty acid differs from that of a simple sugar since the various steps in beta oxidation constitute a spiral rather than a cycle. The beta oxidation spiral is diagrammed in Figure 13.2.

As mentioned earlier, the condensation of two molecules of acetyl CoA into acetoacetyl CoA is a normal event. However, when the needs of the body are being met almost exclusively by oxidation of fats, as in diabetes or starvation, the liver is apparently unable to complete the oxidation of fatty acids, and additional acetoacetyl CoA is produced. The acetoacetyl CoA is changed to acetoacetic acid, which can be decarboxylated to acetone or reduced to β-hydroxybutyric acid:

$$\underset{\text{Acetoacetyl CoA}}{CH_3\overset{\overset{O}{\|}}{C}-CH_2\overset{\overset{O}{\|}}{C}-S-CoA} \xrightarrow{\text{liver cells}} \underset{\text{Acetoacetic acid}}{CH_3\overset{\overset{O}{\|}}{C}-CH_2-\overset{\overset{O}{\|}}{C}-OH} + H-S-CoA$$

$$CO_2 + \underset{\text{Acetone}}{CH_3\overset{\overset{O}{\|}}{C}-CH_3} \qquad \underset{\beta\text{-hydroxybutyric acid}}{CH_3\overset{\overset{OH}{|}}{C}HCH_2\overset{\overset{O}{\|}}{C}-OH}$$

These three compounds are called ketone bodies, and their accumulation results in a condition known as ketosis. The effects of ketosis were described in Section 8.13.

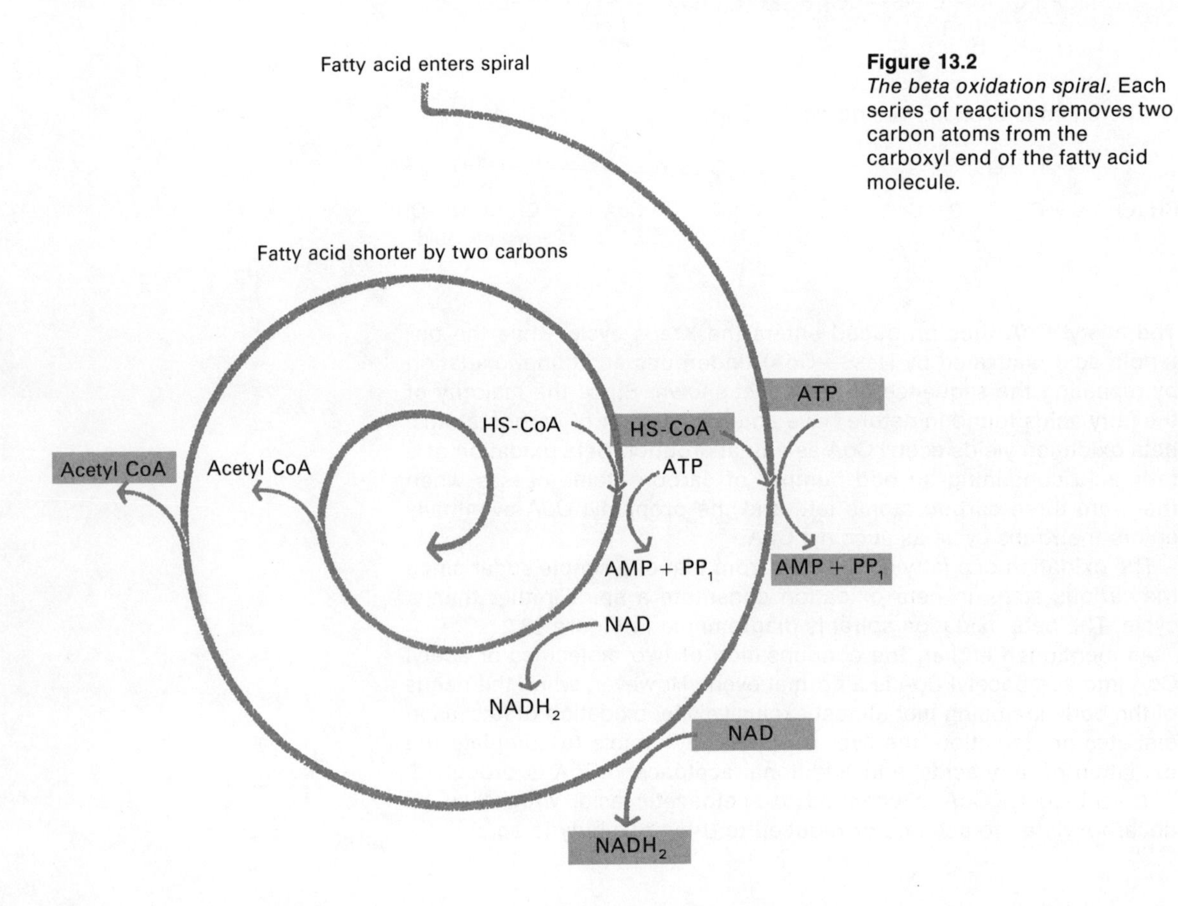

Figure 13.2
The beta oxidation spiral. Each series of reactions removes two carbon atoms from the carboxyl end of the fatty acid molecule.

APPLICATION OF PRINCIPLES

1. The hump on a camel's back is mostly depot fat. Assuming complete hydrolysis of the fat and its subsequent oxidation to CO_2 and water, calculate the number of kilocalories that would be released from 25 kg of camel fat.
2. If the camel's hump were made of glycogen rather than fat, would its oxidation require more or less oxygen? Explain.
3. Name the compound that would be present after the following fat has been completely hydrolyzed:

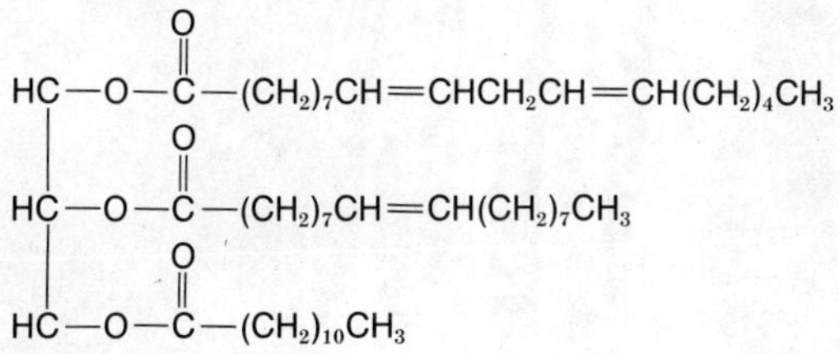

$$
\begin{array}{l}
\overset{\displaystyle O}{\|} \\
HC-O-C-(CH_2)_7CH{=}CHCH_2CH{=}CH(CH_2)_4CH_3 \\
|\overset{\displaystyle O}{\|} \\
HC-O-C-(CH_2)_7CH{=}CH(CH_2)_7CH_3 \\
|\overset{\displaystyle O}{\|} \\
HC-O-C-(CH_2)_{10}CH_3
\end{array}
$$

4. Fats are not digested in the stomach despite the presence in gastric fluid of a lipase. One reason for its lack of activity is the fact that gastric lipase has an optimum pH of about 7. Even if gastric lipase could work at a lower pH, fats would still pass through the stomach virtually unaltered. Why?
5. Hydrolytic rancidity is more likely to occur in butter than in peanut oil. Give two reasons for this.
6. The saponification number of safflower oil is the same as that of lard, yet its iodine number is more than twice as great. What does this tell you about the two lipids?
7. In what respects does the absorption of a digested fat differ from the absorption of digested starch?

Dressed beef carcasses hang in a refrigerated cooler. (Courtesy Swift Fresh Meats Company, Chicago, Illinois)

14

PROTEINS

LEARNING OBJECTIVES

1. Distinguish among the primary, secondary, and tertiary structures of a protein.
2. Give examples of how temperature, radiations, acids, and bases affect the primary, secondary, and tertiary structures.
3. Draw the structural formula for an amino acid above, at, and below its isoelectric point, and explain how amino acids act as buffers.
4. Using structural formulas of two different amino acids, show the formation of a peptide bond between them.
5. Trace the digestion of a protein, including the names and specific chemical action of each proteolytic enzyme.
6. Outline the urea cycle in humans.
7. Summarize the possible metabolic fates of amino acids.
8. Explain why an adequate intake of protein is such a critical part of sound nutrition.

KEY TERMS AND CONCEPTS

coagulation	isoelectric point
conjugation	peptide bond
deamination	primary structure
denaturation	prosthetic group
disulfide linkage	secondary structure
endopeptidase	tertiary structure
enzyme	transamination
incomplete protein	zymogen

If molecules were ranked according to their biological importance, it seems very likely that proteins would be high on the list. Among the reasons for their importance is the discovery by molecular biologists that the genetic information which enables a species to perpetuate itself is essentially a set of coded instructions for manufacturing protein molecules. It would not be an exaggeration to say that the identity and characteristics of an organism are determined by its protein.

14.1 OCCURRENCE, COMPOSITION, AND CHARACTERISTICS OF PROTEIN

Some idea of the distribution and functions of protein molecules may be obtained from the following partial list:

Proteins in the Blood

The protein present in blood serum—albumen—helps to maintain the osmotic pressure balance between tissue cells and extracellular fluid. The clotting process involves a number of proteins, including prothrombin and fibrinogen. Oxygen and carbon dioxide are transported by attachment to the hemoglobin molecule which is mostly protein, and certain other proteins—the gamma globulin type—enable an organism to build up immunity against infectious diseases.

Structural Proteins

The horns, nails, hair, cartilage, muscles, as well as the nonmineral parts of teeth and bones are proteins.

Catalytic Proteins

Enzymes are biochemical catalysts found in every living cell. Without them, the cell's chemical reactions would take place so slowly that life in its present forms would be impossible. All enzymes are proteins.

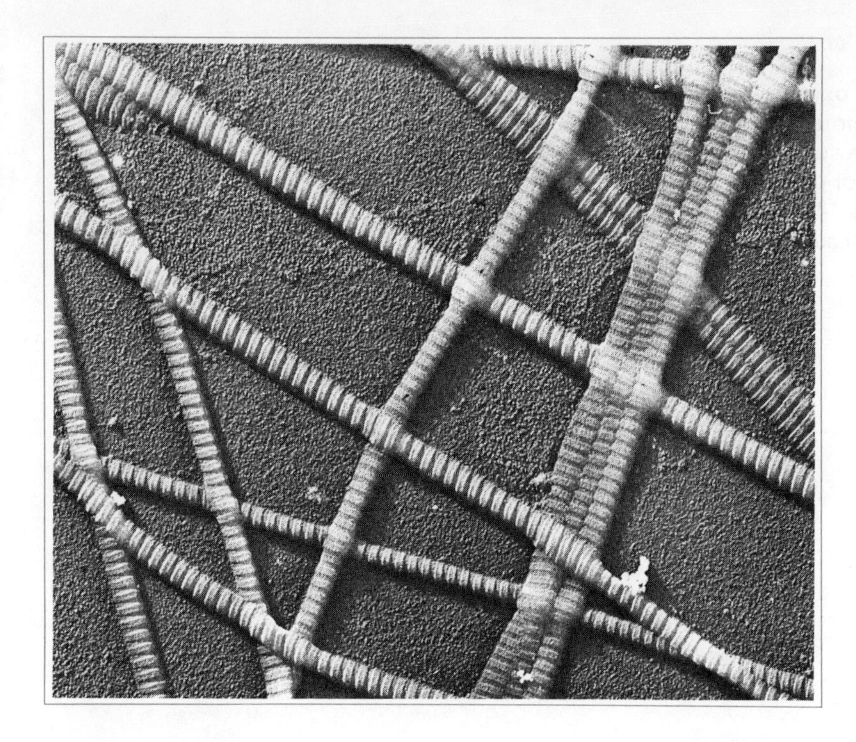

Collagen fibers (magnified 60,000 times). Collagen is a protein that is found in connective tissues such as cartilage and muscle. (Photograph by Dr. Jerome Gross, Massachusetts General Hospital)

Regulatory Proteins

Hormones are secretions of the endocrine glands, such as the pituitary, which exert a regulatory effect on body processes. Although hormones are a structurally diverse group, a number of them, including insulin (which regulates carbohydrate storage) can be classified as proteins.

Poisonous Proteins

Poisonous protein-containing substances which are injected during the sting or bite of an animal are called *venoms.* Protein is an active part of the venoms of rattlesnakes, some spiders, and many insects.

In the *toxin* group is one of the most powerful poisons yet discovered—the protein produced by the bacterium *Clostridium botulinum.* Botulinus toxin is produced by the anaerobic growth of the organism in certain improperly preserved foods, and ingestion of the toxin is nearly always fatal.

Antibiotics

Among the secretions of some bacteria and fungi are substances, called antibiotics, which prevent the growth of other microorganisms. The antibiotics are another chemically diverse group, but several, including gramicidin and tyrocidin, are proteins.

Composition

All proteins contain carbon, oxygen, nitrogen, hydrogen, and sulfur, and many contain one or more additional elements. The structural proteins are especially rich in sulfur, while the proteins of the nucleus contain significant amounts of phosphorus. The average composition of proteins is given in Table 14.1. Some proteins also contain iron, copper, magnesium, manganese, or zinc.

TABLE 14.1 Elemental Composition of Protein

Carbon	51–55%	Hydrogen	6.7–7.3%
Oxygen	21–24%	Sulfur	0.3–2.2%
Nitrogen	15–18%	Phosphorus	0.0–1.5%

Characteristics

Proteins are high-molecular-weight substances varying in molecular weight from about 6,000 to over 2,500,000. The high molecular weight is an indication of the large size of these molecules, and of the complexity. The molecular weights of some representative proteins are shown in Table 14.2.

TABLE 14.2 Molecular Weights of Some Typical Proteins

Insulin (human)	6,000
Myoglobin	17,000
Pepsin (digestive enzyme)	24,000
Egg albumin	48,000
Hemoglobin (horse)	68,000
Serum albumin (human)	69,000
Catalase (cellular enzyme)	250,000
Fibrinogen	400,000

Because of their high molecular weights, proteins tend to form colloidal rather than true solutions, and are retained by a dialysis membrane. Their solubility, however, appears to be more closely related to structure and function than to molecular weight. The serum proteins, for example, range in molecular weight from about 60,000 to 400,000, yet all are quite soluble. The soluble proteins are generally good buffers and, as such, are an essential part of the homeostatic mechanism. Many become insoluble as a result of heating or marked changes in pH. All proteins are hydrolyzed by prolonged contact with strong acids or bases. The hydrolysate always contains amino acids and sometimes includes a mineral, lipid, carbohydrate, or nucleic acid. There are twenty common amino acids, but they are not all found in every protein. For example, egg albumin contains only nineteen of the twenty, and human insulin only sixteen.

14.2 THE AMINO ACIDS

All except two of the amino acids obtained by protein hydrolysis can be represented by the general formula:

$$R-\underset{NH_2}{\overset{H}{\underset{|}{\overset{|}{C}}}}-C\begin{matrix}\nearrow O \\ \searrow OH\end{matrix}$$

where R is a hydrogen, an alkyl group, chain, or ring. The remaining two amino acids can be considered as substituted amines. Since the amino group is attached to the α-carbon (the one next to the carboxyl group), they are all α-amino acids. When R is any group except hydrogen, the α-carbon becomes asymmetric and the molecule exists in both the D and L forms (see glyceraldehyde, Section 12.3). All of the amino acids except one (glycine) contain an asymmetric carbon and have been isolated from proteins only in the L form. Table 14.3 shows the structures of the twenty naturally occurring amino acids. Note that two of the amino acids, lysine and arginine, have a second amino group and are thus diamino acids. Additionally, glutamic and aspartic acid are dicarboxylic acids. Conversion of the dicarboxylic acids to amides produces glutamine and asparagine, whose structures are not shown.

14.3 PROPERTIES OF THE AMINO ACIDS

Physical Properties

The amino acids are crystalline solids having relatively high melting points, and most are readily soluble in water. This combination of characteristics suggests that the amino acids should be strongly polar and capable of existing as ions in polar solvents such as water. The presence of dipolar ions, or "zwitterions," in amino acid solutions has been confirmed. The concept of the zwitterion was first discussed in connection with acids and bases in Chapter 8. Zwitterions are important to the biochemical reactions we will be studying.

Zwitterions and Buffering

The amino group on the carbon adjacent to a carboxyl group is a fairly strong base and withdraws H^+ ions from water as it forms an NH_3^+ ion. The decreased concentration of H^+ ion in the solution increases the dissociation of the carboxyl group and results in a dipolar ion or zwitterion. For each amino acid there is a pH, called the *isoelectric point* (pI), at which its molecules all exist as zwitterions. An amino acid at its isoelectric point will not migrate in an electric field because its net charge is zero. For the monoaminomonocarboxylic acids the pI is between 5.02 and 7.59.

TABLE 14.3 Naturally Occurring Amino Acids

Name	Symbol	Structural Formula
Glycine	gly	$\boxed{H}-\overset{H}{\underset{NH_2}{C}}-C(=O)OH$
L-Alanine	ala	$\boxed{CH_3}-\overset{H}{\underset{NH_2}{C}}-C(=O)OH$
L-Serine	ser	$\boxed{HO-CH_2}-\overset{H}{\underset{NH_2}{C}}-C(=O)OH$
L-Cysteine	cySH	$\boxed{H-S-CH_2}-\overset{H}{\underset{NH_2}{C}}-C(=O)OH$
L-Cystine	(cyS-Syc) or $(cyS)_2$	$\boxed{S-CH_2}-\overset{H}{\underset{NH_2}{C}}-C(=O)OH$; $\boxed{S-CH_2}-\overset{H}{\underset{NH_2}{C}}-C(=O)OH$ (S—S linked)
L-Threonine	thr	$\boxed{CH_3-\overset{OH}{\underset{H}{C}}}-\overset{H}{\underset{NH_2}{C}}-C(=O)OH$
L-Valine	val	$\boxed{H-\overset{CH_3}{\underset{CH_3}{C}}}-\overset{H}{\underset{NH_2}{C}}-C(=O)OH$
L-Leucine	leu	$\boxed{H-\overset{CH_3}{\underset{CH_3}{C}}-CH_2}-\overset{H}{\underset{NH_2}{C}}-C(=O)OH$

TABLE 14.3 (Continued)

Name	Symbol	Structural Formula
L-Isoleucine	ile or ileu	$CH_3-CH_2-CH(CH_3)-CH(NH_2)-COOH$
L-Methionine	met	$CH_3-S-CH_2-CH_2-CH(NH_2)-COOH$
L-Phenylalanine	phe	$C_6H_5-CH_2-CH(NH_2)-COOH$
L-Tyrosine	tyr	$HO-C_6H_4-CH_2-CH(NH_2)-COOH$
L-Tryptophan	try	$(\text{indole: } C_6H_4-NH-CH=C)-CH_2-CH(NH_2)-COOH$
L-Histidine	his	$(\text{imidazole: } H-C=C, N, N-H, C-H)-CH_2-CH(NH_2)-COOH$
Dicarboxylic Acids		
L-Aspartic acid	asp	$HOOC-CH_2-CH(NH_2)-COOH$
L-Glutamic acid	glu	$HOOC-CH_2-CH_2-CH(NH_2)-COOH$

TABLE 14.3 (Continued)

Name	Symbol	Structural Formula
Diamino Acids		
L-Lysine	lys	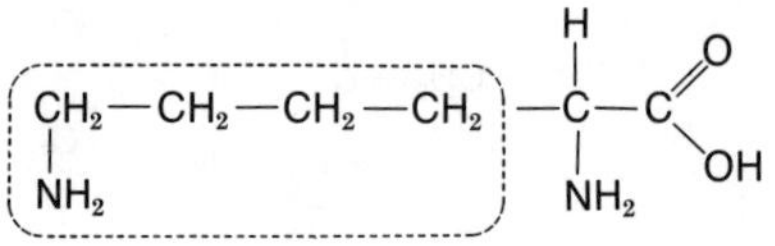 $CH_2(NH_2)-CH_2-CH_2-CH_2-C(H)(NH_2)-C(=O)OH$
L-Arginine	arg	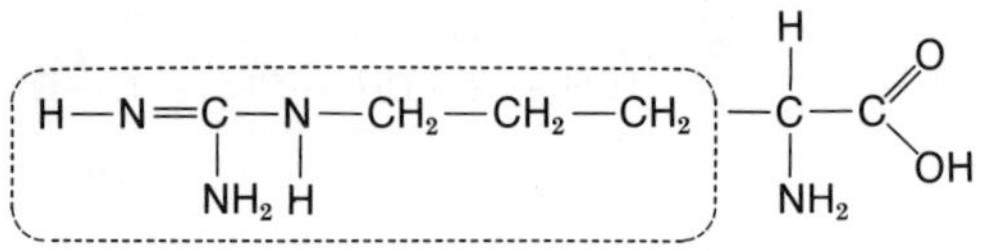 $H-N=C(NH_2)-N(H)-CH_2-CH_2-CH_2-C(H)(NH_2)-C(=O)OH$
N-Substituted Amino Acids		
L-Proline	pro	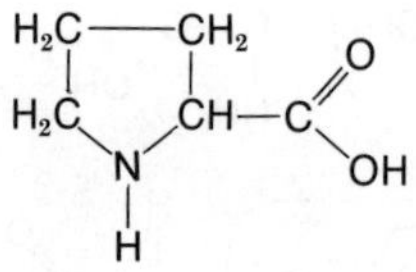 H_2C-CH_2; H_2C, $CH-C(=O)OH$; $N-H$ (ring)
L-Hydroxyproline	hyp	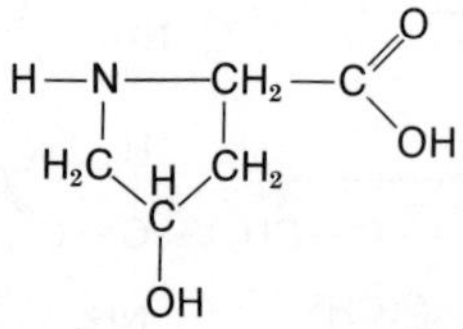 $H-N-CH_2-C(=O)OH$; H_2C, CH_2; $C(H)-OH$ (ring)

One consequence of the zwitterion structure is that solutions of amino acids are buffers. This buffering action is illustrated in Figure 14.1. The carboxyl ion is capable of accepting a proton and is therefore a Brönsted-Lowry base. When additional hydronium ions are added to an amino acid solution at its isoelectric point, the carboxyl ions which are present combine with them and a change of pH is prevented. The addition of base is counteracted by the NH_3^+ group which, in donating its proton, acts as an acid.

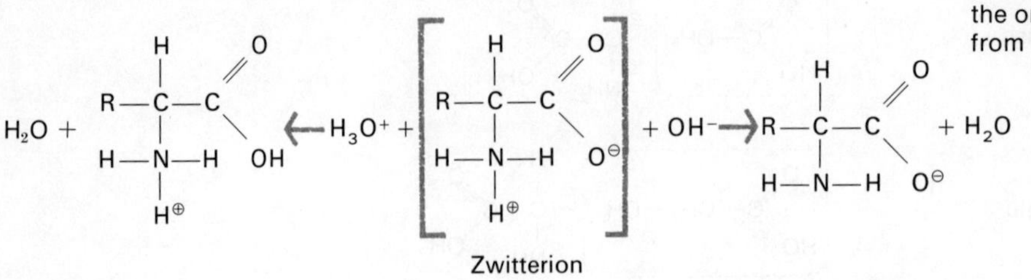

$$H_2O + R-C(H)(NH_3^{\oplus})-C(=O)OH \leftarrow H_3O^+ + \left[R-C(H)(NH_3^{\oplus})-C(=O)O^{\ominus}\right] + OH^- \rightarrow R-C(H)(NH_2)-C(=O)O^{\ominus} + H_2O$$

Zwitterion

Figure 14.1
The buffering action of an amino acid. The reaction on the left takes place when acid is added to the zwitterion, while the one on the right results from the addition of base.

Peptide Bond Formation

All amino acids are capable of forming *peptide bonds* by dehydration synthesis. A peptide bond is an amide linkage between the amino group of one amino acid and the carboxyl group of a second. Figure 14.2 illustrates the formation of a peptide bond between glycine and alanine. Note that two structures are possible: one in which there is a free amino group adjacent to the methyl group, and another in which this amino group is part of the amide linkage.

The molecule in (a) is called glycylalanine, and that in (b) is alanylglycine. It is common practice to use the three-letter abbreviations shown in Table 14.3 to show the sequence of the amino acids joined by peptide linkages; thus, (a) can be represented by gly-ala and (b) by ala-gly. Note that the symbol of the amino acid having an uncombined NH_2 group is written first. This is referred to as the N-terminal amino acid, and the one with the free carboxyl group is designated as the C-terminal.

When two amino acids are joined by a peptide bond, the resulting molecule is called a dipeptide. The fusion of three amino acid molecules produces a tripeptide, and chains varying in length from four to about 50 amino acid residues are classified as polypeptides. The boundary between polypeptides and proteins is rather vague. Insulin, for example, consists of 51 amino acid residues and is generally regarded as a very small protein, while the hormone ACTH, which contains 39 residues, is considered to be a very large polypeptide (not a protein).

The ability of an amino acid to bond through either its amino or carboxyl group gives rise to an astounding number of molecular permutations. Table 14.4 shows that six different sequences are possible for

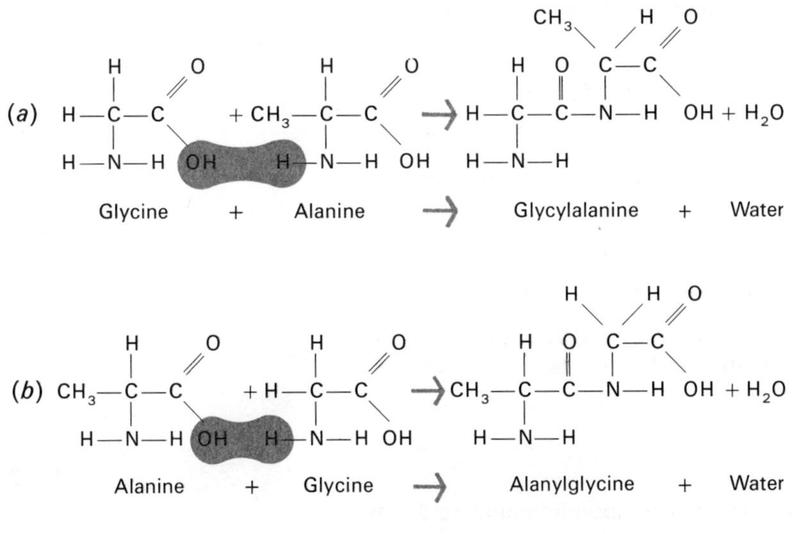

(*a*) Glycine + Alanine → Glycylalanine + Water

(*b*) Alanine + Glycine → Alanylglycine + Water

Figure 14.2
Peptide bond formation. Glycine and alanine may be joined in either of two ways: (*a*) to form glycylalanine and (*b*) to form alanylglycine.

a tripeptide composed of three different amino acids, but that four residues can be arranged in 24 different ways.

TABLE 14.4 Possible Arrangements for Molecules Composed of Three and Four Different Amino Acids

Three amino acids	Four amino acids	
	ala-gly-ser-thr	ser-ala-gly-thr
	ala-gly-thr-ser	ser-ala-thr-gly
ala-gly-ser	ala-ser-gly-thr	ser-gly-ala-thr
ala-ser-gly	ala-ser-thr-gly	ser-gly-thr-ala
gly-ala-ser	ala-thr-gly-ser	ser-thr-ala-gly
gly-ser-ala	ala-thr-ser-gly	ser-thr-gly-ala
ser-ala-gly		
ser-gly-ala	gly-ala-ser-thr	thr-ala-gly-ser
	gly-ala-thr-ser	thr-ala-ser-gly
6	gly-ser-ala-thr	thr-gly-ala-ser
combinations	gly-ser-thr-ala	thr-gly-ser-ala
	gly-thr-ala-ser	thr-ser-ala-gly
	gly-thr-ser-ala	thr-ser-gly-ala
	24 combinations	

Considering all of the possible combinations for a protein containing hundreds of amino acid residues, it is rather remarkable that a given cell produces only those proteins for which there is a specific cellular function. The specificity of protein configurations will be discussed later in connection with enzymes.

Transamination

As the name suggests, *transamination* is the transfer of an amino group from one carbon chain to another one. The donor of the amino group is an amino acid, and the recipient is an α-keto acid such as pyruvic or oxaloacetic. The transamination reaction between glutamic and pyruvic acids is

$$\begin{array}{c} COOH \\ | \\ CH_2 \\ | \\ CH_2 \\ | \\ H-C-NH_2 \\ | \\ COOH \end{array} + \begin{array}{c} CH_3 \\ | \\ C=O \\ | \\ COOH \end{array} \xrightleftharpoons{\text{Transaminase}} \begin{array}{c} COOH \\ | \\ CH_2 \\ | \\ CH_2 \\ | \\ C=O \\ | \\ COOH \end{array} + \begin{array}{c} CH_3 \\ | \\ H-C-NH_2 \\ | \\ COOH \end{array}$$

$$\text{Glutamic acid} + \text{Pyruvic acid} \rightleftharpoons \alpha\text{-ketoglutaric acid} + \text{Alanine}$$

The keto acids are readily available as intermediates of the Krebs cycle, and their conversion into amino acids provides a link between carbohydrate and protein metabolism.

Deamination

The conversion of an amino acid into its corresponding keto acid may also take place by a process called deamination. *Deamination* is the removal of the amino group as NH_3. A typical example is that of glutamic acid:

$$\begin{array}{c} COOH \\ | \\ CH_2 \\ | \\ CH_2 \\ | \\ H-C-NH_2 \\ | \\ COOH \end{array} + H_2O \xrightarrow[\ ]{NAD^+ \quad NADH^+ + H^+} NH_3 + \begin{array}{c} COOH \\ | \\ CH_2 \\ | \\ CH_2 \\ | \\ C=O \\ | \\ COOH \end{array}$$

This is an anaerobic process which occurs in all tissues, especially the liver. The keto acid product of this reaction becomes an intermediate of the Krebs cycle, while the ammonia, which is toxic, is usually disposed of via the urea cycle, which will be discussed shortly.

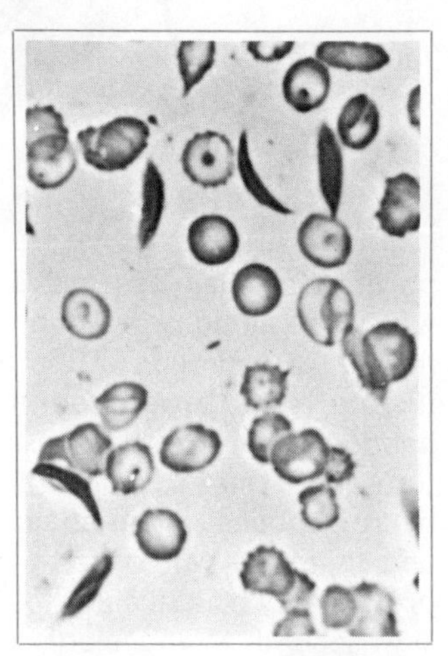

Blood cells from a person with sickle cell anemia. (Photograph by A. M. Winchester, University of Northern Colorado)

14.4 STRUCTURE OF PROTEINS

The word *structure* as applied to proteins has two different meanings. It refers either to the stabilizing forces or bonds which are active within the molecules, or to the three-dimensional structure which results from these bonds. Since neither can exist independently of the other, the difference is a matter of semantics. In the discussion that follows, both aspects of structure will be described.

Primary Structure

The primary structure of a protein refers to the covalent bonds of the peptide linkages. The order or sequence of the amino acids in the protein is a consequence of these bonds, as is the potential for development of subsidiary bonds. And the biological effectiveness of a protein molecule appears to depend upon the sequence of its amino acids. For example, a hemoglobin molecule with decreased oxygen-carrying capacity results when one of its polypeptide chains contains a valine molecule in a position normally occupied by glutamic acid. The resulting anemia, characterized by curling of the red blood cells in reduced oxygen concentrations, is called sickle cell anemia and is invariably fatal.

The covalent nature of the peptide bond makes it quite stable. Although it can be broken by enzymatic or chemical hydrolysis, it is virtually unaffected by changes in pH, solvent, or salt concentration.

The carbon and nitrogen atoms which form the covalent bond, as well as the carbonyl oxygen and the alpha hydrogen, all lie in essentially the same plane. The R groups project at right angles to this plane and are available for the formation of bonds which also play a part in the overall structure (Figure 14.3).

Sickle-cell anemia is a genetically transmitted disease that affects red blood cells. The disease results from the presence of the amino acid valine in a position normally occupied by glutamic acid in the polypeptide of hemoglobin. This defective hemoglobin is designated HbS. It behaves normally in arterial blood where the P_{O_2} is high, but precipitates within the red blood cells when the P_{O_2} is lowered. The precipitated HbS causes the cells to assume a crescent shape (sickle). These deformed cells often block capillaries, cutting off the supply of oxygen to the tissues. Sickle-cell anemia is usually fatal, and is restricted almost exclusively to persons of African descent.

Secondary Structure

The secondary structure of a protein is the result of hydrogen bonding, which can occur in either of two ways. In some proteins the

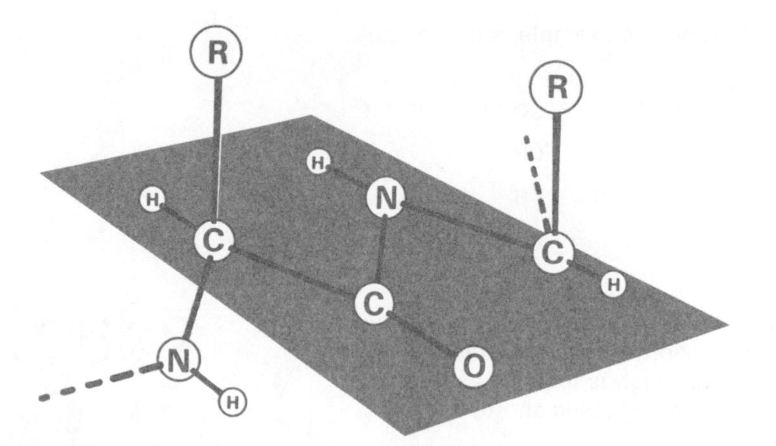

Figure 14.3
Geometry of the peptide bond. The carbon and nitrogen atoms that form the bond, as well as the carbonyl oxygen and the hydrogen from the amino group, all lie in essentially the same plane. The R groups project at right angles from the plane.

polypeptide chain assumes a spiral configuration, called an α-helix, in which the carbonyl oxygen of each amino acid residue is bonded to the amine hydrogen of the third amino acid farther along the chain (Figure 14.4). Viewed from the C-terminal end, the α-helix is a right-handed spiral in which 3.6 amino acid residues form one complete turn. The α-helix is disrupted by the presence of either proline or hydroxyproline in the chain because they lack the amide hydrogen and therefore cannot form hydrogen bonds. Where they occur, the protein may be kinked or folded.

A second arrangement, called the β-configuration or "pleated-sheet" structure, does not involve the formation of a helix, but rather the linkage of several parallel polypeptide chains by hydrogen bonds

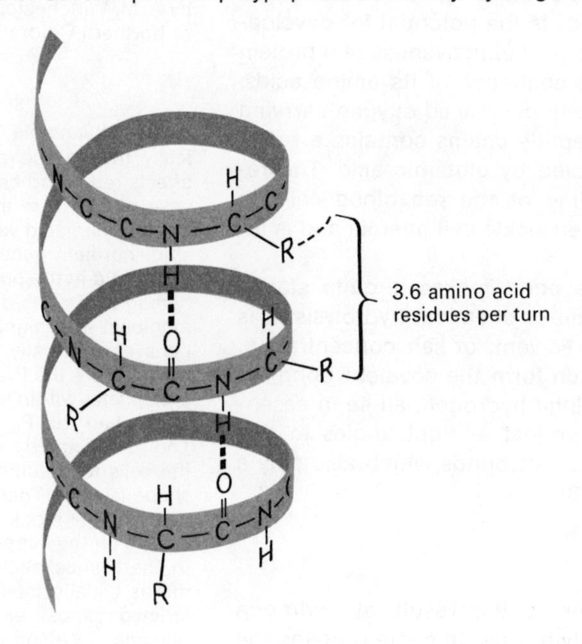

Figure 14.4
The alpha helix. Dotted lines between amino and carbonyl groups indicate hydrogen bonds. The R groups project at right angles to the axis of the helix.

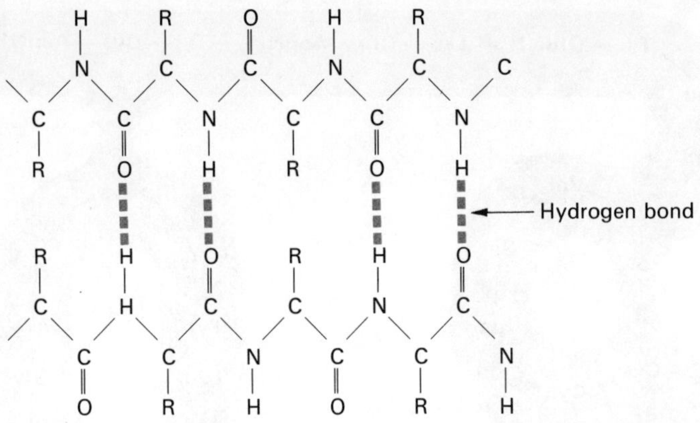

Figure 14.5
The beta configuration, or pleated-sheet structure, of protein. Parallel polypeptides are linked by hydrogen bonds.

(Figure 14.5). Fibrinogen is an example of a protein in the β-configuration.

The nature of the hydrogen bond means that the secondary structure of protein is rather easily disrupted by changes in pH, temperature, solvent, or salt concentration.

Tertiary Structure

The tertiary structure involves interactions between the R groups of the amino acids. In some instances these interactions result in the folding of the α-helix into a globular shape where the hydrophobic R groups are hidden in the folds. Such proteins have marked water solubility. Among these globular proteins are insulin, hemoglobin, and egg albumin.

The interactions between R groups may be classified as disulfide linkages, salt linkages, and hydrophobic attractions.

The disulfide linkage is a covalent bond between two cysteine residues. Because of its covalent nature, it is unaffected by changes in pH, solvent, or salt concentration, but it is subject to rupture by reduction. A disulfide bond may occur between two cysteine residues that are close to each other in the same polypeptide, or between cysteine residues in different polypeptides. The structure of the insulin molecule clearly shows the effect of disulfide bonds (Figure 14.6). Insulin also illustrates the relationship between structure and biological activity because reduction of the disulfide bonds modifies the tertiary structure and makes the molecule inactive.

The R groups other than those of cysteine are responsible for both salt linkages and hydrophobic attractions. Hydrophobic attractions result when two similar hydrophobic R groups lie close to each other. Their mutual repulsion of the aqueous solvent becomes a force driving them together. Interactions of this type are common between residues such as the phenyl ring of phenylalanine.

The salt linkage is an electrostatic attraction between the free ion-

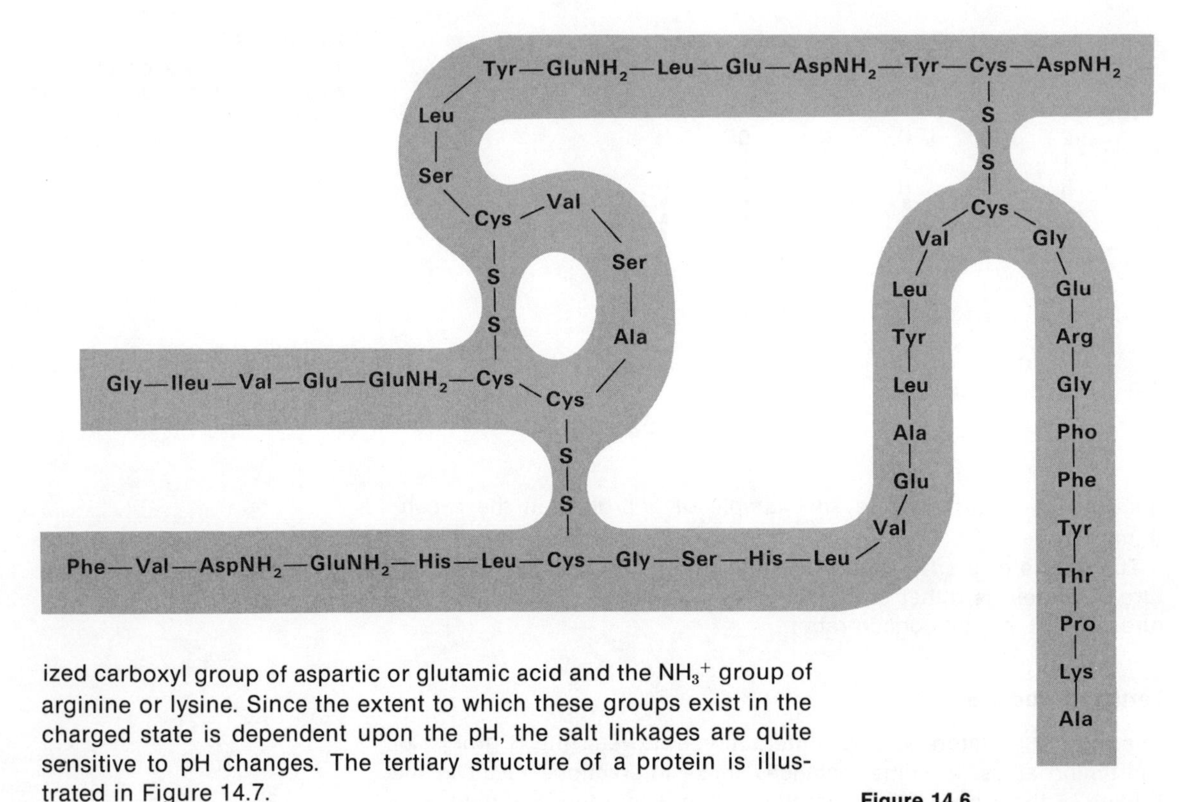

Figure 14.6
The beef insulin molecule. Insulin consists of two polypeptide chains held together by disulfide linkages.

ized carboxyl group of aspartic or glutamic acid and the NH_3^+ group of arginine or lysine. Since the extent to which these groups exist in the charged state is dependent upon the pH, the salt linkages are quite sensitive to pH changes. The tertiary structure of a protein is illustrated in Figure 14.7.

14.5 DENATURATION OF PROTEINS

Denaturation is sometimes thought to be synonomous with coagulation; that is, both terms refer to any change which results in precipitation of a protein. Protein molecules, however, can undergo changes which significantly alter their biological effectiveness but do not result in precipitation. In many cases the change is reversible and biological activity can be restored. The implication is that only the secondary or

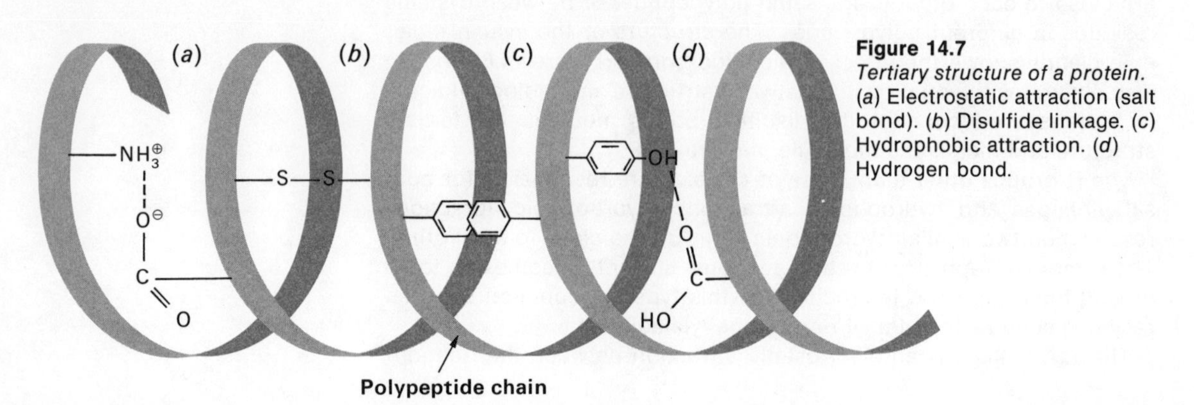

Figure 14.7
Tertiary structure of a protein. (*a*) Electrostatic attraction (salt bond). (*b*) Disulfide linkage. (*c*) Hydrophobic attraction. (*d*) Hydrogen bond.

tertiary structure has been disturbed, since it is most unlikely that the polypeptide chain, once hydrolyzed, would reassemble itself in the original sequence. Therefore, *denaturation* will be the term used to indicate any alteration of the secondary or tertiary structure, and *coagulation* to indicate precipitation.

Denaturation of protein is brought about by a variety of reagents or conditions, some of the most important of which are discussed below:

pH

Changes in pH have their greatest effect on hydrogen bonds and salt linkages. The way in which salt linkages are affected by pH was described earlier. The extent to which pH influences the coiling of the α-helix is exemplified by the behavior of polylysine—a small polypeptide containing only lysine residues. The R group of every lysine residue contains an amino group which, at a pH below its isoelectric point, is in the NH_3^+ form. The coiling of the polypeptide is prevented by repulsions between the positively charged groups, and in solution polylysine is in the extended configuration. If the pH of the solution is adjusted to the basic side of the isoelectric point, the positive charges on the NH_3^+ group disappear and the molecule assumes a helical shape.

The effect of pH changes is also illustrated by the curdling of casein, the protein of milk. Each protein, like the amino acids from which it is made, exists as a zwitterion at its isoelectric point. The pI for casein is 4.7. At this pH the (+) and (−) charges are equal and the molecules tend to aggregate into lattice arrangements like those of ionic compounds. The formation of the lattice creates a unit so large that it can no longer be accommodated in the solvent, and it precipitates. Since fresh milk has a pH of about 6.6, acid must be used to precipitate the casein. When casein curdles naturally, as in the making of cheese, the acid comes from bacterial growth.

Heat

Heating a substance causes increased molecular vibration which tends to disrupt hydrogen bonds and salt linkages. If the heating has been gentle, the bonds may re-form when the substance is cooled. Vigorous heating usually results in coagulation, as in the boiling of an egg.

Radiation

Ultraviolet light and x-rays are forms of radiant energy. They are similar to heat but have higher energies. Apparently proteins of cancer cells are more easily denatured by radiation than those of normal cells; therefore, radiation is used in an attempt to denature cancer cells before they spread to other healthy tissue. Unfortunately, radiation is not always an effective treatment in this dread disease, and may even produce cancer.

Organic Solvents

Solvents such as acetone and the common alcohols are capable of forming hydrogen bonds of their own which compete with those of protein. Contact between alcohol and cellular proteins of bacteria results in coagulation. For this reason, alcohol is swabbed on the skin before taking a blood sample or giving injections. It is worth noting that prolonged contact with alcohol is required to kill most bacteria, and so the usual 5-second scrubbing with alcohol is only a token gesture toward antisepsis.

Salts of Heavy Metals

Soluble compounds containing ions such as Pb^{2+}, Hg^{2+}, and Ag^{+} are called heavy-metal salts. They are extremely toxic to most organisms, probably because of their ability to disrupt the salt linkages of protein and to cause coagulation. Taken internally, they coagulate the proteins in the cells lining the digestive tract. A person who accidentally swallows any of the heavy-metal salts should be given protein such as eggs, milk, or cheese to eat. This added protein is readily coagulated, using up the toxic ions and sparing the tissue proteins. If the coagulated protein remains in the stomach, digestion will once again free the metal ions. For this reason an emetic should be given to induce vomiting.

Alkaloidal Reagents

The alkaloids are a group of naturally occurring chemicals extracted from plants such as *Rauwolfia*. Tannic, tungstic, and picric acids are used in the extraction process and so are called alkaloidal reagents. The negative ions of these acids will precipitate protein when the protein bears a net positive charge, and this occurs on the acid side of the isoelectric point. Tungstate ions are sometimes used in clinical laboratories to precipitate serum proteins prior to glucose measurements on the blood, although this method of preparing protein-free serum is being replaced by techniques in which the protein is removed by dialysis. Tannic acid is sometimes sprayed on severely burned tissue to coagulate the protein, which prevents loss of serum and also lessens the chances of infection.

Reducing Agents

Reducing agents have their greatest effect on the disulfide bonds which are most numerous in fibrous proteins such as cartilage and hair. In "permanent" wave treatments the disulfide bonds are disrupted by a mild reducing agent such as thioglycollate ion, the hair is wound around curlers, and an oxidizing agent is then used to establish the bonds in new positions.

14.6 PROTEIN CLASSIFICATION

Proposals for classifying proteins have been based on such aspects as function, occurrence, solubility, and composition. The latter two are widely accepted and are included here.

Classification According to Composition

Hydrolysis of a protein always yields some amino acids, but the proteins which yield *only* amino acids are referred to as simple proteins, while the others which yield additional products are said to be conjugated. *Conjugation* is the bonding of a polypeptide chain to a compound which is not an amino acid. The added compound is called a *prosthetic group.* The following classification is based on the nature of the prosthetic group:

Simple Proteins

These yield only amino acids on hydrolysis.

Conjugated Proteins

These yield amino acids plus a prosthetic group.

1. Chromoproteins. The prosthetic group is a colored compound as, for example, the heme structure of hemoglobin.
2. Phosphoproteins. The prosthetic group is phosphoric acid. Casein is a phosphoprotein.
3. Glycoproteins. The prosthetic group is a carbohydrate. A glycoprotein found in saliva is mucin, which functions as a lubricant to aid in swallowing food.
4. Lipoproteins. The prosthetic group is a fatty acid or other lipid. It has been suggested that digested lipids circulate through the blood in the form of lipoproteins.
5. Nucleoproteins. The prosthetic group is a nucleic acid. The histones of the nucleus as well as some viruses are nucleoproteins.

Classification According to Solubility

Biochemists are in general agreement with the idea of classifying proteins according to solubility, but some classifications contain more categories than others. The classification given here incorporates the major divisions found in such schemes.

Albumins

Soluble in water and aqueous media, including fairly concentrated salt solutions. Albumins are found in blood serum, egg white, and milk, and are readily coagulated by heat.

Globulins

Insoluble in water, but soluble in dilute solutions of electrolytes. Included in this category are the α-, β-, and γ-globulins of human serum and ovalglobulin. Globulins are also coagulated by heat.

Albuminoids

This category is also referred to as the *scleroproteins,* and includes the proteins which form the outer covering of animals. They are not found in the plant kingdom. The albuminoids are insoluble unless hydrolyzed.

Prolamines

Their name derives from their high content of the amino acid proline. They are insoluble in water and absolute ethanol but dissolve in 70-80% ethanol. They are obtained from cereal grains such as wheat and corn.

Histones

Soluble in water and dilute acids, but insoluble in dilute NH_4OH. They contain large amounts of arginine and are quite basic, and are associated with the nucleic acids.

14.7 COLOR REACTIONS OF PROTEIN

Our understanding of protein structure and function has been increased by the discovery that proteins, and some amino acids, react with specific reagents to produce colored products. For example, these color reactions make it possible to locate proteins and amino acids following chromotography or electrophoresis. The most widely used of the color tests are described in the following sections.

Biuret Test

When a peptide or protein is added to a solution of dilute copper sulfate in strong base, the pale blue color changes to violet. The color is similar to that produced by biuret—hence the name of the test. The reaction apparently depends on the presence of the peptide linkage, and so is not given by a completely hydrolyzed protein.

The biuret test for proteins is so named because the biuret molecule contains arrangements of atoms that resemble the peptide bonds found in proteins:

$$NH_2-\underset{\underset{O}{\|}}{C}-NH-\underset{\underset{O}{\|}}{C}-NH_2$$

Biuret

Millon's Test

Millon's reagent is a solution containing a mixture of mercuric nitrate and mercuric nitrite. Heating a protein in the presence of Millon's reagent produces a white precipitate which, if the protein contains either tyrosine or tryptophan, will upon additional heating turn brick red.

Xanthoproteic Reaction

Concentrated nitric acid reacts with protein to produce a yellow precipitate, which accounts for the yellow stains where nitric acid has contacted the skin. The addition of excess NaOH produces an orange color when an aromatic structure is present, as in tyrosine, tryptophan, and phenylalanine.

Hopkins-Cole Test

In this test the protein is mixed with a solution of glyoxylic acid and then carefully poured over concentrated sulfuric acid, being careful that they do not mix. If an indole group is present, as in tyrosine, a purple color appears at the interface.

Ninhydrin Reaction

This is a quantitative reaction that is specific for α-amino acids. The intensity of the color is directly proportional to the concentration of amino acid. Heating an amino acid in ninhydrin reagent (triketohydrindene hydrate) produces free ammonia, which forms a purple complex with the ninhydrin.

Test for Sulfur-Containing Amino Acids

The disulfide and sulfhydryl groups can be converted into inorganic sulfide by boiling in the presence of an alkali. The addition of lead(II) acetate precipitates black lead(II) sulfide.

14.8 DIGESTION AND ABSORPTION OF PROTEINS

The digestion of proteins begins in the stomach. As food is swallowed, a hormone called *gastrin* is released. When gastrin stimulates the cells of the stomach lining (gastric mucosa), the parietal cells secrete hydrochloric acid, and the chief cells produce a zymogen called *pepsinogen*. A zymogen is a protein converted into an active enzyme by a specific change in its structure. Pepsinogen is converted into the enzyme *pepsin* by reaction with the hydrochloric acid. A portion of the pepsinogen molecule is removed by acid hydrolysis, which apparently exposes the active site of the enzyme.

This reaction deserves comment for two reasons: it illustrates the mechanism by which a number of zymogens are converted into active enzymes, and, since the enzymes are produced in an inactive form, it helps to explain how cells can produce digestive enzymes without themselves being digested. The discovery that other cells in the lining of the digestive tract produce a viscous secretion that resists digestion explains how the cells are protected from the activated enzymes.

Pepsin is an endopeptidase that appears to preferentially attack peptide bonds involving the amino group of either phenylalanine or tyrosine. Hydrolysis of a protein by pepsin produces a mixture of polypeptides with molecular weights between 600 and 3,000. This mixture is called *peptone.* Peptone is a common ingredient in media for the growth of bacteria, since most bacteria thrive on amino acids and polypeptides.

The contents of the stomach are released through the pyloric valve into the duodenum of the small intestine, where it stimulates the secretion of *enterokinase* and *aminopeptidase,* and at the same time triggers the release of the hormone *secretin* into the blood. Under the influence of secretin the pancreas adds *trypsinogen, chymotrypsinogen,* and *procarboxypeptidase* to the intestinal contents, which are now alkaline.

The conversion of trypsinogen into *trypsin* is accomplished by enterokinase in the following reaction:

$$\text{Trypsinogen} \xrightarrow{\text{enterokinase}} \text{Trypsin} + \text{Octapeptide}$$

As in the case of pepsin, the active site appears to be exposed as a result of the removal of a portion of the polypeptide chain. Trypsin completes the zymogen conversions by catalyzing the following reactions:

$$\text{Chymotrypsinogen} \xrightarrow{\text{trypsin}} \text{Chymotrypsin}$$

and

$$\text{Procarboxypeptidase} \xrightarrow{\text{trypsin}} \text{Carboxypeptidase}$$

Trypsin is an endopeptidase which acts specifically on peptide bonds involving the carboxyl group of either lysine or arginine. Chymotrypsin is also an endopeptidase, but it preferentially attacks peptide bonds on the carboxyl side of either tyrosine or phenylalanine. Both carboxypeptidase and aminopeptidase are exopeptidases and attack peptide bonds involving the C-terminal and N-terminal amino acids, respectively. Figure 14.8 shows the action of the proteolytic enzymes on a hypothetical polypeptide.

Although both pepsin and trypsin attack specific sites in a protein molecule, they both exhibit general proteolytic action. This is accounted for by the fact that almost all proteins contain significant quantities of the amino acids whose bonds these enzymes specifically attack.

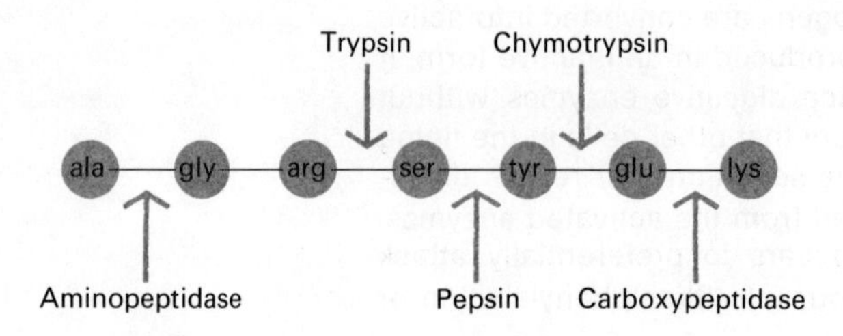

Figure 14.8
The action of the proteolytic enzymes on a hypothetical polypeptide. Aminopeptidase splits off the C-terminal amino acid, carboxypeptidase attacks the bond of the N-terminal amino acid, and the other three enzymes act on specific internal peptide linkages.

At the completion of the digestive process, the protein has been completely hydrolyzed. The amino acids are absorbed through the villi of the small intestine into the portal vein, which transports them to the liver.

14.9 METABOLISM OF AMINO ACIDS

Since it is not the proteins themselves, but rather the products of their digestion which are absorbed into the body, the metabolic fate of a protein becomes the fate of its amino acids. The manner in which an amino acid is used depends upon such factors as the current protein requirements of the body, availability of other amino acids, and the structure of the amino acid in question. Figure 14.9 diagrams the major pathways for the use of amino acids which are discussed in the following section.

Protein Synthesis

An organism must continuously synthesize new protein because its own proteins are constantly being degraded and replaced. The replacement rate varies with the location of the protein—the rate being highest for those in the serum and lowest for structural proteins. The cells maintain pools of amino acids that are used in this synthesis, and these pools are replenished primarily from proteins which have been digested.

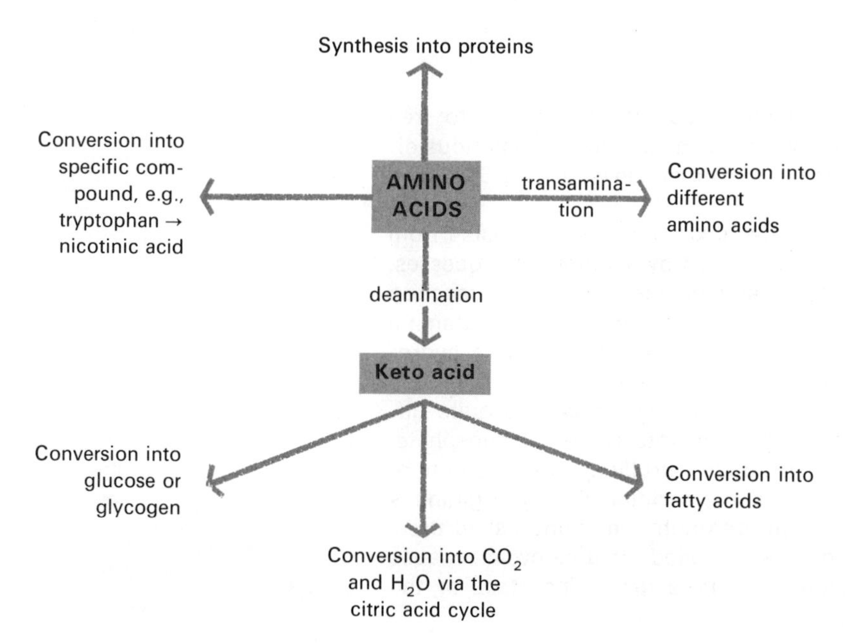

Figure 14.9
Metabolic pathways of amino acids.

Transamination

This reaction was described in Section 14.3. Transamination provides a pathway for the conversion of α-keto acids into amino acids. Since the metabolism of carbohydrates is the major source of keto acids, transamination makes possible the interconversion of proteins and carbohydrates. More than half of the amino acids are glycogenic, i.e., capable of being converted into carbohydrate.

Conversion into a Specific Compound

Although the pathways are often long and complicated, a number of amino acids can be transformed into specific compounds. Tyrosine, for example, can be converted into melanin—one of the major pigments of the skin. Tryptophan becomes the vitamin niacin, which is found in the coenzymes NAD and NADP, while cysteine is changed into taurine, which combines with cholic acid to produce taurocholic acid, a bile salt.

Deamination

The removal of the amino group produces an α-keto acid which may be utilized in reverse glycolysis to generate either glucose or glycogen. The keto acids may also be degraded via the Krebs cycle to CO_2 and H_2O with the energy being trapped as ATP. A third alternative is the conversion of the keto acids into saturated fatty acids. This reaction was discussed in detail in Chapter 13.

14.10 THE UREA CYCLE

In a deamination reaction the amino group is converted into free ammonia. The accumulation of ammonia has generally deleterious effects on most organisms. In man it is toxic at a level of only 5 mg% in blood.

The removal of ammonia and other toxic nitrogenous wastes from the body fluids of animals is accomplished by a variety of processes. In fish and most aquatic animals the ammonia is converted by reaction with glutamic acid into the less toxic glutamine. When glutamine comes into contact with the gills or similar membranes, it is hydrolyzed, freeing the ammonia which diffuses outward.

In humans and most terrestrial creatures the process is quite different. Ammonia is converted by the liver into carbamyl phosphate, which reacts with ornithine to produce citrulline. The addition of another NH_2 group changes citrulline into arginine. Finally, arginine is hydrolyzed into ornithine and a nitrogen-rich compound called urea, which is excreted. The entire process is called the urea cycle because the ornithine is recycled to produce more urea. The steps of the process are summarized below:

Step 1. $$CO_2 + H_2O + NH_3 \xrightarrow[\text{carbamyl phosphate synthetase}]{2\ ATP \qquad 2\ ADP} \begin{array}{c} NH_2 \\ | \\ C{=}O \\ | \\ O \\ | \\ HO{-}P{=}O \\ | \\ OH \end{array}$$

carbamyl phosphate

Step 2.

$$\begin{array}{c} NH_2 \\ | \\ C{=}O \\ | \\ OPO_3H_2 \end{array} + \begin{array}{c} NH_2 \\ | \\ (CH_2)_3 \\ | \\ H{-}C{-}NH_2 \\ | \\ COOH \end{array} \longrightarrow H_3PO_4 + \begin{array}{c} NH_2 \\ | \\ C{=}O \\ | \\ N\ \ H \\ | \\ (CH_2)_3 \\ | \\ H{-}C{-}NH_2 \\ | \\ COOH \end{array}$$

Carbamyl phosphate + Ornithine ⟶ Phosphoric acid + Citrulline

Step 3. Citrulline + Aspartic acid ⟶ [Arginosuccinate]
⟶ Arginine + Fumaric acid

Step 4.

$$\begin{array}{c} NH_2 \\ | \\ C{=}NH \\ | \\ N{-}H \\ | \\ (CH_2)_3 \\ | \\ H{-}C{-}NH_2 \\ | \\ COOH \end{array} + H_2O \longrightarrow \begin{array}{c} NH_2 \\ | \\ (CH_2)_3 \\ | \\ H{-}C{-}NH_2 \\ | \\ COOH \end{array} + \begin{array}{c} NH_2 \\ | \\ C{=}O \\ | \\ NH_2 \end{array}$$

Arginine + Water ⟶ Ornithine + Urea

Damage to the liver, as in cirrhosis resulting from alcoholism, prevents carbamyl phosphate synthesis, permitting the accumulation of ammonia, which often causes death.

In humans ammonia may be changed into ammonium salts by reaction with phosphate or sulfate ions, but the quantity of ammonia which can be detoxified by this process is limited because the continuous loss of these ions would result in drastic pH changes.

Conservation of moisture is an important aspect of the metabolism of birds, insects, and reptiles. In these organisms urea is changed into uric acid, which is excreted along with a minute quantity of water. The formation of uric acid consumes more energy than is required in the synthesis of urea, but is compensated by the reduction in water losses.

Uric acid production in man and other primates results only from the catabolism of nucleic acids. If the uric acid accumulates, it produces a disease called gout, which is characterized by swelling of the joints, especially of the legs.

14.11 PROTEINS AND HEALTH

One of the most essential elements of good nutrition is a constant supply of protein to replace that which is being degraded. Since proteins, unlike fats and carbohydrates, cannot be stored, this means the protein must be eaten daily. The recommended allowance is 1.0 gram of protein per kilogram of body weight. For a 75-kilogram (165-lb) man this would amount to about 3 ounces. Because protein-rich foods such as meat, fish, eggs, and cheese contain as much as 70% water, a person would have to eat between 1/3 and 3/4 lb of such foods to obtain his daily requirement.

When proteins are completely burned in a calorimeter, they yield approximately 4 calories per gram. However, the body is able to extract only about 70% of this energy when it processes protein. By contrast, the burning of carbohydrates in the body is about 95% efficient. This greater inefficiency of the protein furnace has implications for persons who wish to lose weight. Replacement of a significant portion of the dietary carbohydrate by protein will allow a person to eat approximately the same weight of food but to obtain from it fewer calories. Assuming that the caloric intake has already been adjusted to the point where it sustains the body processes but does not contain excess calories, a high-protein, low-carbohydrate diet should result in weight loss.

Although most of the amino acids can be synthesized from related structures (e.g., alanine from pyruvic acid by transamination), there is strong evidence that at least eight amino acids cannot be produced by man. These essential amino acids are listed in Table 14.5. It has been suggested that the essential amino acids cannot be synthesized in sufficient quantities because their structures include rather exotic groups that are not available in the body in significant amounts.

Proteins in Common Foods

	grams
Baked beans, with pork in tomato sauce, 1 cup	13.0
Bread, white, 1 slice	2.4
Chicken, roasted, thigh or drumstick	27.4
Corn flakes, 1 cup	1.8
Cottage cheese, 1 cup	30.0
Egg, 1 medium	6.4
Hamburger, 1 lean patty, 3 × 3 × 1/4″	8.5
Ice cream, chocolate, 1/2 cup	3.1
Milk, whole, 8 oz	7.0
Sirloin steak, medium, 4 × 3 × 1/2″	20.4
Soybeans, cooked, 3/4 cup	17.6
Swiss cheese, 4 × 4 × 1/8″ slice	5.5
Tuna fish, canned, drained 3/4 cup	30.5

TABLE 14.5 The Essential Amino Acids[1]

Isoleucine	Phenylalanine
Leucine	Threonine
Lysine	Tryptophan
Methionine	Valine

[1] The amino acids listed above are essential for man. Rats require arginine and histidine in addition to those above.

Many of the dietary proteins, especially those from plant sources, are deficient in one or more essential amino acids and are called *incomplete proteins.* Among these are zein (from corn), which is low in tryptophan and lacking in lysine, and gliadin (from wheat), which also contains no lysine. When a population is dependent upon plants for their primary source of protein, deficiency diseases are widespread. One of these diseases, called Kwashiorkor (which means "the disease of the first-born"), afflicts hundreds of millions of children in Africa and Asia. The disease is so named because its symptoms appear after the child is displaced from his mother's breast by a younger sibling, and the complete protein in his mother's milk is replaced by incomplete protein in a diet that consists largely of grains such as wheat or rice.

The need to increase the world's supply of protein has provoked schemes for filtering sea water to extract the protein-rich microscopic plants and animals called *plankton.* It has also stimulated research aimed at the conversion of petroleum wastes into high-grade protein. In a process developed by French microbiologists the waste oil is used as food by single-celled organisms which are then harvested and eaten much like yeast. It is unlikely that either of these schemes will produce any significant alleviation of the mass starvation which is the legacy of an overpopulated world. It is a distressing commentary on the leaders of the world's most resourceful nations that being first in the exploration of space takes priority over being the first to reduce human suffering.

APPLICATION OF PRINCIPLES

1. Predict the hydrolysis products of the polypeptide shown below following digestion by a mixture of trypsin and chymotrypsin:

   glu-phe-val-gly-arg-pro-ala

2. Draw the structural formula for glycyltyrosylalanine.
3. Why is insulin injected into the body rather than taken orally?
4. Write an equation for the transamination of serine with pyruvic acid.
5. Draw structural formulas for phenylalanine as it would appear below, above, and at its isoelectric point.
6. Is the ability of an amino acid to counteract added acid greater below or above its isoelectric point? Explain.
7. Draw the structural formula of the keto acid that results from the deamination of threonine.
8. When proteins are metabolized for energy, the amino group is removed as NH_3 and excreted. What significance does this fact have in the amount of energy that can be obtained from the oxidation of protein as compared with energy obtained from oxidation of carbohydrates?
9. Would a mixture of corn and wheat meet the protein requirements of the body? Explain.

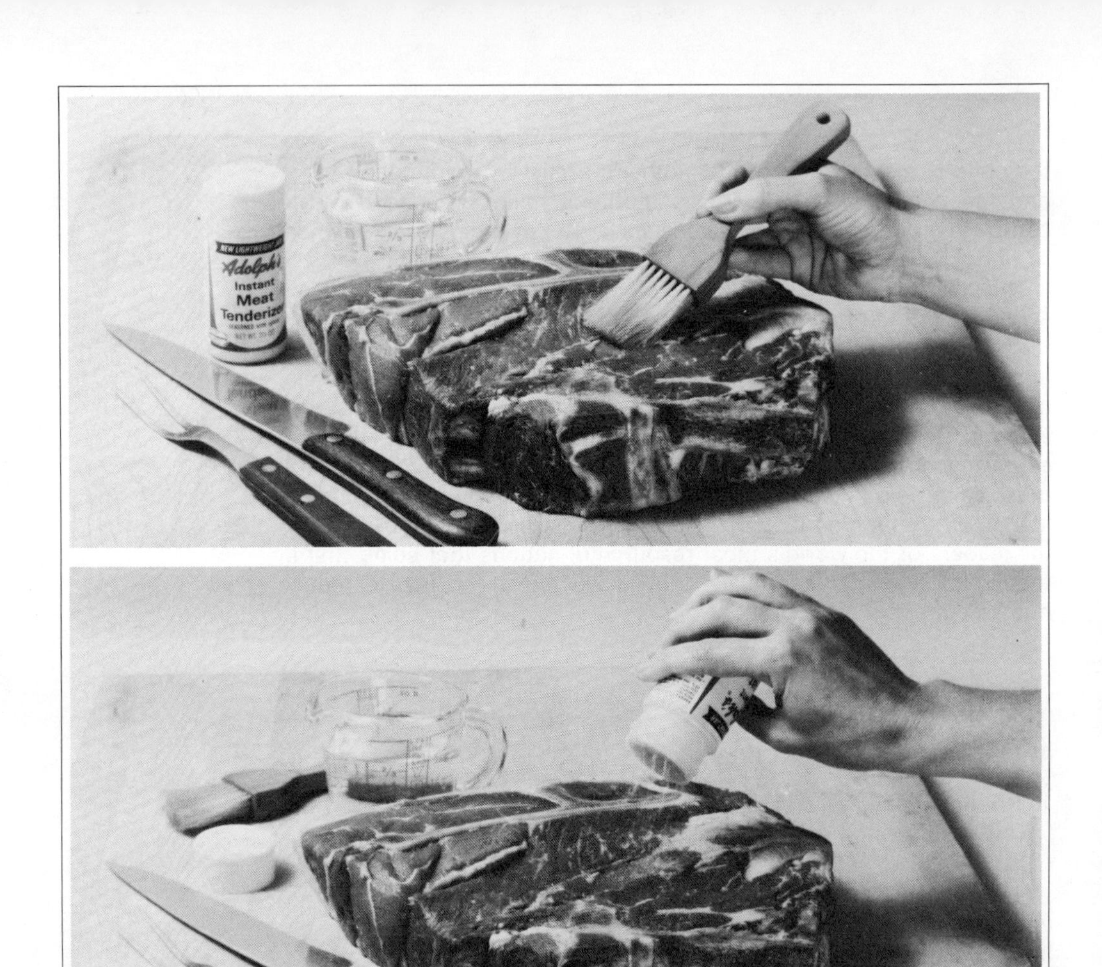

Three steps for tenderizing meat with the product called Instant Meat Tenderizer, containing a natural food enzyme, papain. (Courtesy Adolph's Ltd., Burbank, California)

15

ENZYMES

LEARNING OBJECTIVES

1. Describe the conversion of a zymogen into the active form of the enzyme.
2. Explain and give examples of the procedure for naming enzymes.
3. Describe the current model that is used to explain how an enzyme functions.
4. List the major factors that affect enzyme activity and briefly describe and explain the effect of each factor.
5. Distinguish between competitive and noncompetitive inhibition.
6. Propose a classification scheme for enzymes based on the type of reaction they catalyze.
7. Describe the applications of enzyme technology to medicine and industry.

KEY TERMS AND CONCEPTS

apoenzyme	holoenzyme
activator	induced-fit hypothesis
active site	inhibition
coenzyme	noncompetitive inhibition
competitive inhibition	substrate
enzyme	turnover number

It was pointed out in Chapter 9 that a catalyst changes the rate of reaction by lowering the activation energy, and thus the reaction temperature. For example, sucrose can be hydrolyzed by boiling in the presence of strong acid, but in living organisms the reaction is catalyzed by enzymes and proceeds at much lower temperatures. Since all of the enzymes that have been isolated and studied have proved to be proteins, the properties of enzymes can be explained using the properties of proteins we have just discussed.

15.1 CHARACTERISTICS OF ENZYMES

Enzymes may be defined as protein catalysts, although many enzymes are actually conjugated proteins. Such enzymes are called *holoenzymes,* and consist of an inactive protein, or *apoenzyme* portion, and a nonprotein prosthetic group. The prosthetic group may consist of a loosely attached organic structure called a *coenzyme,* or it may be a metal ion. The coenzyme, like the apoenzyme, is inactive by itself. Each of the coenzymes studied thus far has included a vitamin.

Zymogens

Many of the enzymes appear to be produced in an inactive form and become active only after some structural modification. The inactive forms of the enzymes are called *zymogens,* and their names usually end in *-ogen.* For example, trypsinogen is the inactive form of the digestive enzyme trypsin. Trypsinogen is activated by the enzyme enterokinase which removes a portion of the zymogen and apparently exposes the active site of the enzyme.

Substrates

The substance which is transformed by the action of an enzyme is called its *substrate.* Many enzymes act only on a specific substrate, while others affect a number of compounds. For example, the enzyme that hydrolyzes sucrose is incapable of hydrolyzing lactose, but trypsin is able to catalyze the hydrolysis of a number of proteins. The action of trypsin is actually quite specific in that it attacks the peptide bonds that include the carboxyl group of either lysine or arginine. However, since most proteins contain these amino acids, the proteolytic activity of trypsin is rather broad.

15.2 HOW ENZYMES ARE NAMED

When the first enzymes were named, the only concession to a systematic approach was to terminate the name in "*-in*" to show that it was a protein. Thus we have such names as ptyalin for an enzyme in the saliva which attacks the amylose and amylopectin forms of starch, and trypsin, which hydrolyzes polypeptides. In a later, and more orderly, procedure the name of the enzyme was derived from the name of its substrate and the suffix "*-ase*" was added. This changed the name of ptyalin to amylase and that of trypsin to peptidase.

The current practice is to name an enzyme in such a way as to indicate both the substrate and the type of reaction that the enzyme causes it to undergo. An example is phosphohexose isomerase, which converts glucose-6-phosphate into fructose-6-phosphate, and the reverse. Despite this systematic approach, many of the old names persist. The best we can do for trypsin, for example, is to call it an endopeptidase, a name which recognizes that it acts on peptide bonds in the interior of a polypeptide chain. This contrasts with an exopeptidase, which attacks the peptide bond of either the C-terminal or N-terminal amino acid.

15.3 ENZYME CLASSIFICATION

Enzymes are classified according to the type of reaction they catalyze. Six general categories have been recognized.

Hydrolases are hydrolytic enzymes and include the digestive enzymes. Among the subcategories are the carbohydrases, proteases, esterases, etc. Maltase, for example, cleaves maltose into two molecules of glucose.

Isomerases are enzymes that change a structure into one of its isomers. The isomerase of glucose-6-phosphate was described earlier.

Oxidoreductases cause oxidation-reduction reactions. Subdivisions include dehydrogenases, oxidases, catalases, etc. Lactic dehydrogenase, for example, removes two hydrogens from lactic acid to produce pyruvic acid.

Transferases bring about the transfer of a chemical group from one molecule to another. The action of transaminase was described earlier.

Lyases cause the nonhydrolytic addition or removal of a chemical group. Since covalent bonds are affected, the carbon chain may be broken or extended. Carbonic anhydrase is a lyase which combines CO_2 and H_2O to form H_2CO_3.

Ligases (or synthetases) cause the linkage of two molecules and the simultaneous disruption of a high-energy phosphate bond. RNA synthetase is involved in the synthesis of proteins.

Important Types of Oxidoreductases

(A) Dehydrogenase. Causes the removal of hydrogen atoms (electrons) from a substrate and their subsequent transfer to any acceptor other than molecular oxygen.

$$AH_2 + B \xrightarrow{\text{dehydrogenase}} A + BH_2$$

(B) Oxidase. Transfers hydrogen atoms (electrons) to molecular oxygen.

$$AH_2 + O_2 \xrightarrow{\text{oxidase}} A + H_2O$$

15.4 MECHANISM OF ENZYME ACTION

Since all proteins consist of amino acids, but only a fraction of the proteins in an organism have the ability to catalyze chemical reactions, it is logical to assume that enzymatic activity is in some way related to the arrangement of the amino acids. The arrangement of the amino acids, in turn, determines the physical shape of the protein molecule, including the types and locations of charged groups which project from the surface. Studies of enzyme-catalyzed reactions suggest that the substrate is attracted by this specific pattern of charges and forms a temporary bond with the enzyme. Since many substrate molecules are extremely small in comparison with proteins, the area where attachment takes place must involve only a small portion of the protein's surface, referred to as the *active site.* When enzymes are produced as zymogens, the active site appears to be obscured, and is uncovered by hydrolysis of a portion of the polypeptide. A conceptualized view of the bonding between enzyme and substrate is shown in Figure 15.1. Because of the apparently close correspondence between the shape of the substrate and that of the active site, chemists often use the analogy of a lock and key to describe the union of enzyme and substrate.

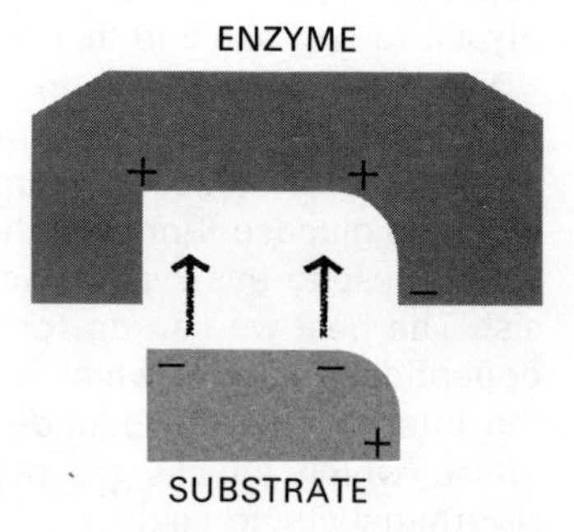

Figure 15.1
Conceptualized view of bonding between enzyme and substrate.

Uncatalyzed reactions are largely the result of random collisions between reacting species. Many of these collisions are not fruitful because the reactants are not oriented properly for reaction to take place. In a catalyzed reaction the substrate may be attracted to the active site of the enzyme where it is held in such a way as to increase the chances of reaction. The reaction of the substrate may alter its structure in such a way that it no longer exactly fits the active site, resulting in its release from the enzyme. The overall reaction can be visualized as

$$\text{Enzyme} + \text{Substrate} \longrightarrow [\text{Enzyme} - \text{Substrate complex}] \longrightarrow \text{Enzyme} + \text{Altered substrate}$$

One of the most amazing aspects of enzymatic reactions is the rate at which the enzyme-substrate complex is formed and dissolved. One way of expressing this rate is called the *turnover number* and is defined as the number of molecules of substrate affected by one molecule of enzyme in one minute at optimum conditions. Calculations have indicated that some enzymes have turnover numbers of over one million! The optimum conditions mentioned in the definition will be explained in one of the following sections.

Turnover Numbers of Representative Enzymes

Catalase	2×10^8
Glutamic dehydrogenase	3×10^4
Sucrose invertase	1×10^6

The theory of enzyme activity just outlined assumes that there is an exact complementarity between the structure of the substrate and the active site of the enzyme. It appears in some cases that an exact fit is not necessary, and that contact between enzyme and substrate alters the configuration of the active site to produce a better match. This is known as the *induced-fit hypothesis* and is diagrammed in Figure 15.2.

Many enzymatic reactions fail to take place unless a specific coenzyme is present. A number of these coenzymes may act by attaching

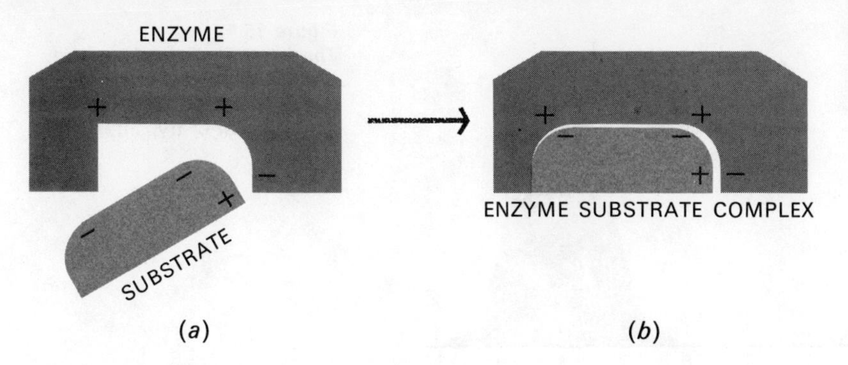

Figure 15.2
The induced-fit hypothesis. (*a*) Substrate contacts enzyme. (*b*) Active site of enzyme is modified so that the fit is improved.

themselves to the peptide chain and becoming the active site of the enzyme. The coenzymes NAD and FAD, for example, have structures which can accommodate hydrogens as they are removed from substrates such as lactic acid. A hypothetical representation of the role of a coenzyme is shown in Figure 15.3.

15.5 FACTORS AFFECTING ENZYME ACTIVITY

The structure of a protein may be affected by changes in pH, temperature, solvent, or electrolyte concentration. Since enzymes are proteins having specific configurations, it follows that these same factors should have an effect upon their activity as well. Changes in solvent or electrolyte concentrations usually result in denaturation, while changes in pH or temperature have generally less drastic effects. The following discussion will consider the variable effects of pH and temperature, as well as substances which specifically activate or inhibit enzymes.

pH

Changes in pH are reflected in the charges of the R groups which are responsible for the secondary and tertiary structure, and probably for the charges on the active site. Since the exact distribution of charges which makes the active site most efficient occurs only at a specific pH, raising or lowering the pH reduces the efficiency until the active site

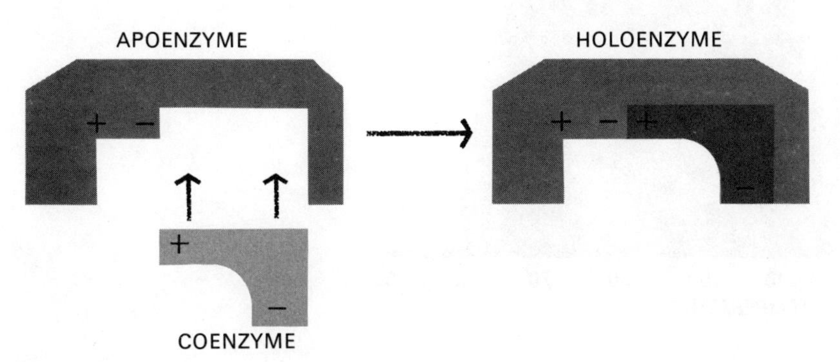

Figure 15.3
The possible role of a coenzyme. The coenzyme may become part of the functional group or active site of the enzyme.

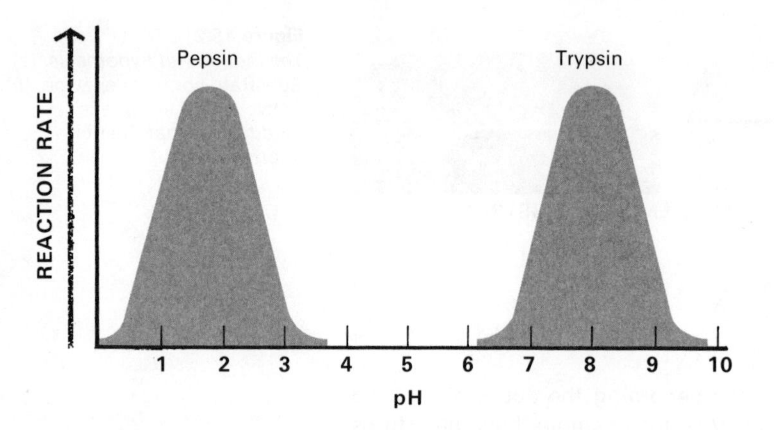

Figure 15.4
The effect of pH on the activity of two proteolytic enzymes. The optimum pH for pepsin is 1.5–2.0; that of trypsin is 8.0.

no longer is able to attract and bind the substrate. Figure 15.4 shows the effect of pH on the activity of trypsin and pepsin.

Temperature

The optimum temperature for the enzymes found in the body is about 37°C. As the temperature is increased above this point, the activity increases slightly and then abruptly falls off. The decrease in activity results from denaturation of the protein by heat. Unless the enzyme has been irreversibly denatured, which usually occurs for most enzymes at about 80°C, most of the activity can be restored by cooling (Figure 15.5).

Below the optimum temperature the activity of an enzyme gradually decreases until, at about 0°C, it is essentially inactive. Temperatures below 0°C do not result in any appreciable denaturation of the en-

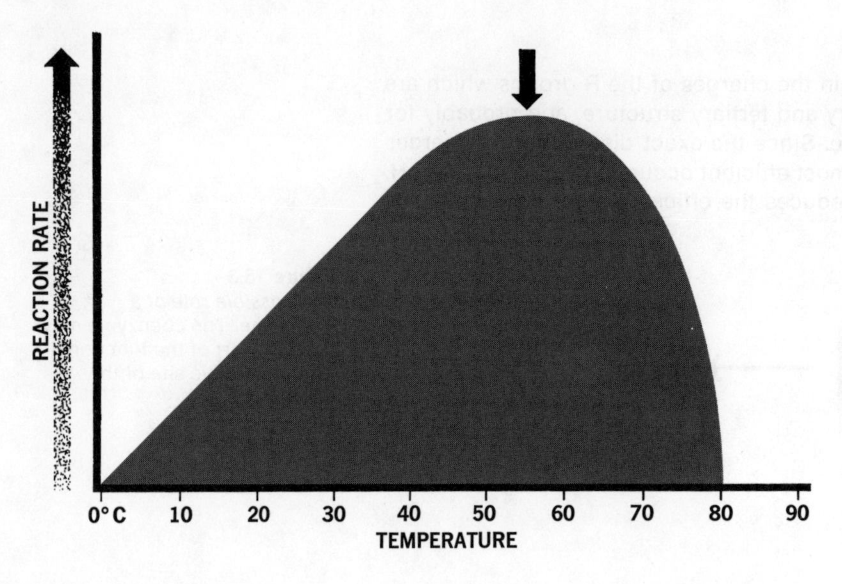

Figure 15.5
The effect of temperature on enzyme activity. The arrow indicates the temperature at which denaturation becomes accelerated.

zymes of most organisms, producing instead a kind of reversible state of suspended animation. This fact permits foods to be preserved against bacterial spoilage by freezing.

Activators

A number of enzymes function only in the presence of specific inorganic ions, called *activators.* Salivary amylase requires the chloride ion, while rennin will not curdle milk unless calcium ions are present.

Inhibitors

Substances which reduce or eliminate the activity of enzymes are called *inhibitors.* The cyanide ion, for example, is poisonous because it interferes with respiratory enzymes such as cytochrome oxidase, precipitating the iron which is an essential part of the enzyme. Inhibitors may also work by preventing the coenzyme from joining with the apoenzyme, or by occupying the active site so that the enzyme cannot bind to the substrate. Blockage of the enzyme may be either competitive or noncompetitive.

Competitive inhibition occurs when a molecule which resembles the substrate, but cannot act like it, becomes attached to the active site. This type of inhibition can be recognized because it is only temporary, and increasing the concentration of the normal substrate displaces the inhibitor and restores enzyme activity. The action of sulfa drugs on bacteria is a classic example of competitive inhibition.

Some types of disease-producing bacteria require folic acid for growth, and synthesize it by a pathway which begins with *para*-aminobenzoic acid (PABA). Sulfanilamide, on the other hand, has a structure similar to PABA but cannot be used for production of folic acid:

H—N—H / COOH — *para*-aminobenzoic acid

H—N—H / SO_2NH_2 — Sulfanilamide

The enzyme of the bacteria is unable to distinguish between the two compounds and binds readily to both. The introduction of large amounts of sulfanilamide into the body causes most of the active sites to bind to sulfanilamide, instead of PABA, interfering with folic acid synthesis and preventing bacterial growth while the body defenses work to overcome the infection.

In *noncompetitive inhibition* the inhibitor bonds more strongly to the active site, and increasing the substrate concentration has little effect. Many noncompetitive inhibitors are toxic to living organisms

Some of the More Effective Sulfa Drugs

Sulfanilamide

H_2N—(benzene ring)—SO_2—NH_2

Sulfathiazole

H_2N—(benzene ring)—SO_2—N(H)—C(S—CH=CH—N)

Sulfapyridine

H_2N—(benzene ring)—SO_2—N(H)—(pyridine ring, N)

Sulfaguanidine

H_2N—(benzene ring)—SO_2—N(H)—C(=N—H)—NH_2

Sulfadiazine

H_2N—(benzene ring)—SO_2—N(H)—(pyrimidine ring, N, N)

because they interfere with the normal transmission of nerve impulses, which involves the enzyme cholinesterase. When a nerve impulse reaches a junction between two nerve cells, a substance called acetylcholine is produced by the cell which is carrying the impulse. Acetylcholine diffuses across the junction and stimulates the receptor protein of the next cell so that the impulse is continued. The acetylcholine, having done its job, is destroyed by the enzyme cholinesterase so that the cell is once again ready to receive a nerve impulse. The toxin of *Clostridium botulinum* acts as an inhibitor to block the synthesis of acetylcholine, thus effectively blocking transmission of nerve impulses. One symptom of botulism poisoning is paralysis. The insecticide parathion is one of a class of toxic substances called "nerve gases" because they inhibit the functioning of cholinesterase by blocking its active site. Curare, procaine, and a number of other anaesthetics also interfere with nerve impulses by blocking the receptor protein.

15.6 MEDICAL AND INDUSTRIAL APPLICATIONS OF ENZYMES

The discovery that specific enzymes can be used as powerful chemical tools is not a recent one. For centuries the South Sea islanders have tenderized fish and meats by wrapping them in papaya leaves while they were being cooked. Papaya leaves contain a proteolytic enzyme, papain, that tenderizes the food by partial hydrolysis of connective tissue. Within recent years the enzyme has been extracted and purified, and it is now conveniently available to cooks in powdered form.

During the late 1960s powerful proteolytic and lipolytic enzymes extracted from fungi and bacteria became part of the formulation of many laundry detergents. Since clothing is often soiled with blood (protein), grease (lipid), and food (possibly both), it was reasoned that the enzymes would digest these materials and enable the detergent to wash them away. Because of the regional variation in the pH and solutes in water, and the numerous cases of allergic reactions by users of these enzyme preparations, most detergents that contain enzymes have been withdrawn from the market.

Enzymes are also replacing caustic solutions for removing hair from animal hides before the tanning process.

In addition to their industrial importance, enzymes have become invaluable in the field of diagnostic medicine. Under normal conditions cellular enzymes are seldom found in measurable concentrations in the blood serum, cerebrospinal fluid, or urine. However, it has been found that a number of specific enzymes are released into these fluids as a result of a disease condition that causes tissue damage. Sensitive assays have been developed to measure the concentrations of several enzymes whose appearance in the body fluids is related to specific pathological conditions. Among these are SGOT (serum glutamic-oxaloacetic transaminase), SGPT (serum glutamic-pyruvic transaminase), acid phosphatase, and pancreatic amylase. SGOT cata-

Defective Enzymes and Diseases

A number of diseases appear to be caused by metabolic errors. Such a disease is *phenylketonuria,* which is characterized by mental retardation, and results from a deficiency of the enzyme phenylalanine hydroxylase. This enzyme assists in the conversion of phenylalanine to tyrosine. In its absence phenylalanine is changed into phenylpyruvic acid—a phenylketone. Phenylpyruvic acid accumulates in the body and eventually spills over into the urine, hence the name of the disease. If this condition is detected at birth and the infant fed a diet low in phenylalanine for about the first year, there is a high probability that the child will be normal.

lyzes the transfer of an amino group from glutamic acid to oxaloacetic acid. Its appearance in the blood has been correlated with damage to the heart muscle resulting from a blockage of the coronary artery. An increase in the level of SGPT has been linked with infectious hepatitis. This indicator is so sensitive that diagnosis is possible even before clinical symptoms appear. Acid phosphatase is increased in cancer of the prostate and in certain types of bone diseases. (Acid phosphatase catalyzes the hydrolysis of phosphoric acid esters at low pH.) Although traces of amylase are normally present in the blood, the level rises significantly in disease of the pancreas.

APPLICATION OF PRINCIPLES

1. What name would you suggest for an enzyme that catalyzes the removal of hydrogen atoms from lactic acid?
2. Enzymes that have been inactivated by heating often regain their activity upon cooling. Would you expect an enzyme that has been inactivated by heating in the presence of dilute hydrochloric acid to regain its activity after it has been cooled and the acid removed? Justify your answer.
3. Carbon dioxide is able to attach to the hemoglobin molecule and prevent it from transporting oxygen. The carbon monoxide can be displaced by high concentrations of oxygen. Is this an example of competitive or noncompetitive inhibition?
4. Meats that contain large amounts of connective tissue can be tenderized by the action of proteolytic enzymes. Is there a danger that these enzymes could harm the stomach lining when the food is eaten? Explain.
5. The turnover number of catalase, the enzyme that assists the breakdown of hydrogen peroxide into oxygen and water, is 2×10^8. What weight of hydrogen peroxide could be decomposed in one hour by one molecule of catalase?
6. Since each enzyme acts on a specific substrate, how can the general proteolytic activity of trypsin be explained?
7. If you were going to isolate and purify a proteolytic enzyme that you would then use in detergents, would you attempt to extract one from the digestive tract of a steer, or from a culture of mold? Explain.

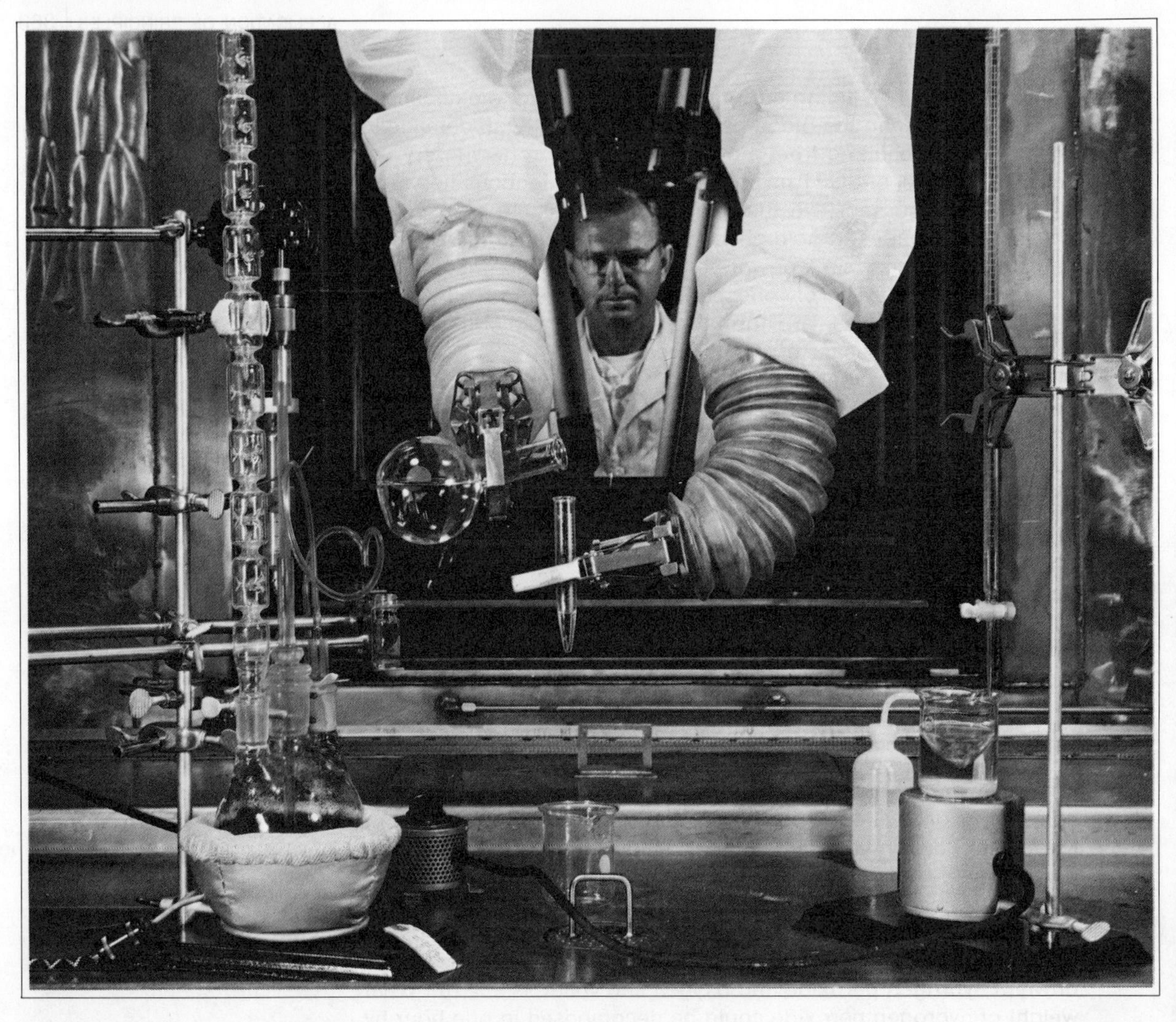

A technician uses a pair of mechanical hands to handle a radioactive liquid. (Courtesy Oak Ridge National Laboratory, Oak Ridge, Tennessee)

16

RADIATION CHEMISTRY

LEARNING OBJECTIVES

1. Compare ordinary chemical changes with those taking place during nuclear decay.
2. Describe several devices commonly used for detection or measurement of radiation and briefly explain how each one works.
3. Describe several ways in which radiation is used for man's benefit.
4. Describe the process of nuclear fission.
5. Describe the acute radiation syndrome.
6. Explain briefly how the daughter element differs from the parent in alpha, beta, and gamma decay.
7. Draw a simple schematic diagram of a nuclear reactor.
8. Describe the interaction of ionizing radiation with matter.
9. Explain what steps can be taken to reduce the risk of damage from radiation exposure.

KEY TERMS AND CONCEPTS

acute radiation syndrome
curie
daughter nucleus
decay
fission
half-life
ionizing radiation
LD_{50}
rad
roentgen
tenth-value thickness
tracer
transmutation

A number of unique and important changes take place within substances as they either absorb or emit high-energy radiation. These changes range from subtle (an almost undetectable rise in temperature) to spectacular (the explosion of an atomic bomb). Man's use of radiation constitutes a paradox in that it may lead to the advancement of civilization or to the destruction of it.

The high-energy radiations involved in these changes may be short waves from the electromagnetic radiation spectrum (see Figure 3.11), streams of particles ejected from the nuclei of radioactive atoms, or charged bits of matter which have been accelerated to high velocities by devices commonly called "atom smashers." The characteristics and involvement of each of these kinds of radiation will be discussed in the following sections.

16.1 ELECTROMAGNETIC RADIATION

Ultraviolet Rays

Approximately 9% of the energy radiated by the sun falls within the 40-4000 Å region and is classified as ultraviolet light. However, oxygen molecules in the earth's atmosphere absorb most of the radiation shorter than 3000 Å and are converted into reactive ozone molecules by the reaction

$$3\ O_{2\,(g)} \xrightarrow{\text{u.v.}} 2\ O_{3\,(g)}$$

Ozone is a poisonous, colorless gas with a distinctive pungent odor and is a powerful oxidizing agent. In the presence of nitrogen dioxide (which is present in most polluted air samples), large amounts of ozone are also created by the following reactions:

$$NO_{2\,(g)} \xrightarrow{\text{u.v.}} NO_{(g)} + O_{(\text{free radical})}$$

$$O_{2\,(g)} + O \longrightarrow O_{3\,(g)}$$

Ozone is irritating to the mucous membranes and is damaging to plants, rubber, and textiles. Ozone is present in both photochemical smog and other types of polluted air.

Radiation in the 2900-3050-Å range is absorbed by tissues and is responsible for tanning the skin and for sunburn, while that in the region of 2660 Å is particularly damaging to nucleic acids. Ultraviolet rays of this wavelength are used to destroy surface bacteria—especially on plastics, rubber, and other materials that would be damaged by heating or by chemical disinfectants.

The retina of the eye is especially sensitive to ultraviolet radiation. For this reason one should never look directly at sunlamps, sterilizing lamps, or the so-called black light devices used with fluorescent posters and signs.

X-Rays

The velocity of x-rays is the same as that of visible light, yet they pass readily through most materials. Although x-rays are generated by our sun and by distant stars, they are almost completely absorbed by the earth's atmosphere. It is not surprising, then, that experiments with the kinds of discharge tubes that produce x-rays had been carried out by scores of other scientists before Wilhelm Röntgen discovered their presence by accident.

Modern x-ray generators do not resemble Röntgen's primitive cathode ray tube in appearance, yet there is a basic similarity. The cathode is a tungsten filament in an evacuated glass tube. The filament is electrically heated and emits electrons which are attracted toward a tungsten anode by the difference in charge and the potential difference. As the electrons strike the anode, some of the energy is transformed into x-rays, the remainder being converted into heat which must be dissipated. In older tubes the anode is embedded in a heavy piece of copper that conducts the heat away (Figure 16.1), but modern generators employ a spinning anode (Figure 16.2) which

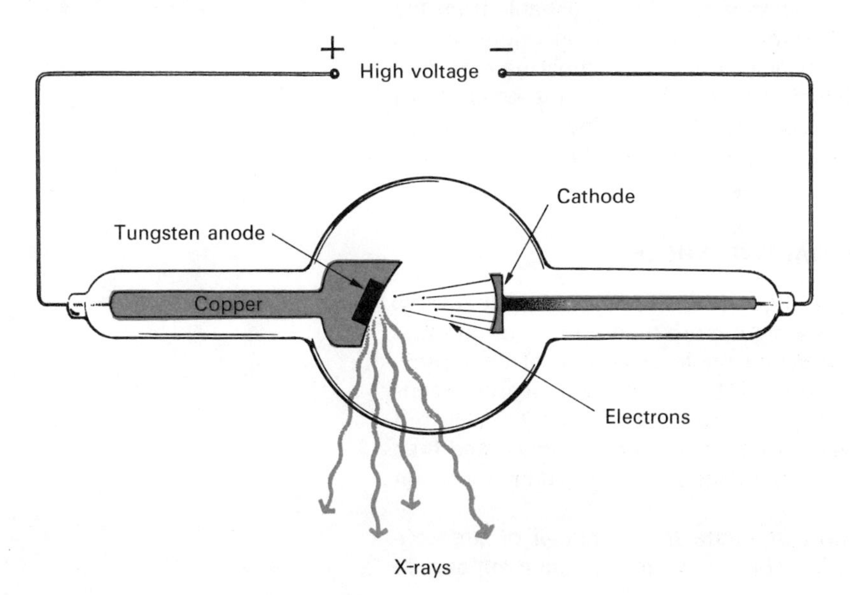

Figure 16.1
An early model x-ray tube. Electrons from the heated cathode strike the tungsten anode, and radiation in the x-ray region is produced. Heat is dissipated by the copper.

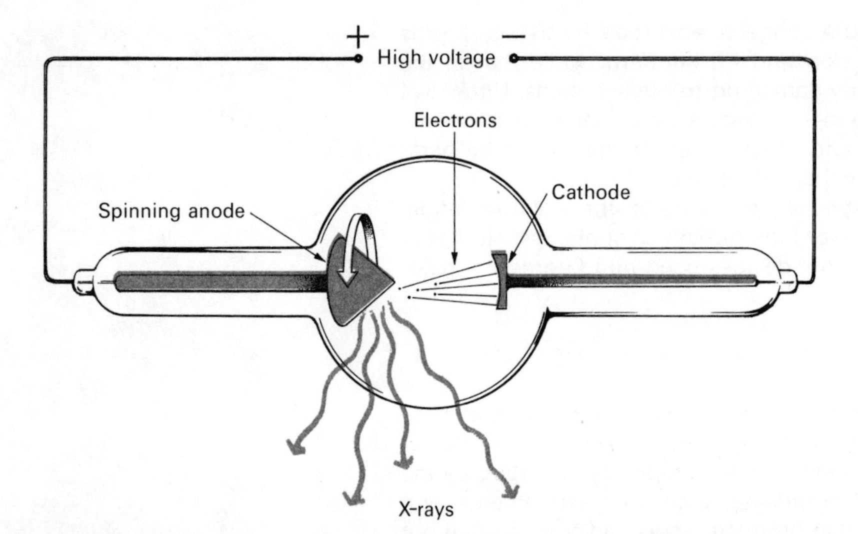

Figure 16.2
A spinning-anode x-ray tube. This type of tube can be operated at high voltages since the anode surface is constantly turning and heat build-up is avoided.

allows the tube to be operated at a higher potential difference. The higher potential difference, in turn, produces x-rays having greater energy. For example, the radiation from a 5-kV (5000 volts) machine has a wavelength of approximately 3 Å, while the 0.01 Å waves from a 1000-kV generator have several hundred times as much energy. Because of their higher energy and greater penetrating ability, the extremely short x-rays are referred to as "hard" radiation.

Gamma Rays

The overlap of adjacent regions of the electromagnetic spectrum means that the longest gamma radiation is indistinguishable from the shortest x-rays. There is a difference, however, in their origins: while x-rays are normally generated in discharge tubes, most gamma rays originate in the nuclei of radioactive atoms. Because the emission of gamma radiation is usually preceded or accompanied by the ejection of a subatomic particle from the unstable nucleus, the two events will be considered together in the following section.

16.2 EMISSIONS FROM RADIOACTIVE NUCLEI

During the 75 years following Becquerel's discovery of radioactivity, scientists have observed and cataloged more than a score of rays and particles in connection with unstable nuclei. Because of their rarity and complexity, most of them are of interest only to research scientists. However, alpha, beta, and gamma rays described by Rutherford in 1899 are most often involved in radiation-related changes and thus deserve detailed study. Their characteristics are summarized in Table 16.1.

Alpha, beta, and gamma rays originate in the nuclei of unstable atoms. The exact cause of this instability is unknown, but it appears to

TABLE 16.1 Characteristics of Alpha, Beta, and Gamma Radiations

	Symbols	Charge	Relative Mass (amu)	Velocity Range (cm/sec)	Penetration (cm in air)	Identity or Nature
Alpha	α $^4_2He^{2+}$	+2	4.00260	1.6×10^9 to 3.2×10^9	2.5–8.5	Helium nucleus or ion
Beta	β $_{-1}^{0}e$	−1	0.00055	9.6×10^9 to 2.6×10^{10}	30–3000	Electron
Gamma	γ hν	0	0	3×10^{10}	>3000	Electro-magnetic radiation

be related to the ratio between neutrons and protons in the atom and to the arrangement of the protons in the nucleus. In general, it can be said that the greatest stability is associated with a neutron/proton ratio of 1:1 (as in $^{12}_{6}C$ and $^{16}_{8}O$), and that a higher ratio increases the likelihood of instability, as shown by the fact that all of the elements with a n/p ratio greater than 1.54:1 are radioactive. Unusual stability is also associated with filled nuclear shells (not the same as electron shells), and atoms with 2, 8, 20, 50, and 82 protons are especially stable.

A nucleus which emits either an alpha or beta particle is changed into a new or "daughter" nucleus. The original nucleus is spoken of as the "parent," and the emission process is called *decay* or *disintegration.*

Alpha Particle Emission

In the case of alpha decay, the mass number of the daughter nucleus is four units, and the atomic number two units less than in the parent nucleus. The overall change that accompanies alpha decay is shown by the equation

$$^{\text{Mass number}}_{\text{Atomic number}}\text{Symbol of parent element} \longrightarrow {}^4_2He^{2+} + {}^{\text{Mass number-4}}_{\text{Atomic number-2}}\text{Symbol of daughter element}$$

and is illustrated by the following examples:

$$^{222}_{86}Ra \longrightarrow {}^4_2He^{2+} + {}^{218}_{84}Po$$

$$^{226}_{88}Ra \longrightarrow {}^4_2He^{2+} + {}^{222}_{86}Rn$$

Note that nuclear reactions can be represented by equations similar to those used for chemical reactions. The mass and atomic numbers are shown for each reactant and product, and a nuclear equation is balanced when the sum of the mass numbers of the reactants equals the sum of the mass numbers of the products, and the sum of the atomic numbers on the left equals the sum of the atomic numbers on the right.

Radon-222 and radium-226 do not decay at the same rate although both decay by alpha emission. The rate of disintegration for a given isotope is proportional to the number of atoms of that isotope that are present. A constant fraction of the radioactive atoms disintegrate during a given time period—thus the number of unstable atoms decreases very rapidly at first and then more slowly with the passage of time. Since the rate of decay is constantly changing, it is more meaningful to describe the rate of change of a radioisotope in terms of its half-life. A *half-life* is defined as the time required for the decay of half of the radioactive atoms present in a sample. The half-life of radon-222 is 3.82 days, while that of radium-226 is 1620 years.

Beta Particle Emission

Beta particle emission is slightly more complex. A beta particle is indistinguishable from an electron, yet it originates in the nucleus. An essentially true explanation of a beta particle is that it is combined with a proton and masquerades as a neutron. The emission of a beta particle may be thought of as occurring in two steps: first, the separation of the neutron into a proton and an electron, and second, the ejection of the electron. Thus beta decay produces a daughter nucleus whose atomic number is greater than that of its parent by one unit, but whose mass number is unchanged. The process is summarized by the equation

$$_{\text{Atomic number}}^{\text{Mass number}}\text{Symbol of parent element} \longrightarrow \beta + {}_{\text{Atomic number}+1}^{\text{Mass number}}\text{Symbol of daughter element}$$

Some examples of beta decay are

$$^{14}_{6}C \longrightarrow {}^{0}_{-1}e + {}^{14}_{7}N$$

and

$$^{3}_{1}H \longrightarrow {}^{0}_{-1}e + {}^{3}_{2}He$$

Gamma Ray Emission

Sometimes the daughter nucleus produced by the alpha or beta decay is in an excited state, much like the excited state of an orbital electron. When the nucleus adjusts to a lower energy level, it emits photons of electromagnetic energy called gamma rays. Gamma ray emission does not produce a new daughter element, and may occur simultaneously with, or following, the emission of an alpha or beta particle. Some of the more common radioactive isotopes are shown in Table 16.2.

TABLE 16.2 Radioactive Isotopes of Biological Importance

Isotope Symbol	Name	Emissions	Half-Life
$^{3}_{1}H$	Tritium	Beta	12.26 years
$^{14}_{6}C$	Carbon-14	Beta	5570 years
$^{24}_{11}Na$	Sodium-24	Beta, gamma	15.0 hours
$^{32}_{15}P$	Phosphorus-32	Beta	14.3 days
$^{42}_{19}K$	Potassium-42	Beta, gamma	12.4 hours
$^{59}_{26}Fe$	Iron-59	Beta, gamma	46 days
$^{60}_{27}Co$	Cobalt-60	Beta, gamma	5.27 years
$^{90}_{38}Sr$	Strontium-90	Beta	28 years
$^{99m}_{43}Tc$	Technicium-99^{m}	Gamma only	6.1 hours
$^{131}_{53}I$	Iodine-131	Beta, gamma	8.05 days
$^{198}_{79}Au$	Gold-198	Beta, gamma	2.69 days
$^{226}_{88}Ra$	Radium-226	Alpha, beta, and gamma	1620 years

The superscript ^{m} after the mass number indicates an excited state of this isotope.

16.3 CHEMICAL VERSUS NUCLEAR CHANGE

It was pointed out earlier that the rate of nuclear decay is independent of the number of radioactive atoms present. Several other differences between ordinary chemical changes and the nuclear changes that accompany radioactivity are summarized in Table 16.3.

TABLE 16.3 A Comparison of Chemical Reactions and Radioactive Decay Processes

Chemical Reaction	Radioactive Decay
1. The reaction involves orbital electrons.	1. The process involves nuclear particles.
2. New elements *are not* produced.	2. A new element *is* produced.
3. Rate of reaction is affected by temperature, concentration, catalysts, etc.	3. Rate of change is independent of external factors.
4. Reactions involve relatively small amounts of energy.	4. Extremely large amounts of energy may be produced.

16.4 RADIATION EFFECTS AND COUNTERMEASURES

The way in which matter is acted upon by radiation depends upon the type and energy of the radiation and the chemical structure of the matter. Radiation effects range from a small increase in molecular energy to extreme changes in molecular structure. These molecular

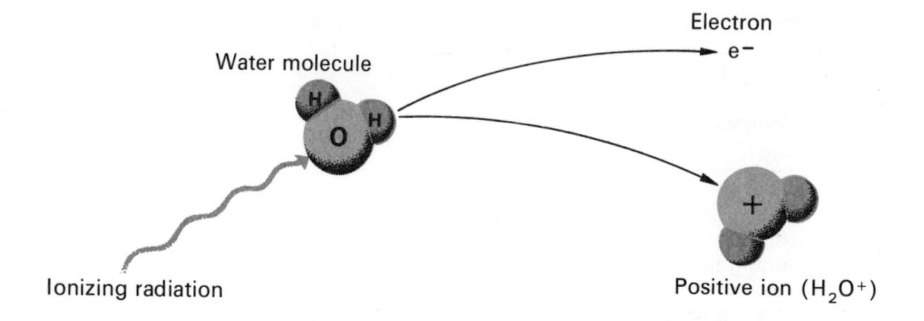

Figure 16.3
Formation of an ion pair from a water molecule. Ion pairs can be formed from atoms as well as from molecules.

rearrangements are reactions that usually involve ions or free radicals. Radiation which creates ion pairs or free radicals in its passage through matter is called *ionizing radiation.* Alpha, beta, and gamma radiation are all in this category. The formation of an ion pair is diagrammed in Figure 16.3.

Alpha and beta particles, although less penetrating than gamma rays, have considerable mass and charge and produce more ion pairs as they pass through matter. The ion pairs and free radicals formed in water are of special importance because living cells are mostly water. Studies of irradiated water have shown the presence of the free radicals H• and OH•, the ions HO^+ and HO_2^+, and relatively high concentrations of hydrogen peroxide. Many of the chemical changes that take place in irradiated materials may be due to reactions of the OH• radical (which can cause oxidation, addition at a double bond, and removal of a hydrogen atom from organic molecules), or to the strongly oxidizing hydrogen peroxide. Such reactions are capable of producing profound changes should they occur within a living cell. The metabolism of the cell may be slightly impaired, or it may be so disrupted that the cell dies. Among the irradiated survivors the mutation rate is much higher than that of untreated cells.

Tissues vary widely in their sensitivity to irradiation. Rapidly growing tissues are more susceptible to radiation damage than those that grow more slowly. Lymphocytes, epithelial cells, and the basal cells of the testis and ovary are especially sensitive, while muscle and nerve cells are extremely resistant. Tissues of the developing embryo are so sensitive that a pregnant woman should avoid all except the most necessary x-ray examinations.

The quantitative aspects of radiation chemistry are complicated by the use of several different units for the expression of radiation intensity. One unit, the curie, is not appropriate for x-rays because it indicates the rate at which radioactive atoms decay. One *curie* amounts to 37 billion disintegrations per second—the number of disintegrations observed to take place in one gram of pure radium during that time. Because this represents a very large amount of radiation, the units millicurie (37 million disintegrations) and microcurie (37 thousand) are frequently used. Since the curie is merely the number of disintegrations, without regard to either the nature or effect of the radiations, it is of questionable value as a unit of radiation intensity.

A more useful unit than the curie is the *roentgen* (r), which gives some indication of the energy of the radiation. A roentgen is the quantity of ionizing radiation required for the production of one electrostatic unit of charge (equivalent to roughly two billion ion pairs) in a cubic centimeter of air at STP.

Despite this seemingly large number, one roentgen actually represents a very small amount of energy. Since a single alpha particle produces as many as 100,000 ion pairs in air, a roentgen is roughly equivalent to the energy delivered by just 20,000 alpha particles. However, since every ion represents a potential chemical reaction, a single roentgen is still a significant amount of radiation.

Although it is a more useful unit than the curie, the roentgen also has its shortcomings: it represents the amount of radiation which is absorbed by air, but not the amount which would be absorbed by other materials. In other words, the roentgen does not take into account the fact that materials vary widely in the extent to which they absorb radiation. A different unit called the rad (radiation absorbed dose) makes allowance for these differences. A *rad* is an amount of ionizing radiation which results in the absorption of 100 ergs by each gram of absorbing material. The erg is a unit of energy so small that it takes about 40 million of them to equal one calorie. For most ionizing radiations one rad is about the same as one roentgen.

Organisms subjected to large amounts of ionizing radiation may be damaged or killed. Human beings who have received whole-body irradiation at a dose level of 100 r or greater generally display a series of symptoms known as *acute radiation syndrome,* or simply radiation sickness. These symptoms include nausea with vomiting, a reduction of circulating lymphocytes, a decrease in the total white blood cell count, diarrhea, fatigue, and easy bruising. Changes in the blood may be detected following doses as low as 50 r.

Based upon present information, approximately half of the human beings who are exposed to whole-body irradiation of 400 r would be expected to die. This measure of radiation sensitivity in which half of the treated subjects die is called the LD_{50} (50% lethal dose). Organisms vary greatly in their sensitivity to radiation, as shown in Table 16.4.

Barring unusual accidents it is unlikely that a person would be exposed to enough radiation to cause radiation sickness: the average

TABLE 16.4 Radiation Sensitivity of Various Organisms[1]

Organism	LD_{50} (roentgens)	Organism	LD_{50} (roentgens)
Rabbit	300	Goldfish	750
Man	400	Algae	4000–8500
Mouse	400	Bacteria	5000–130,000
Monkey	450	Viruses	100,000–200,000
Rat	600	Protozoa	3,000,000

[1] McGraw-Hill *Encyclopedia of Science and Technology.*

yearly dosage from such sources as fallout, cosmic rays, television sets, and medical x-rays totals less than 1 rad. Nevertheless, every precaution should be taken to reduce radiation exposure because of the possibility of mutations occurring.

Potential damage resulting from exposure to radiation can be reduced by any or all of the following measures: (1) reducing the duration of exposure, (2) the use of shielding, (3) increasing the distance from the source of radiation, and (4) ingestion of protective chemicals. The effectiveness of a shielding material in reducing the level of radiation that passes through it is expressed in terms of a unit called the tenth-value thickness. A *tenth-value thickness* is the number of inches of material that will reduce the intensity of γ-rays to one-tenth their initial level. Table 16.5 gives tenth-value thicknesses for common shielding materials.

TABLE 16.5 Tenth-Value Thicknesses of Common Shielding Materials

Material	Tenth-Value Thickness (inches)
Steel	4.5
Concrete	15
Earth	22
Water	32
Wood	60

Reduction of potential radiation damage by ingestion of protective chemicals is still largely experimental. Administration of chemicals containing the sulfhydryl group prior to exposure approximately doubles the radiation resistance of an organism. Unfortunately, the amount of chemical required to provide this protection produces toxic side effects. The best countermeasures are still time, distance, and shielding.

16.5 DETECTION AND MEASUREMENT OF RADIATION

Devices that detect or measure ionizing radiation fall roughly into three categories: photographic, electronic, and electroscopic. Despite their design differences, there is a basic similarity to all such devices: they record the ionizations produced by radiation. The following is a brief description of representative devices from the different categories.

Photographic Devices

The Film Badge

Photographic film consists of a sheet of cellulose coated with silver bromide. Radiation striking the silver bromide drives an electron from

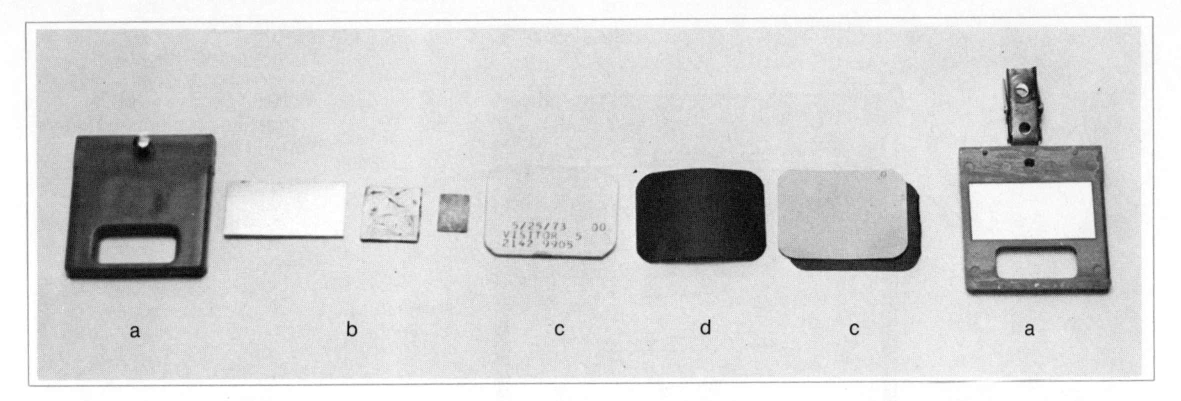

Components of a film badge: (*a*) plastic housing, (*b*) metal shields, (*c*) film wrappers, (*d*) film. (Photograph by J. Asdrubal Rivera)

a Br^- ion. This electron is captured by an Ag^+ ion, reducing it to atomic silver. During the development process, other silver atoms are deposited around the first one, producing a tiny black spot on the film. The amount of darkening is directly proportional to the intensity of the radiation.

A film badge consists of two or three layers of film of different sensitivity covered with paper and enclosed within a cadmium shield about 1 mm thick. There are several openings in the front of the shield and some means of attaching it to the clothing. The badge is worn by persons who work around radiation to provide a record of their exposure. New badges are issued at intervals and the film in the old badge is processed and evaluated.

The cadmium shield absorbs both alpha and beta particles and weak gamma rays. Beta particles and weak gammas pass through the openings while strong gammas penetrate the shield. This makes it possible to determine both the type and intensity of radiation to which the worker has been exposed during a given period.

The Cloud Chamber

The cloud chamber is a glass container saturated with alcohol vapor. The chamber has a transparent top and a black bottom and rests on a container of Dry Ice® (Figure 16.4). The atmosphere in the chamber becomes supersaturated with alcohol vapor as the chamber is cooled by the ice. Radiation passing through the chamber collides with molecules of gas and produces ions. A tiny droplet forms around each ion. The process is similar to the formation of visible trails (contrails) of condensation by an airplane operating at high altitudes.

A camera records the tracks and an analysis of the pictures provides information about the interactions of radiation with matter.

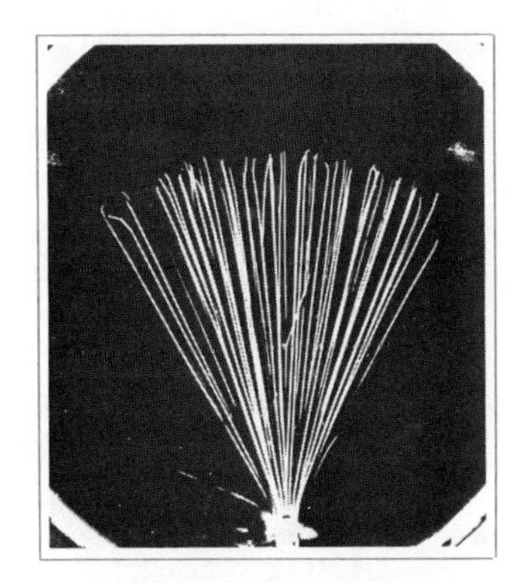

Photograph of a cloud chamber showing tracks of alpha particles. (Courtesy J. K. Boggild, Niels Bohr Institute, Copenhagen, Denmark)

Electroscopic and Electronic Devices

The Ion Chamber

One type of ion chamber, called a pocket dosimeter, is an air-filled tube about the same size and shape as a fountain pen. Running

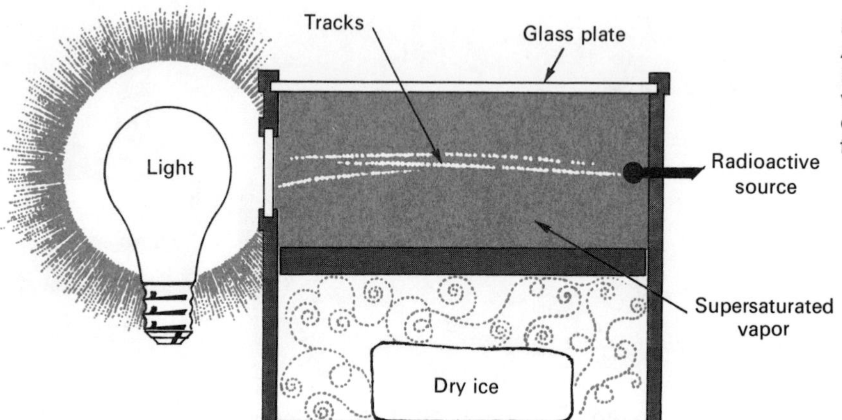

Figure 16.4
A cloud chamber. The tracks of radioactive particles become visible because water droplets condense on the ions that are formed by them.

through the center of the tube is a flexible metal-coated fiber (Figure 16.5). Both the fiber and the chamber walls are positively charged. Radiations passing through the air in the chamber produce ions and free electrons. The free electrons are attracted to the fiber, reducing its positive charge and causing it to bend toward the wall. The movement of the fiber is proportional to the radiation intensity. The position of the fiber can be seen at any time by simply holding the dosimeter to the light and looking through the viewing window. A calibrated scale makes it possible to read the radiation exposure directly.

The pocket dosimeter, unlike the film badge, provides an immediate indication of radiation exposure, but is responsive only to fairly penetrating radiations.

Geiger-Müller Counter

The heart of the Geiger-Müller counter (often called a Geiger counter) is a slender argon-filled tube whose glass walls are exceedingly thin. In the center of the tube is the anode—a wire which is positively charged—while the metal-coated inside wall acts as a cathode (Figure 16.6).

A 1500-volt potential difference is maintained between the anode and the cathode. Ionizing radiation which penetrates the wall strikes an atom of argon, freeing an electron and forming a positively charged ion of argon. The free electron is attracted by the anode and moves toward it, colliding with other argon atoms in its path, and freeing more electrons. When these electrons reach the anode, they cause a temporary reduction in its positive charge and thus changes the potential difference between it and the cathode. These changes

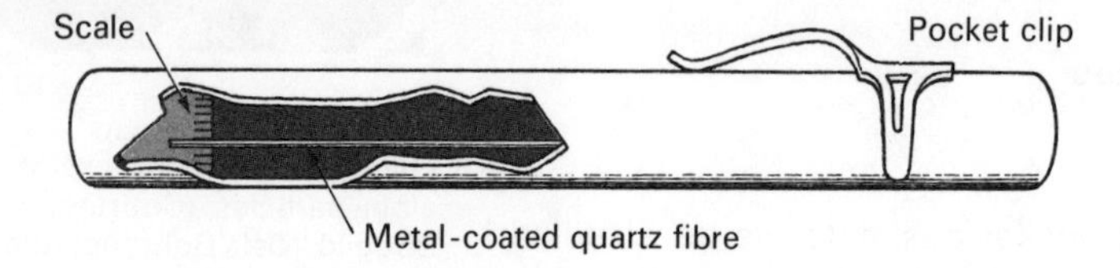

Figure 16.5
A pocket dosimeter. The deflection of the quartz fiber is proportional to the amount of ionizing radiation that has passed through the chamber.

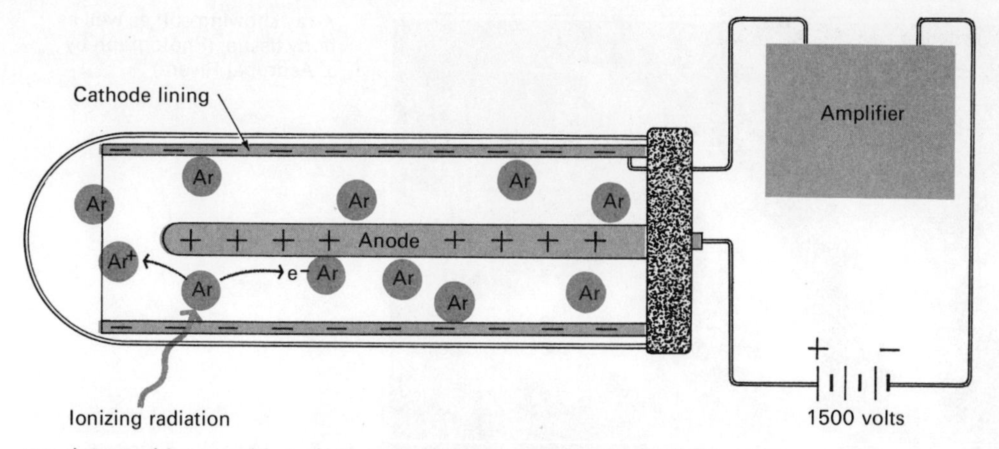

Figure 16.6
A diagram of a Geiger-Muller counter. Ionizing radiation enters the tube and dislodges an electron from an atom of argon. As this electron moves toward the anode, it collides with other argon atoms, causing them to lose electrons. The cascade of electrons striking the anode lowers the potential difference between it and the cathode.

are detected by sensitive electronic circuits and converted into a current which simultaneously drives a counter and a speaker. The speaker gives an audible indication that radiations are being received.

Since alpha particles are unable to penetrate the glass tube, Geiger counters normally do not respond to them. It is possible, however, to change the construction of the apparatus so that alpha rays can be counted.

16.6 THE USES OF RADIATIONS

It is probable that the military applications of radiation technology have had a greater impact on our lives than any other aspect of science; yet there are few people living who have not also been affected in some way by other, less spectacular, uses of radiations. Some philosophers have suggested that scientists are trying to make amends for unleashing the destructive power of radiation by discovering as many beneficial uses as possible. Be that as it may, the applications of radiation chemistry are certainly numerous, and rank among the most ingenious and imaginative of man's efforts.

X-Rays

Medical Diagnosis

Soft x-rays are absorbed to varying degrees by bone, teeth, cartilage, and other tissues, making it possible to produce a photographic record of internal structures. The portion of the body to be examined is positioned so that x-rays passing through it strike a cellulose film coated with silver bromide and create a latent image. When the film is processed, the body parts that are most opaque to the passage of x-rays appear as light areas, while the less absorptive areas are darker.

Compounds such as barium sulfate, which are almost completely opaque to x-rays, may be swallowed or administered by enema to

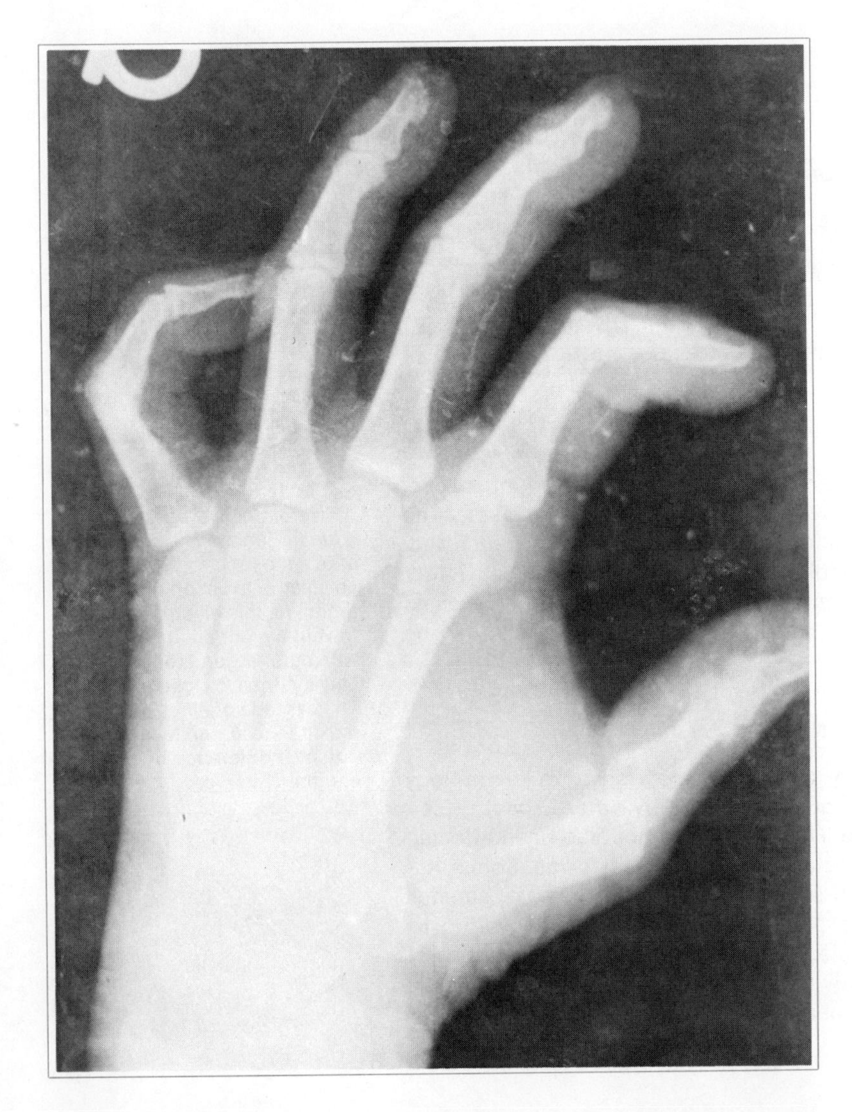

X-ray showing soft as well as bony tissue. (Photograph by J. Asdrubal Rivera)

produce additional contrast between tissues of the gastrointestinal tract and other soft tissues. Although barium ions are extremely toxic, barium sulfate is virtually insoluble in body fluids, and thus is harmless. Blood vessels may also be examined after injection of a radioopaque dye. X-rays used for diagnostic purposes usually fall within the 40-60-kV range.

While an x-ray plate provides only a stop-motion view of internal structures, a process called fluoroscopy allows continuous x-ray examination. The x-rays which emerge from the body strike a sheet of thick glass coated with a fluorescent compound, producing a glowing image of the body parts. The opposite side of the glass is coated with an x-ray-absorbing substance to protect the viewer.

Therapy

Very soft x-rays (~5 kV) have been used to treat skin conditions such as acne, keloid scars, eczema, and plantar warts. Hard x-rays, with their greater penetrating ability, are employed in the treatment of deep, and often inoperable, tumors. In such treatments the point of entry of the x-ray beam is constantly varied in order to minimize damage to the overlying tissue. Cervical cancer and Hodgkin's disease respond rather well to irradiation, although the remission in the latter case is not permanent.

Prevention of Transplant Rejection

Within the last decade skilled surgeons have astounded the world by transplanting organs from one person to another. In most cases the transplanted organ eventually ceases to function properly because it is attacked by chemicals called antibodies. Antibodies are produced by the body as a defense mechanism against foreign materials. The rejection can be postponed if the organs responsible for antibody production are subjected to continued large doses of x-radiation. The patient is now placed in a critical position: if the irradiation treatments do not continue, the transplant will be rejected; and if they do, he may either suffer from radiation sickness or die from some minor bacterial

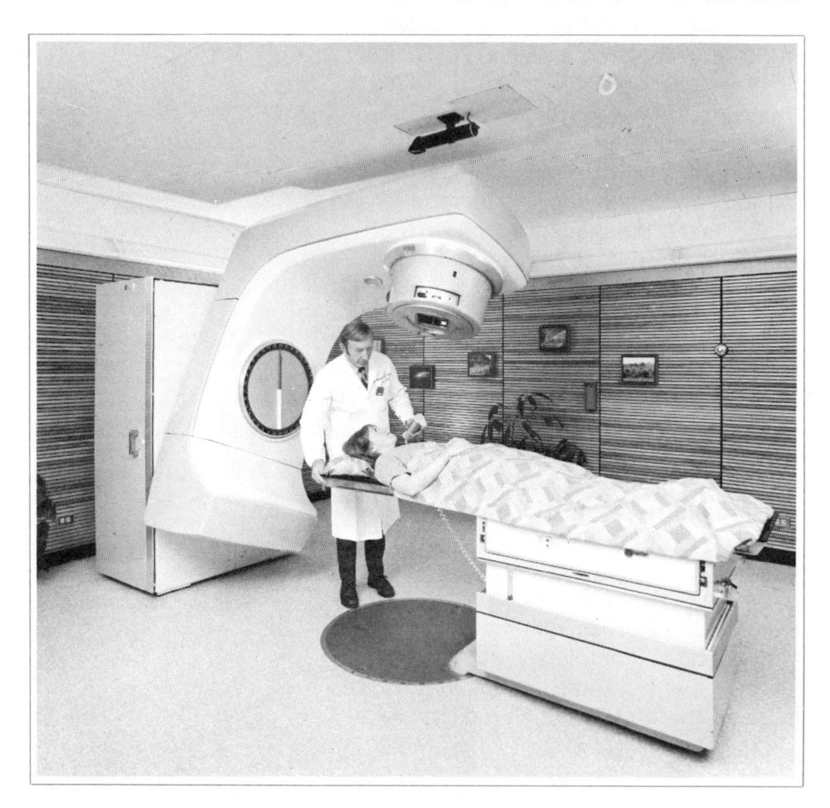

A device for irradiation of deep-seated tumors. The apparatus rotates about the patient so that the point of entry of the radiation is constantly changing but the focal point of the beam remains the same. (Courtesy Varian Associates, Palo Alto, California)

or viral infection that his body is unable to overcome. Until this dilemma has been resolved, organ transplants should be viewed only as a last resort.

Modification of Materials

Chemists have discovered that subjecting a substance to x-rays or gamma rays can result in chemical structures which cannot be produced by any other method. For example, polyethylene can be converted from a flexible, low-strength material easily deformed by heat, to a tough, resilient substance capable of withstanding repeated autoclaving. These properties, coupled with polyethylene's low chemical reactivity, make the irradiated material preferable even to glass for some laboratory apparatus.

Gamma Rays

Eradication of Pests

The growth rate of the world's population is so phenomenal that the food supply is barely able to keep pace. Man must wage unrelenting war against insects which threaten to destroy or consume it. Control of these insects has been primarily by the use of chemicals—a practice which may have serious long-term effects on other species. A most promising alternative to toxic chemicals is a form of insect population control. Large numbers of male insects are subjected to gamma irradiation—a treatment that renders them sterile. The sterile males are then released to breed in their natural environment. Since the eggs are fertilized by sterile sperm, they do not hatch. This method of insect control has already greatly decreased the damage done by the screwworm which infests cattle, and shows promise for controlling a number of other pests.

Preservation of Food

Even when man is successful in thwarting his insect foes, he still loses much of his harvest, especially meat, because of spoilage. Freezing has significantly lengthened the storage life of many foods, but this technique is available to only a small fraction of the world's population. At room temperatures meat is spoiled by bacterial action, and vegetables such as potatoes and onions begin to sprout if stored too long. Gamma irradiation can be used to kill tissue in sprouting vegetables and has been successful in extending the storage life of potatoes, beans, onions, beets, carrots, and turnips. The preservation of meat by gamma irradiation, however, has not been completely successful. Because of the high dosage required, the taste and appearance of most irradiated meats are affected, and there may be some decrease in the nutritional value as well.

Results of an experiment to determine the effectiveness of radiation as a means of preventing vegetables from sprouting. (Courtesy Research Branch, Agriculture Canada)

Cancer Treatment

Gamma rays may be preferred to x-rays for the treatment of deep tumors because of their greater penetrating ability and because they do not require bulky and expensive generators. The radioactive isotope of cobalt, cobalt-60, emits both beta and gamma radiation, but the betas may be screened out by plastic or aluminum. The cobalt-60 source is sometimes referred to as a cobalt "bomb," and is less hazardous to use than radium or other gamma sources because it does not become incorporated into the body if accidently ingested.

Radioisotopes

Leak Detection and Flow Meters

Leaks in hidden pipelines may be located by mixing a radioisotope with the material being transported and monitoring the surface for areas of high radioactivity. It is also possible through the use of radioisotopes to measure the rate at which gases or liquids move through a pipe. The isotope is added to the flowing stream at spaced intervals, and the number of radioactive pulses passing a monitoring station in a given time indicates the flow rate.

Chemical Reaction Mechanisms and Metabolic Pathways

The fact that the radioactive atoms of a given element have the same chemical properties as those of its stable isotopes means that both will be metabolized by a living organism in exactly the same way.

Chemists have made use of this fact to trace the pathways of complex chemical reactions such as photosynthesis.

A compound normally metabolized by an organism is prepared in such a way that one of its constituent elements is radioactive. A compound containing a radioactive atom is said to be "tagged" or "labeled," and the radioactive atom is called a *tracer*. The labeled compound is made available to the organism by its normal route. The subsequent appearance of the tracer in another compound implies a metabolic link between the two.

Melvin Calvin was awarded the Nobel Prize for Chemistry in 1961 for studies in which he used carbon dioxide tagged with carbon-14 to work out the complex series of reactions by which green plants produce glucose. Studies with tracers have also revealed the fate of iron, phosphorus, and calcium in the body.

Diagnosis and Therapy

An interesting and useful aspect of metabolism is that certain elements concentrate in specific tissues: gold-198 in the liver; chromium-51 in red blood cells; cesium-137 in muscle, and iodine-131 in the thyroid. Phosphorus-32, barium-140, calcium-45, strontium-90, and radium-226 all accumulate in bone. Some of these isotopes have long half-lives and emit powerful radiations so that their introduction into the body would be extremely dangerous. However, a few of them are used to test the functioning of a specific organ. The test of thyroid function is such an example.

Iodine is converted by the thyroid into thyroglobulin—a compound which controls the basic metabolic rate. In the test of thyroid function the patient drinks a dilute solution of sodium iodide which has been tagged with iodine-131, a beta and gamma emitter having a half-life of 8 days. Twenty-four hours later a sensitive counter scans the thoracic area and measures the percentage of radiation that is localized in the thyroid. The thyroid of a person whose metabolic rate lies within the normal range will accumulate between 15 and 45% of the total iodine-131. An uptake of more than 45% is evidence of hyperthyroidism, while a figure below 15% denotes insufficient thyroid activity. Higher concentrations of iodine-131 are used in treating cancer of the thyroid. In this way the tumor is irradiated from within, and the short half-life of the isotope minimizes the hazard.

Fats labeled with iodine-131 are used to measure the degree of fat absorption through the intestinal wall. Tagged methyl iodide is put into the bloodstream to measure circulation times.

Compounds labeled with phosphorus-32 are used to treat myelogenous leukemia (too rapid multiplication of white blood cells). The treatment does not cure the disease but reduces its severity and progress. White blood cells have a higher percentage of phosphorus than most other cell types, and it may be for this reason that the treatment is moderately successful.

Malignant brain tissue accumulates phosphorus-32—the concentration reaching 100 times that in normal brain cells. When a labeled phosphorus compound is injected into an artery leading to the brain, the radiation defines the size and location of the tumor.

Radioactive gold in colloidal form can be injected into the pleural cavity to reduce pain and fluid accumulation resulting from lung cancer. Gold-198 is also used for treatment of chronic leukemia.

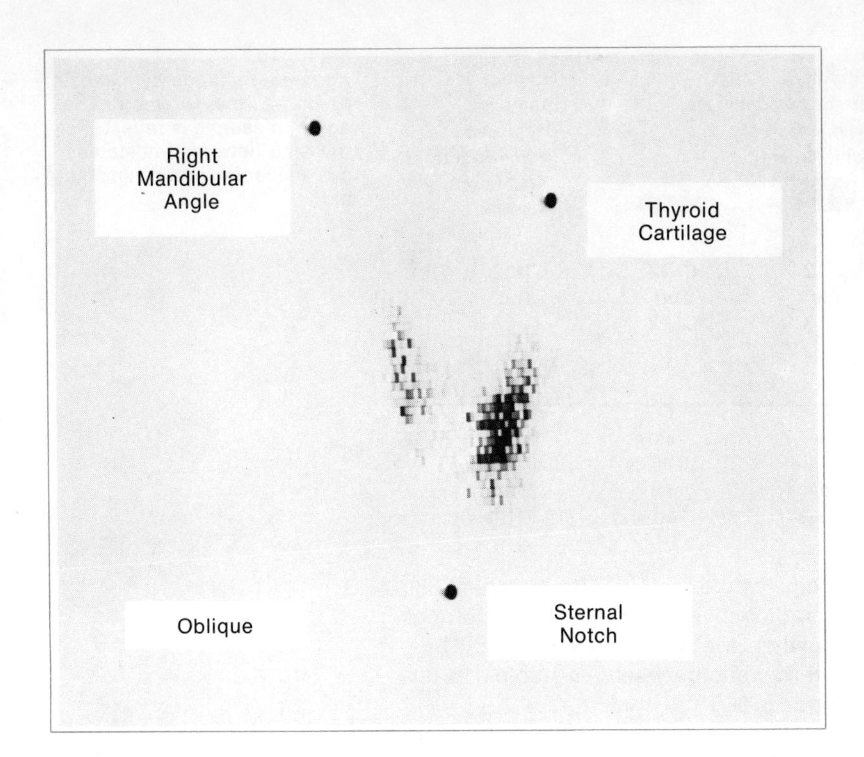

A two-dimensional plot showing the localization of radioactive iodine in the thyroid gland. (Photograph by J. Asdrubal Rivera)

Age Determination

The full value of an archaeological discovery is often not realized unless the age of the artifacts is known. In many instances age determination is little more than an educated guess, but the procedure known as carbon-14 dating has provided a high degree of accuracy for materials of biological origin such as wood, leather, and textiles.

The distribution of the isotopes of carbon today is such that approximately one carbon atom in every trillion is radioactive. This may sound like a very small ratio, but in one gram of carbon there would be about 50 billion such atoms. Carbon-14 emits a beta particle and decays into stable nitrogen-14. However, carbon-14 is generated in the atmosphere by neutron capture in the reaction

$$^{14}_{7}N + ^{1}_{0}n \longrightarrow ^{14}_{6}C + ^{1}_{1}H$$

An important assumption of the dating method is that the rate of this regeneration is essentially the same as it was thousands of years ago, so that the ratio of radioactive to total carbon may be considered constant.

During the lifetime of an organism, growth and repair processes cause a continuous interchange of its carbon with that of the atmosphere, but at death the interchange ceases. The carbon-14 in the organism decays but is not replaced. The half-life of carbon-14 is 5570 years, which means that after the passage of this period of time, only

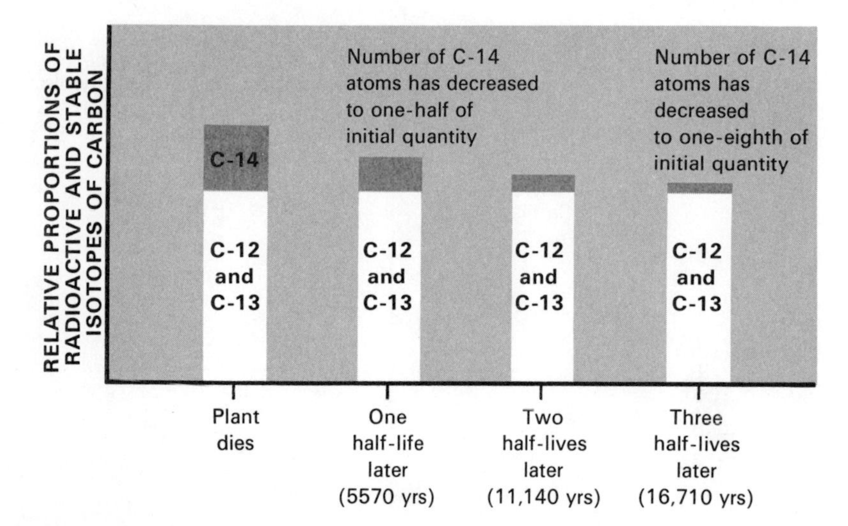

Figure 16.7
Age determination by means of carbon-14 measurement. The age of a sample is reflected in the ratio between the stable and unstable carbon remaining in it.

half of the original radioactive atoms would remain. After another 5570 years the ratio between the nonradioactive and radioactive isotopes would be four times as great as when the organism was alive. Thus, the age of a sample is reflected in the ratio between the stable and unstable carbon remaining in it (Figure 16.7).

As an example of the accuracy of this method, wood from the tomb of Sesostris III has been dated at 3800 ±50 years. The ruler is known to have died about 1800 B.C., which would make the tomb close to 3800 years old. It should be noted that the use of carbon-14 dating is limited to samples whose age does not exceed 50,000 years—samples older than this will contain so few radioactive carbon atoms that it becomes impossible to count them with an acceptable degree of accuracy.

16.7 TRANSMUTATION

Transmutation is the conversion of one element into another by the introduction of a particle into its nucleus. The process may either occur naturally or be directed by man. The conversion of nitrogen into carbon-14 mentioned above is an example of natural transmutation. The first recorded artificial transmutation took place in 1919 when Ernest Rutherford introduced an alpha source into a cloud chamber containing nitrogen and noticed tracks branching off from those left by the alpha particles. The branches were caused by protons produced in the reaction

$$^{14}_{7}N + ^{4}_{2}He \longrightarrow \underbrace{\left[^{18}_{9}F\right]}_{\text{Theoretical}} \longrightarrow ^{17}_{8}O + ^{1}_{1}H$$

In 1932 Chadwick demonstrated the existence of the neutron by bombarding beryllium with alpha rays. The reaction is

$${}^{9}_{4}\text{Be} + {}^{4}_{2}\text{He} \longrightarrow \left[{}^{13}_{6}\text{C}\right] \longrightarrow {}^{12}_{6}\text{C} + {}^{1}_{0}n$$

Theoretical

The development of high-energy particle accelerators such as cyclotrons and linear accelerators has made it possible to modify nuclear structure almost at will. Chemists can, for example, fulfill the dreams of the alchemists by converting other elements to gold—but only at an expense far greater than the value of the gold. Transmutation is employed primarily for the production of useful radioisotopes—for example, the cobalt-60 which is used as a source of gamma rays. Cobalt-60 is produced by the reaction

$${}^{59}_{27}\text{Co} + {}^{1}_{0}n \longrightarrow {}^{60}_{27}\text{Co}$$

Another application of transmutation is the production of synthetic elements (elements which have not been found in nature), a process which leads to our increased understanding of matter. The latest in the series of synthetic elements is tentatively called hahnium (symbol Ha). The name is tentative because the Russians claim prior discovery and have suggested the name Kurchatovium. It was produced in 1970 by the nitrogen bombardment of californium:

$${}^{249}_{98}\text{Cf} + {}^{15}_{7}\text{N} \longrightarrow {}^{260}_{105}\text{Ha} + 4\ {}^{1}_{0}n$$

16.8 NUCLEAR WEAPONS AND ATOMIC POWER

Atomic Fission

The most widely publicized use of radiation is based on a form of transmutation observed by Hahn, Meitner, Frisch, and Strassman in 1938. They bombarded atoms of uranium-235 with neutrons and noticed that the capture of a neutron was accompanied by the appearance of smaller atoms such as barium, strontium, and krypton. Their investigations led to the discovery that the uranium nuclei were being split into two atoms of approximately equal size, and that the splitting was accompanied by the release of additional neutrons. Among the reactions now known to take place are

$${}^{235}_{92}\text{U} + {}^{1}_{0}n \longrightarrow \begin{cases} {}^{103}_{42}\text{Mo} + {}^{131}_{50}\text{Sn} + 2\ {}^{1}_{0}n \\ {}^{90}_{38}\text{Sr} + {}^{144}_{54}\text{Xe} + 2\ {}^{1}_{0}n \\ {}^{141}_{56}\text{Ba} + {}^{92}_{36}\text{Kr} + 3\ {}^{1}_{0}n \end{cases}$$

The division of the nucleus into two small nuclei resembles the biological process of cell division and the name *fission* is used for both. Uranium-233 and plutonium-239 are also capable of atomic fission.

Far more important than the identity of the fission products was the discovery that the larger uranium atom contained more stored poten-

Portion of a letter from Albert Einstein to President Franklin D. Roosevelt regarding the potential use of atomic fission for military purposes:

Sir: Some recent work by E. Fermi and L. Szilard, which has been communicated to me in manuscript, leads me to expect that the element uranium may be turned into a new and important source of energy in the immediate future. Certain aspects of the situation seem to call for watchfulness and, if necessary, quick action on the part of the administration. I believe, therefore, that it is my duty to bring to your attention the following facts and recommendations.

In the course of the last four months it has been made probable—through the work of Joliot in France as well as Fermi and Szilard in America—that it may become possible to set up nuclear chain reactions in a large mass of uranium, by which vast amounts of power and large quantities of new radium-like elements would be generated. Now it appears almost certain that this could be achieved in the immediate future.

This new phenomenon would also lead to the construction of bombs, and it is conceivable—though much less certain—that extremely powerful bombs of a new type may thus be constructed. A single bomb of this type, carried by boat or exploded in a port, might very well destroy the whole port together with some of the surrounding territory. However, such bombs might very well prove to be too heavy for transportation by air.

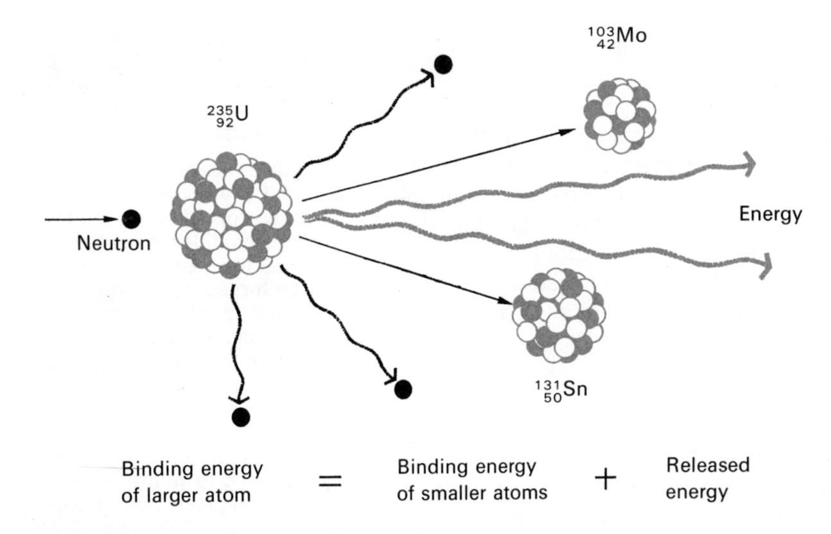

Figure 16.8
Fission of an atom of uranium-235. The binding energy of the fission products is less than that of the uranium atom.

tial energy (binding energy) in its nucleus than the combined binding energies of the smaller nuclei, and that energy equal to this difference was released when the uranium atom split (Figure 16.8).

It was pointed out by other scientists, among them Albert Einstein, that the additional neutrons produced in the reaction might also be captured by other uranium nuclei, setting off a chain reaction, and that the combined energy released by the chain reaction would far exceed anything ever observed. The main features of a chain reaction are shown in Figure 16.9.

During the next few years, a massive technological effort made the atomic bomb a reality. The essential parts of an atomic bomb are shown in Figure 16.10. The fissionable material is cast into pieces so small that the majority of neutrons would escape from them without being captured. When the bomb is detonated, a conventional explosive drives the pieces together and the larger fissionable mass becomes capable of sustaining a chain reaction. Temperatures in excess of one million degrees Celsius are produced within a millionth of a second, generating a destructive pressure wave in the heated air. The blast also generates high-intensity ionizing radiations whose absorption converts many stable isotopes into their radioactive counterparts. This radioactive debris, which is called fallout, eventually settles to the earth where it may constitute a hazard to life for many years. Strontium-90 and cobalt-60 are especially dangerous because they have such long half-lives. Strontium-90 presents a special hazard because of its chemical similarity to calcium (both elements are in the same family). Like calcium, it is incorporated into bony tissue, from which location it bombards the sensitive bone marrow with beta particles. Evidence suggests that such radiation may lead to a cancer of the bone, called leukemia.

Figure 16.9
Diagrammatic representation of a chain reaction taking place in uranium-235.

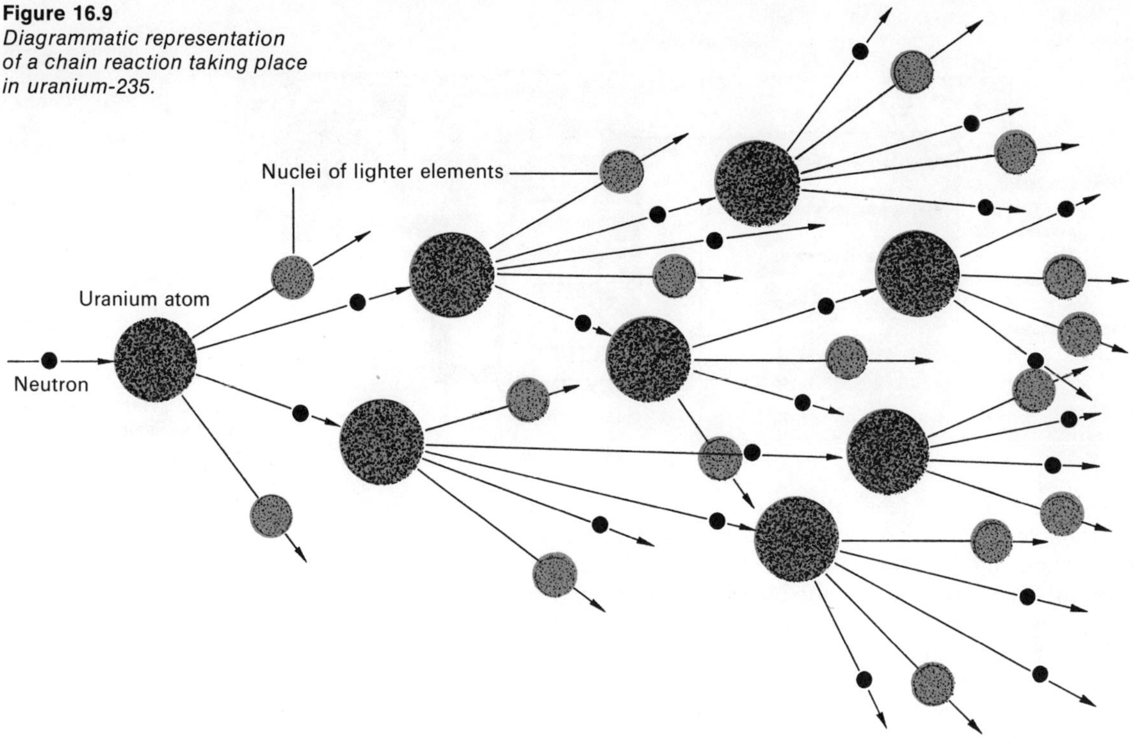

Nuclei of lighter elements

Uranium atom

Neutron

Nuclear Reactors

A nuclear reactor is a device in which a controlled fission reaction yields electrical energy. The essential difference between a reactor and an atomic bomb is that the fission proceeds at a much lower rate in the reactor. This is accomplished by absorbing most of the neutrons with a nonfissionable material so that the number of nuclei undergoing fission at any time will be very small, thus releasing energy at a manageable rate.

F = Fissionable material

N = Neutron source such as mixture of radium and beryllium

◗ = Explosive charges

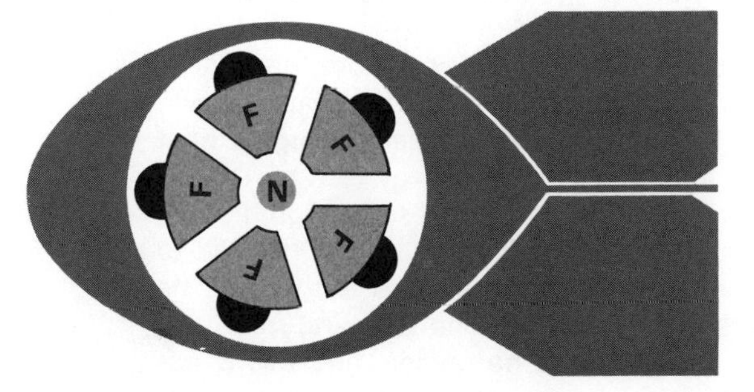

Figure 16.10
Diagram of an atomic bomb. The fissionable material is cast into pieces too small to sustain a chain reaction. The bomb is detonated by driving these pieces together.

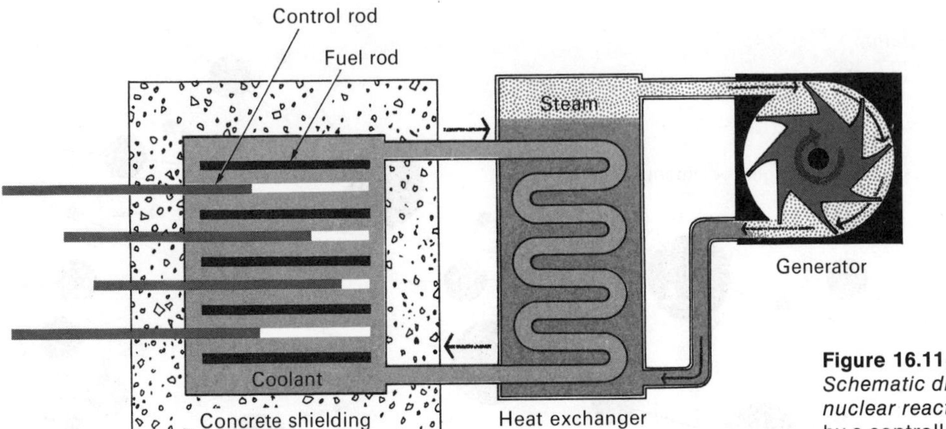

Figure 16.11
Schematic diagram of a nuclear reactor. Heat produced by a controlled fission reaction is used to power a generator. The reactor can be shut down by inserting the control rods between the fuel rods.

The major components of a reactor are shown in Figure 16.11. The fissionable material, or fuel, is formed into rods that are spaced throughout the reactor core. The fuel is either uranium-235 or plutonium-239. Control rods made of neutron-absorbing substances such as boron steel or cadmium are placed between the fuel rods. When the control rods are fully inserted, the reactor is shut down. When the control rods are partly withdrawn, neutrons are captured by the fuel and heat is generated. A coolant, usually water or molten sodium, circulates through the core and carries the heat to a heat exchanger where it is used to generate electricity in the conventional way. Thick concrete shielding encloses the core to protect the operating personnel against radiation.

It is difficult to imagine any major application of technology that would not have some undesirable feature, and nuclear reactors are no exception. Although they have enabled us to conserve our dwindling reserves of fossil fuels, and have reduced air pollution, they have also created problems in the disposal of radioactive wastes in the form of spent fuel, and are threatening to produce a biological problem through thermal pollution. Thermal pollution is an increase in the temperature of a body of water such that ecological changes take place. Manufacturing processes and nuclear reactors are typical causes of thermal pollution.

In a typical reactor the material that absorbs heat from the core is recirculated in the same way as the water that cools an automobile engine. In an automobile engine the hot water passes through a heat exchanger, called a radiator, and transfers its energy to a current of moving air. The coolant in a reactor also passes through a heat exchanger, transferring its energy to water. The water is converted to steam and used to run the generators. Following its trip through the generator, the steam condenses to hot water. The temperature of the condensate must be lowered before it can be reused, and this is accomplished by a secondary heat exchanger using cooler water. Because of the enormous amount of heat involved, large quantities of water are required. The source of this water may be a lake, stream, or even the ocean. When the water is returned to its source, it is invariably warmer than before.

Studies indicate that the types and numbers of organisms in a body of water may be significantly altered by temperature changes. The behavior of many organisms is also temperature dependent. For example, the feeding activity of trout is significantly reduced when the temperature of their habitat is raised above 65°F.

The generation of electric power by nuclear fission is clearly a major step in the conservation of our resources, but until the problems of thermal pollution and radioactive waste disposal are solved, it will remain a mixed blessing.

APPLICATION OF PRINCIPLES

1. What are the atomic and mass numbers of the daughter element produced by the beta decay of $^{131}_{53}I$?
2. What are the atomic and mass numbers of the daughter element produced by the alpha decay of $^{226}_{88}Ra$?
3. The symbol $^{99m}_{43}Tc$ represents the excited state of an isotope of technicium-99. This excited form emits gamma radiation as it adjusts to a less energetic state. The half-life of this transition is 6.1 hours. How long will it take for a sample containing 10 billion atoms of technicium-99^{m} to decay so that fewer than 10 million technicium-99^{m} atoms remain?
4. Why is it a good idea for a person whose work involves frequent handling of radioisotopes to wear both a film badge and a pocket dosimeter?
5. Both sodium-24 and iron-59 emit beta radiation. Both elements are important constituents of blood—sodium in the form of sodium ion, and iron as a component of hemoglobin. Consult Table 16.2 and suggest a reason for adding sodium-24 rather than iron-54 to the blood for the measurement of blood volume.
6. Why is it unlikely that the cooling water discharged from a nuclear reactor would be radioactive?
7. Consult Table 16.5 and determine the thickness of wood that would have the same shielding effect as 5 inches of concrete.
8. A biologist wishes to know whether the preformed cholesterol that is obtained by eating certain fatty materials is the same cholesterol that is sometimes deposited on the inner walls of blood vessels. How might the use of radioisotopes help to solve this problem?

APPENDIX A

MEASUREMENTS AND THEIR MANIPULATION

A measurement system is a collection of standards that can be used to assign numerical values to various aspects of matter, such as length or mass. For convenience every measurement system includes standards of different sizes for measuring the same thing. For example, in the British system, length may be measured in inches, feet, miles, etc., while the metric units of length include millimeters, meters, and kilometers. If you are not familiar with the metric system, it may at first seem awkward or confusing. However, the relationships among the units in the metric system are simpler, and more logical, than those in the British system. And once you have mastered the prefixes that are used to modify the basic units in the metric system, you will find that the conversion of a measurement into its equivalent in some other unit is surprisingly easy.

The "Factor-Label" Method

Conversions within either system, or between systems, are simplified by the use of a procedure called the "factor-label" method. Despite its impressive name, this procedure can be used successfully by persons who feel that their mathematical abilities are limited. It is based on three principles that were derived from mathematics, but which are stated here in nonmathematical terms.

First Principle

An equal sign (=) separates two factors which have the same value. Thus $8 = 2 \times 4$ means that 8 has the same value as the product of 2 and 4. The expression 1 foot = 12 inches means that the distance represented by one of the measurement standards known as a foot is the same as the distance corresponding to 12 of the inch standards. Statements of equality between two measurement standards are called *conversion factors.* One foot = 12 inches is a conversion factor.

Second Principle

Anything divided by itself (or its equivalent) is equal to 1 (unity).

Examples of this principle are $\frac{8}{8} = 1$, $\frac{8}{2 \times 4} = 1$, $\frac{12 \text{ inches}}{1 \text{ foot}} = 1$, and $\frac{8 \text{ apples}}{8 \text{ apples}} = 1$. Note that the result obtained by dividing 8 apples by 8 apples is not 1 apple, but simply 1.

Third Principle

Multiplying a quantity by unity does not change its value. Since $\frac{8}{8}$ = unity, we can multiply any quantity by $\frac{8}{8}$ without altering its value.

The value of 8 apples is likewise unchanged by the operation $8 \text{ apples} \times \dfrac{12 \text{ inches}}{1 \text{ foot}}$.

Just how these principles are applied in converting a measurement in one unit to its equivalent in a different unit will become evident from the following examples.

Let's begin with the statement of a problem. How many feet are there in 48 inches? What we are seeking is the value of the unknown factor in the relationship

$$48 \text{ inches} = ?\ \text{feet.}$$

The conversion factor that relates feet and inches is

$$12 \text{ inches} = 1 \text{ foot}$$

According to the second principle, this conversion factor can be written in either of two ways: $\dfrac{12 \text{ inches}}{1 \text{ foot}}$ or $\dfrac{1 \text{ foot}}{12 \text{ inches}}$. Remember that each of these expressions has a value of unity, and that we can multiply any quantity by unity without changing its value. We can determine the value of the unknown factor in our initial statement by simply multiplying 48 inches by the proper conversion factor (which has a value of unity). Recall that a conversion factor can be written in either of two ways. If we choose the former, in which inches appears in the denominator, our initial statement becomes

$$48 \text{ inches} \times \frac{1 \text{ foot}}{12 \text{ inches}} = ?\ \text{feet}$$

The second principle tells us that inches divided by inches is unity, leaving us with

$$48 \times \frac{1 \text{ foot}}{12} = ?\ \text{feet}$$

The value of the unknown factor is obtained by completing the operation $48 \div 12$. Thus we see that 48 inches equals 4 feet. Here are several more examples of this method.

Sample problem A.1 The effective dose of a certain hallucinogen is 50 μg. How many milligrams is this?

Solution Since there are 1000 μg in 1 milligram, the conversion factor is 1000 $\mu\text{g} = 1$ mg. The relationship to be solved is

$$50\ \mu\text{g} = ?\ \text{mg}$$

The conversion factor is written so that μg appears in the denominator.

$$50\ \mu\text{g} \times \frac{1 \text{ mg}}{1000\ \mu\text{g}} = ?\ \text{mg}$$

The answer is 0.050 mg.

Sample problem A.2 The average volume of urine excreted in 24 hours is about 1200 ml. How many liters is this?

Solution There are 1000 ml in 1 liter. The conversion factor is 1000 ml = 1 liter, and it is written so that the ml appears in the denominator.

$$1200 \text{ ml} \times \frac{1 \text{ liter}}{1000 \text{ ml}} = ? \text{ liter}$$

The answer is 1.200 liters.

In some instances you may not know the relationship that could be used for a conversion factor. For example, suppose you were trying to determine the number of milligrams in 1 1/2 ounces. Two conversion factors are needed: one to change ounces to grams, and another to change grams to milligrams. The solution, using two conversion factors, is shown below.

$$1\ 1/2 \text{ oz} \times \frac{28.35 \text{ g}}{1 \text{ oz}} \times \frac{1000 \text{ mg}}{1 \text{ g}}$$

The answer is approximately 42,500 mg.

Temperature Conversions

The relationship between Fahrenheit and Celsius temperatures is °C = 5/9 (°F − 32°). This relationship can also be stated as °F = 9/5°C + 32°. Memory is a frail crutch and should not be made to bear more than is absolutely necessary. Instead of memorizing a pair of formulas, you may prefer to make use of the way in which the two scales were constructed in carrying out conversions between them. Remember that the common reference point is the freezing point of water. The temperature at which water boils can also be used for reference. A comparison of these reference points is shown below.

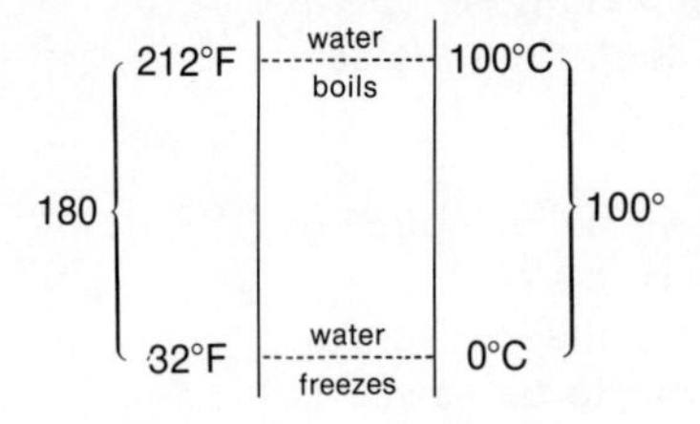

Note that between the two points there are 100 degrees on the Celsius scale and 180 degrees on the Fahrenheit scale. Thus each Celsius degree covers the same temperature change as 1.8 Fahrenheit degrees.

Sample problem A.3 Room temperature averages about 20°C. What is this temperature on the Fahrenheit scale?

Solution The difference between the lower reference point and the given temperature on the Celsius scale is 20 degrees. Since each Celsius degree is equal to 1.8 Fahrenheit degrees, a difference of 20 Celsius degrees equals 36 Fahrenheit degrees. Recall that this temperature is also above the lower reference point on the Fahrenheit scale, so the Fahrenheit temperature is $32° + 36°$, or 68°.

Sample problem A.4 Milk is pasteurized by heating it to 145°F. What is this temperature on the Celsius scale?

Solution A temperature of 145°F is 113° above the lower reference point. This is 113/1.8, or 62.8, degrees above the lower reference point on the Celsius scale. Since the Celsius reference point is 0°, the Celsius temperature is $0° + 62.8°$, or 62.8°.

APPENDIX B

SELECTED ANSWERS

Chapter 1

1. (b) 142 cg (e) 127 mm (h) 100 ml
4. 600 cc
7. −78°C or 196°K
9. Density = 3.5 g/cc and specific gravity = 3.5
12. 8 mm
13. 600 calories

Chapter 2

2. (a) chemical (d) physical (g) physical
5. Evaporate the water and test the residue for the properties associated with NaOH – melting point, for example.
8. Glycerine and water boil at different temperatures. Set up a distillation apparatus with each liquid. The temperature in the flask containing pure water will rise gradually to about 100°C, where it will stay until all of the liquid has been distilled. The temperature in the flask containing the mixture will stay for awhile at one temperature, but will move up to a different temperature eventually.
9. (c) mixture (e) mixture (water + magnesium hydroxide)
   (g) mixture

Chapter 3

2.

Mg	$1s^2 2s^2 2p^6 3s^2$
Cl	$1s^2 2s^2 2p^6 3s^2 3p_x^{\;2} 3p_y^{\;2} 3p_z^{\;1}$
S	$1s^2 2s^2 2p^6 3s^2 3p_x^{\;2} 3p_y^{\;1} 3p_z^{\;1}$
Ar	$1s^2 2s^2 2p^6 3s^2 3p^6$

3. Because it violates Hund's Rule. There are 2 electrons in the $2p_y$ orbital, but the $2p_z$ orbital is vacant.
5. :Si· (with one dot below) :P· (with one dot below) :Al· Ca:
9. 142
14. Dalton's model declared that all atoms of a given element were alike in all respects. The neutron was discovered in connection with isotopes, and the discovery showed that all atoms of a given element did not have the same mass.

Chapter 4

2. With an electronegativity difference of only 1.0 the bond will probably be relatively nonpolar. The compound should have low water solubility, low melting and boiling points, and be a poor conductor of electricity when in solution.
4. (c)

```
   xx   oo   xx
  xClx°P°xClx
  x  xox  x  x
   xx   ox   xx
       xClx
       x  x
        xx
```

5. (a) $1s^2 2s^2 2p_x^{\;1} 2p_y^{\;1} 2p_z^{\;1}$. Since its outer energy level contains 5 electrons, nitrogen will tend to gain three more electrons to conform

to the Octet Rule. This would give the nitride ion 7 protons and 10 electrons, or a charge of 3−.
(b) $1s^2 2s^2 2p^6 3s^2$. Magnesium would tend to lose its 2 valence electrons, leaving it with a nuclear charge of 12^+, but only 10 electrons. The magnesium ion should have a charge of 2+.
9. Ionically bonded. One atom of the Group IIA element with 2 atoms from Group VIIA.
12. Oxygen, 2; nitrogen, 3; carbon, 2; and beryllium, 2.

Chapter 5

4. (b) Iron(III) oxide or ferric oxide (e) boron trifluoride (h) magnesium nitrate
5. (a) $(NH_4)_2SO_4$ (e) Ca_3N_2 (g) SF_6
6. (b) 132.1
7. (c) Al = 15.8%, S = 28.1%, and 0 = 56.1%
9. (c) Not a type reaction.
10. (d) $H_2O + Ca_3(PO_4)_2$
12. (b) $2\ Na_2O_2 + 2\ H_2O \rightarrow 4\ NaOH + O_2$
(f) $3\ Mg(OH)_2 + 2\ H_3PO_4 \rightarrow Mg_3(PO_4)_2 + 6\ H_2O$

Chapter 6

3. $Na_2O + H_2O \rightarrow 2\ NaOH$
5. Combine 3 volumes of 10% peroxide with 7 volumes of distilled water.
7. Sugar is readily transported through the intestinal walls into the bloodstream, but positively charged ions such as the Na^+ ion from salt are not. The presence of a large number of sodium ions in the lumen of the intestine creates an osmotic pressure differential, causing water to flow from the tissues into the lumen and resulting in severe dehydration.
12. No, because the suspended particles in a colloidal dispersion generally do not settle out on standing.

Chapter 7

2. 152 torr
4. Smaller. The pressure of the expired air was due to the combined pressures of the gases and the water vapor. If the water vapor is removed, the remaining gases will exert a lower pressure than before, and the volume will decrease.
7. 138 liters
8. Collect it by displacement of air. Place a bottle upright on the desk. Run the delivery tube to the bottom of the bottle. The less dense air will be forced out as the bottle fills with HCl.
11. Approximately 2.0 g per liter.

Chapter 8

2. 0.11 *M*
4. $(NH_4)_2SO_4 + 4\ H_2O \rightarrow 2\ NH_4OH + 2\ H_3O^+ + SO_4^{2-}$
5. Solution B is more acidic. Its hydronium ion concentration is 10 times that of solution A.
8. (b) $HCOO^-$ (d) ClO_4^-
12. Acidic. The amino group is a stronger base than water and has removed protons from hydronium ions, converting the amino group to $—NH_3^+$. In a basic solution the carboxyl group (a weaker base than the amino group) would have donated a proton and formed a carboxylate ion.

Chapter 9

2. It will cause a 9-fold increase in reaction rate.
3. (b) Exergonic
6. (a) Endergonic. (b) 40 kcal (c) None. The useful energy is applied to the activation of more HI.

Chapter 10

2. (b)

```
                H
                |
            H—C—H
        H       |       H      H     H   H
        |       |       |      |     |   |
     H—C——————C——————C——————C——————C—C—H
        |       |       |      |     |   |
        H       |       H      |     H   H
            H—C—H     H—C—H
                |           |
                H       H—C—H
                            |
                            H
```

3. (a) $2\ C_6H_6 + 15\ O_2 \rightarrow 12\ CO_2 + 6\ H_2O$
6. 2-methylpentane has the higher boiling point. This isomer is less branched and thus has a greater surface area. Since the van der Waals forces are proportional to surface area, the substance will have greater intermolecular attractions, and thus a higher boiling point.
7. (b) 2-iodo-2,3,3-trimethylhexane
10. The production of HCl in the reaction indicates that the hydrogen atoms in benzene are being replaced—in other words, a substitution reaction is taking place. This means that the double bonds in benzene are not being attacked.

Chapter 11

1.

```
     H   H   H   H            H   H   H   H          H   OH   H
     |   |   |   |            |   |   |   |          |   |    |
   H-C---C---C---C-OH       H-C---C---C---C-H      H-C---C----C-H
     |   |   |   |            |   |   |   |          |   |    |
     H   H   H   H            H   H   OH  H          H   |    H
                                                       H-C-H
                                                         |
                                                         H
      1-butanol                  2-butanol          2-methyl-2-propanol
```

```
      H      H      H
      |      |      |
    H-C------C------C-OH
      |      |      |
      H      |      H
           H-C-H
             |
             H
      2-methyl-1-propanol
```

2. (d)

```
       H  H     H      H  H  H
       |  |     |      |  |  |
     H-C==C-----C------C--C--C-H
                |      |  |  |
                |      H  H  H
              H-C-H
                |
                H
```

5. (b) There are two hydrogen atoms on the carbon to which the hydroxyl group is attached: it is a primary alcohol.
6. The structure in (d) is an amide.

Chapter 12

2. An overdose of insulin causes the removal of glucose from the blood, and its conversion into glycogen. This lowers the blood sugar level to such an extent that the body can no longer function normally.
3. Galactose can be identified by the fact that it is not fermented by yeasts, while the other two are. Fructose, which is a ketohexose, can be distinguished from glucose (an aldohexose) since ketoses are dehydrated more quickly by hot HCl than aldoses are. Seliwanoff's reagent will produce a bright red color with fructose and a pink color with glucose after the same period of time.
7. They should enter the cycle as α-ketoglutaric acid.
10. The statement is false. Lactose can be hydrolyzed to one mole each of glucose and galactose; thus its oxidation would yield more energy than a mole of glucose.

Chapter 13

2. It would require less oxygen if it were made of glycogen. Glycogen is more oxidized (i.e., contains a higher proportion of oxygen) to begin with.
4. Fats would not be digested to any extent in the stomach because they must be emulsified before there can be effective contact between the fat and an enzyme. Emulsification is brought about by bile salts, and they are released into the intestine.
7. Starch must be completely hydrolyzed before it is absorbed, whereas fats do not have to be. Digested starch is absorbed into the bloodstream directly, while fats are absorbed into the lymphatic system.

Chapter 14

1. Trypsin preferentially attacks the peptide linkage involving the carboxyl group of either lysine or arginine. Its action would yield the dipeptide pro-ala and the pentapeptide glu-phe-val-gly-arg. Chymotrypsin acts on the carboxyl side of phenylalanine, so it would hydrolyze the larger peptide to glu-phe and val-gly-arg.
4.

$$H{-}\underset{\underset{COOH}{|}}{\overset{\overset{CH_2OH}{|}}{C}}{-}NH_2 + \underset{\underset{COOH}{|}}{\overset{\overset{CH_3}{|}}{C}}{=}O \longrightarrow \underset{\underset{COOH}{|}}{\overset{\overset{CH_2OH}{|}}{C}}{=}O + H{-}\underset{\underset{COOH}{|}}{\overset{\overset{CH_3}{|}}{C}}{-}NH_2$$

8. Since the bonds between nitrogen and hydrogen are not broken when proteins are oxidized, the release of energy from a protein molecule is not as complete as the release from a carbohydrate, where *all* bonds are broken.
9. No. Both corn and wheat contain incomplete proteins. The diet would be deficient in lysine.

Chapter 15

1. Lactic dehydrogenase
4. It is doubtful that these proteolytic enzymes would have much activity at the low pH of the stomach. Also, since heating alters enzyme activity, it is likely that most of the enzyme would be inactivated by cooking.
6. Although trypsin specifically hydrolyzes the bond involving the carboxyl group of lysine or arginine, most proteins contain these amino acids and thus are acted upon by trypsin.

Chapter 16

2. Mass number = 222; atomic number = 86.

3. After 6.1 hours the number of excited atoms will be reduced to 5 billion. After a total elapsed time of 12.2 hours the number will be 2.5 billion. At this rate it would take 10 half-lives, or 61 hours.
8. It might be possible to synthesize cholesterol molecules in which carbon-14 atoms occupy key positions in the structure. These radioactive atoms make it possible to trace the cholesterol. If it is deposited on the walls of blood vessels, the level of radiation should be high in these areas.

INDEX

Page numbers for definitions are in boldface type.